UNIVERSITY OF CHICAGO
GRADUATE PROBLEMS IN PHYSICS
With Solutions

JEREMIAH A. CRONIN | DAVID F. GREENBERG
VALENTINE L. TELEGDI

The University of Chicago Press
Chicago and London

The University of Chicago Press, Chicago 60637
The University of Chicago Press, Ltd., London

PREFACE

Since 1947, the Department of Physics at the University of Chicago has required every graduate student to pass a comprehensive examination in the fundamentals of classical and modern physics before allowing the student to begin his thesis research for the Ph.D. This examination, called the "Basic" in the early days, and now known officially as the "Candidacy Exam," is given twice yearly; each time a committee of four faculty members is assigned the task of devising a new test.

We present here a selection of problems from these exams, together with solutions. It is our hope that the collection will be useful to students as a means of both measuring and improving their understanding of the subject matter. This collection will serve this purpose only if it is properly used. Browsing through the solutions will not be of permanent value to the student; only after honest effort should he consult the solutions.

We have tried to make each solution self-contained and coherent, and we have strived to avoid reproducing material found in standard textbooks. It is difficult to know how much knowledge may be assumed on the part of the reader; our general rule has been to assume the level of knowledge expected of a second-year graduate student at the University of Chicago.

The solutions were prepared by two of the authors, Jeremiah A. Cronin and David F. Greenberg. The third author, Professor Valentine L. Telegdi, assisted in putting the solutions to Chapters 1 and 2 into final form. He also reviewed Chapters 6, 9, and 11. All three authors benefited from conversations with colleagues of the University of Chicago. In particular, they would like to thank Dr. Sol Krasner of the Physics Department for his numerous contributions. They are grateful to the Department for its support and encouragement, and to the secretarial staff of the Enrico Fermi Institute for their assistance in the preparation of the manuscript.

Chicago, Illinois
June 1967

J. A. C.
D. F. G.
V. L. T.

The names of those faculty members of the University of Chicago Physics Department who have served on the Candidacy Examination Committees over the years appear below:

EDWARD ADAMS, II	PETER MEYER
SAMUEL K. ALLISON	ROBERT S. MULLIKEN
HERBERT L. ANDERSON	DARRAGH E. NAGLE
NORMAN E. BOOTH	YOICHIRO NAMBU
MORREL H. COHEN	PETER D. NOERDLINGER
RICHARD H. DALITZ	REINHARD OEHME
RUSSELL J. DONNELLY	EUGENE N. PARKER
DAVID H. DOUGLASS	SHERWOOD PARKER
LEOPOLDO M. FALICOV	GEORGE PARZEN
ENRICO FERMI	JAMES C. PHILLIPS
PETER G. O. FREUND	JOHN R. PLATT
HELLMUT FRITZSCHE	MICHAEL PRIESTLEY
RICHARD L. GARWIN	FREDERICK REIF
NORMAN M. GELFAND	CLEMENS C. J. ROOTHAAN
MURRAY GELL-MANN	J. J. SAKURAI
CLAYTON GIESE	MARCEL SCHEIN
MARVIN L. GOLDBERGER	JOHN A. SIMPSON
ROGER H. HILDEBRAND	ROYAL STARK
MARK G. INGHRAM	VALENTINE L. TELEGDI
ULRICH E. KRUSE	EDWARD TELLER
ANDREW W. LAWSON	ROBERT W. THOMPSON
RICCARDO LEVI-SETTI	DEREK A. TIDMAN
WILLIAM LICHTEN	GREGOR WENTZEL
LEONA MARSHALL	S. COURTNEY WRIGHT
BERNDT MATTHIAS	CARL M. YORK
MARIA G. MAYER	WILLIAM H. ZACHARIASEN

CONTENTS

PROBLEMS

MATHEMATICAL PHYSICS

1. A community practices birth control in the following peculiar fashion. Each set of parents continues having children until a son is born; then they stop. What is the ratio of boys to girls in this community if, in the absence of birth control, 51% of the babies born are male?

2. A die consists of a cube which has a different color on each of 6 faces.
(a) How many distinguishably different kinds of dice can be made?
(b) How many different ways are there to make a pair of dice?

3. Each face of a regular octahedron is to be given a different color. If eight different colors are available, how many distinguishable octahedra can be made?

4. In dealing 52 cards, consisting of 4 suits of 13 cards each, among two teams (each team containing two partners), what is the probability that a particular pair of partners obtains at least one complete suit between them?

5. A certain process has the property that, regardless of what has transpired in an interval [0 to t], the probability that an event will take place in the interval [t, ($t + h$)] is λh. Assume that the probability of more than one event is of higher order in h. Determine the probability that at a time t, n events have taken place, passing to the limit of h going to zero. Evaluate the average value of n and the average value of n^2 for the distribution function.

6. There are about 6500 stars visible to the naked eye. Sometimes two stars appear very close together, though upon careful examination no physical connection is found between them. Such a pair is called an optical double star.
(a) Assuming the stars to be distributed at random on the celestial sphere, compute the expectation value of the number of optical double stars with a separation of no more than 1′ of arc.

(b) What is the probability that there are precisely two optical double stars?

(c) Estimate roughly the probability of an optical triple star.

7. Find eigenvalues and normalized eigenvectors of the matrix

$$M = \begin{pmatrix} 0 & 0 & 0 & 1 \\ 0 & 0 & 1 & 0 \\ 0 & 1 & 0 & 0 \\ 1 & 0 & 0 & 0 \end{pmatrix}.$$

8. Let λ_i ($i = 1,2,3$) be the eigenvalues of the matrix

$$H = \begin{pmatrix} 2 & -1 & -3 \\ -1 & 1 & 2 \\ -3 & 2 & 3 \end{pmatrix}.$$

Calculate the sums

(a) $\sum_{i=1}^{3} \lambda_i$ and   (b) $\sum_{i=1}^{3} \lambda_i^2$.

9. Calculate $T = \text{Tr} [e^{i\boldsymbol{\sigma}\cdot\mathbf{a}} e^{i\boldsymbol{\sigma}\cdot\mathbf{b}}]$, where the components of $\boldsymbol{\sigma}$ are the three standard Pauli matrices σ_i for spin $\frac{1}{2}$.

10. Consider a symmetric second-rank tensor **T** with components T_{ik} (i, $k = 1,2,3$).

(a) Show that there exist three invariants, say I_0, I_1, I_2, with respect to coordinate transformations, associated with **T**.

(b) Associate a surface $1 = \sum_{i,k} T_{ik} X_i X_k$ (X_j = Cartesian coordinates) with **T**. Give interpretations of the three invariants in terms of properties of the surface.

11. What are the residues of the following functions at the points indicated?

(a) $\dfrac{e^{az}}{z^5}$ at $z = 0$,   (b) $\dfrac{1}{\sin^3 z}$ at $z = 0$.

12. Calculate

$$\lim_{\epsilon \to 0^+} \int_{-\infty}^{+\infty} \frac{dk}{(k^2 - a^2 - i\epsilon)^3} \quad \text{with} \quad a > 0.$$

13. Evaluate

$$I = \int_{-\infty}^{\infty} \frac{\sin^3 x}{x^3} dx.$$

14. Calculate:

$$\text{(a)}\ I_1 = \int_0^\infty \frac{x\,dx}{e^x - 1}, \qquad \text{and} \qquad \text{(b)}\ I_3 = \int_0^\infty \frac{x^3\,dx}{e^x - 1}.$$

15. Develop $f(x) = \cos(x^2)$ in a Fourier integral.

16. Find $f(t)$ by inverting the Laplace transform

$$\frac{a^2}{p^2 + a^2} = \int_0^\infty e^{-pt} f(t)\,dt.$$

17. Calculate $\displaystyle\int_0^{2\pi} d\phi/(\alpha + \cos\phi)$

(a) when $\alpha > 1$,
(b) when $\alpha = \alpha_0 + i\epsilon$, α_0, ϵ real, $\epsilon > 0$ and $0 < \alpha_0 < 1$, as $\epsilon \to 0$,
(c) when $\alpha = -1$.

18. Evaluate $\displaystyle I_1 = \int_{-\infty}^{+\infty} dx/\cosh x$ and $\displaystyle I_3 = \int_{-\infty}^{+\infty} dx/\cosh^3 x$.

19. Evaluate

$$I = \int_0^{2\pi} d\phi \frac{b + a\cos\phi}{a^2 + b^2 + 2ab\cos\phi}, \qquad |a| \neq |b|.$$

20. The gamma function is defined by

$$\Gamma(x) = \int_0^\infty t^{x-1} e^{-t} dt, \qquad \text{Re}(x) > 0.$$

Show that for $0 < x < 1$,

$$\int_0^\infty t^{x-1} \cos t\,dt = \Gamma(x) \cos\frac{\pi x}{2},$$

$$\int_0^\infty t^{x-1} \sin t\,dt = \Gamma(x) \sin\frac{\pi x}{2}.$$

21. Show that

$$\int_0^\infty \frac{\sinh(ax)}{\sinh(\pi x)} dx = \frac{1}{2} \tan\frac{a}{2}, \qquad -\pi < a < \pi,$$

by integrating $e^{az}/\sinh(\pi z)$ around an appropriate contour.

22. Evaluate by contour integration

$$\int_0^\infty \frac{x^{1/2} dx}{1 + x^2}.$$

Show your contour and all poles and branch cuts in the complex plane.

23. Evaluate the series

$$\sum_{n=1}^{\infty} \frac{(-1)^n}{n^4} = \frac{-7\pi^4}{720}$$

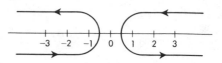

by contour integral techniques. [*Hint:* Use the fact that the function $1/(\sin z\pi)$ has poles along the real axis at $z = 0, \pm 1, \pm 2, \ldots$ Consider the contour above.]

24. Consider the analytic function

$$F(z) = \rho(z) \ln \left[1 - \frac{2z}{a}(1 - \rho(z)) \right]$$

with $\rho(z) = \sqrt{(z - a)/z}$. Here a is real and positive. Choose the branch lines for $\rho(z)$ along the real axis from $-\infty$ to 0 and from a to ∞.
(a) Discuss the Riemann surface of $F(z)$.
(b) Show that there is one sheet where $F(z)$ may be represented in the form

$$F(z) = F(z_0) + (z - z_0) \int_a^{\infty} ds \frac{W(s)}{(s - z)(s - z_0)}$$

and determine $W(s)$.

25. Evaluate

$$\lim_{n \to \infty} \sqrt{n} \int_{-\infty}^{+\infty} \frac{dx}{(1 + x^2)^n}$$

where n is a positive integer.

26. Compute $f(a, b) = \int_0^{\infty} (dx/x) (e^{-xa} - e^{-xb})$.

27. Find the sum of the following infinite series:

$$S = 1 + 2x + 3x^2 + 4x^3 + \cdots \quad \text{for } |x| < 1.$$

28. A generating function $F(x, t)$ of the Hermite polynomial $H_n(x)$ is

$$F(x, t) = e^{x^2 - (t-x)^2} = \sum_{k=0}^{\infty} H_k(x) t^k / k!$$

(a) Express $H_n(x)$ as a contour integral.
(b) Prove that $H_n(x)$ satisfies Hermite's differential equation

$$\frac{d^2 H}{dx^2} - 2x \frac{dH}{dx} + 2nH = 0.$$

(c) Deduce the relation

$$\frac{dH_n}{dx}(x) = 2nH_{n-1}(x).$$

29. A generating function for the Legendre. polynomials $P_l(x)$ is

$$G(x, r) = \frac{1}{(1 - 2xr + r^2)^{\frac{1}{2}}} = \sum_{l=0}^{\infty} r^l P_l(x), \qquad x = \cos\theta, |r| \le 1.$$

Prove that $x P_l'(x) = P_{l-1}'(x) + l P_l(x)$ where $P_l'(x) = d P_l(x)/dx$.

30. Given the Laurent series for $e^{(\mu/2)(z-1/z)}$ as $\sum_{-\infty}^{\infty} A_n z^n$ where $A_n = J_n(\mu)$, obtain an expression for the Bessel function $J_n(\mu)$ as an integral from $-\pi$ to π of a trigonometric function.

31. The function $\phi(x, y)$ is given on the plane $z = 0$. Find, for $z > 0$, a solution $\psi(x, y, z)$ of Laplace's equation that reduces to $\phi(x, y)$ on the plane $z = 0$.

32. Show that

$$K_0(x) = \int_0^{\infty} e^{-x\cosh\phi} d\phi$$

satisfies Bessel's equation of zeroth order and imaginary argument, that is $K_0(x) \equiv J_0(ix)$. Show that $K_0(x)$ has the asymptotic form $De^{-x}/\sqrt{x}$ for very large x; give the value of the constant D.

33. Calculate $\int \mathbf{r} \cdot d\mathbf{A}$ over the surface of a torus.

34. Calculate the volume V of a four-dimensional unit sphere:

$$x_1 = r \sin\phi_2 \sin\phi_1 \cos\phi,$$
$$x_2 = r \sin\phi_2 \sin\phi_1 \sin\phi,$$
$$x_3 = r \sin\phi_2 \cos\phi_1,$$
$$x_4 = r \cos\phi_2.$$

35. Gaseous helium is flowing without turbulence at a velocity v down a pipe and into the atmosphere. Within a very short distance from the end of the pipe, the helium is rapidly diluted to essentially zero concentration.

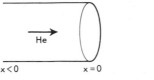

Set up and solve the differential equation for the concentration of air in the pipe as a function of distance from the end of the pipe. Assume equilibrium conditions, neglect wall friction and end effects, assume no temperature difference, and assume that the coefficients of diffusion of O_2 and N_2 into He are the same and equal to D.

36. The equation describing the neutron density in a chain reacting pile is $\nabla^2 n + K^2 n = 0$.

(a) With the boundary condition that the neutron density be finite and positive everywhere, and that it vanish outside the pile, find the radius of a spherical pile for a given value of K.

(b) Now suppose that a thin layer of material of thickness t is added to the surface, and that the neutron density in the layer is described by $\nabla^2 n - \mu^2 n = 0$.

Assume the boundary conditions at the interface are that n and **grad** n are continuous. Demanding that n vanish outside the pile and material layer, find for fixed values of K, μ and t, an expression for the radius of the internal region. Assuming $K \ll \mu$, derive an approximate relation for the difference between the radii without and with the layer.

37. A point source of neutrons on the axis of a long square column of graphite 150 cm on a side emits 10^6 neutrons per sec. Calculate the flux of neutrons at a point on the axis 1 m from the source if the diffusion coefficient of the neutrons is $D = \lambda v/3$, v is their velocity, and $\lambda = 2.8$ cm is the mean free path for scattering. Neglect the effects of slowing down and capture.

2

MECHANICS

1. Derive the form of Stokes' law by dimensional analysis. Assume that this force is independent of the density ρ of the fluid. What happens when this assumption is dropped?

2. A gas bubble from a deep explosion under water oscillates with a period $T \sim p^a d^b e^c$ where p is the static pressure, d the water density, and e the total energy of the explosion. Find a, b, and c.

3. A satellite is put into a circular orbit at a distance R_0 above the center of the earth. A viscous force resulting from the thin upper atmosphere has a magnitude $F_v = Av^\alpha$, where v is the velocity of the satellite. It is noted that this results in a rate of change in the radial distance r given by $dr/dt = -C$, where C is a positive constant, sufficiently small so that the loss of energy per orbit is small compared to the total kinetic energy. Obtain expressions for A and α.

4. A point mass m under no external forces is attached to a weightless cord fixed to a cylinder of radius R. Initially the cord is completely wound up so that the mass touches the cylinder. A radially-directed impulse is now given to the mass, which starts unwinding.

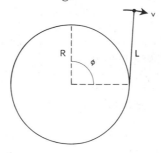

(a) Find the equation of motion in terms of some suitable generalized coordinate,

(b) find the general solution satisfying the initial condition, and

(c) find the angular momentum of the mass about the cylinder axis, using the result of (b).

5. Consider a lawn sprayer consisting of a spherical cap ($\alpha_0 = 45°$) provided with a large number of equal holes through which water is ejected with velocity v_0. The lawn is not uniformly sprayed if these holes are evenly spaced. How must $\rho(\alpha)$, the number of holes per unit area, be chosen to achieve uniform spraying of a circular area? Assume the radius of the sprinkling cap is very much less than the radius of the area to be sprayed, and the surface of the cap is at the level of the lawn.

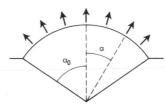

6. Find the differential equation for the contour of a constraining surface on which a point mass will oscillate with a period independent of the amplitude.

7. Three masses (m_1, m_2, m_3), forming the corners of an equilateral triangle, attract each other according to Newton's Law. Determine the rotational motion which will leave the relative position of these masses unchanged.

8. A mass m moves in a circular orbit of radius r_0 under the influence of a central force whose potential is $-km/r^n$. Show that the circular orbit is stable under small oscillations (that is, the mass will oscillate about the circular orbit) if $n < 2$.

9. Two particles move about each other in circular orbits under the influence of gravitational forces, with a period τ. The motion is suddenly stopped at a given instant of time, and the particles are then released and allowed to fall into each other. Prove that they collide after a time $\tau/(4\sqrt{2})$.

10. If r_e and ρ_e are the earth's radius and density, respectively, the corresponding quantities for the moon are $0.275r_e$ and $0.604\rho_e$. A man standing on earth bends his knees, lowering his center of mass 50 cm. Exerting his maximum strength he jumps straight up, raising his center of mass 60 cm above its height at his normal erect posture. How much higher can he jump, in this manner, on the moon?

11. A uniform thin rigid rod of weight W is supported horizontally by two vertical props at its ends. At $t = 0$ one of these supports is kicked out. Find the force on the other support immediately thereafter.

12. Three identical cylinders with parallel axes are in contact with each other on a rough plane, with two cylinders lying on the plane and the third resting on top of them, as in the figure. What is the minimum angle which the direction of the force acting between the cylinders and the plane makes with the vertical?

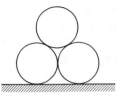

13. A yo-yo rests on a level surface. A gentle horizontal pull (see figure) is exerted on the cord so that the yo-yo rolls without slipping. Which way does it roll and why?

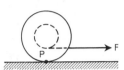

14. A horizontal circular disk of mass M is free to rotate about a vertical axis through a point on its rim. If a dog of mass m walks once around the rim, show that the disk turns through an angle given by the expression

$$\int_0^\pi \frac{4m \cos^2 \gamma \, d\gamma}{(3M/2) + 4m \cos^2 \gamma}.$$

15. A layer of dust is formed h feet thick (h small compared to the earth's radius) by the fall of meteors reaching the earth from all directions. Show, by considering angular momentum, that the change in the length of the day is approximately $(5hd/RD)$ of a day, where R is the radius of the earth, and D and d the densities of earth and dust, respectively. Use a notation in which the initial quantities carry subscript zero, final quantities a subscript 1. The moment of inertia of a sphere about an axis through its center is $(2/5) MR^2$, that of a thin-walled, hollow sphere of mass m and radius R is $(2/3) mR^2$.

16. A simple gyrocompass consists of a gyroscope spinning about its axis with angular velocity ω. The moment of inertia about this axis is C, that about a transverse axis is A. The gyroscope suspension floats on a pool of mercury so that the only torque acting on the gyroscope is one constraining its axis to remain in a horizontal plane. If the gyro is placed at the earth's equator, the angular velocity of the earth being Ω, show that the axis of the gyro will oscillate about the north-south direction; and for small amplitudes of oscillation, find this period. Remember that $\omega \gg \Omega$ is an excellent approximation.

17. The surface of a sphere is vibrating slowly in such a way that the principal moments of inertia are harmonic functions of time:

$$I_{zz} = \frac{2mr^2}{5}(1 + \epsilon \cos \omega t), \qquad I_{xx} = I_{yy} = \frac{2mr^2}{5}\left(1 - \epsilon \frac{\cos \omega t}{2}\right),$$

where $\epsilon \ll 1$. The sphere is simultaneously rotating with angular velocity $\mathbf{\Omega}(t)$. Show that the z-component of $\mathbf{\Omega}$ remains approximately constant. Show also that $\mathbf{\Omega}(t)$ precesses around z with a precession frequency $\omega_p = (3\epsilon\Omega_z/2)\cos\omega t$ provided $\Omega_z \gg \omega$.

18. Three rigid spheres are connected by light, flexible rods with relative masses as shown below:

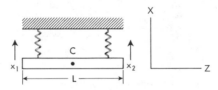

$$m_1 : m_2 : m_3 = 1:2:1$$

Describe all the normal modes of the system and state whatever you can about the relative frequencies.

19. A rigid uniform bar of mass M and length L is supported in equilibrium in a horizontal position by two massless springs attached one at each end.

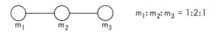

The springs have the same force constant k. The motion of the center of gravity is constrained to move parallel to the vertical X axis. Find the normal modes and frequencies of vibration of the system, if the motion is constrained to the XZ-plane.

20. A particle of mass M hangs from one end of a uniform string of mass m and length L; the other end of the string is fixed. The particle is given a small lateral displacement δ and released from rest. Set up the differential equations and boundary conditions to determine the motion of string and particle. Set up a transcendental equation that determines the natural frequencies, and solve the equation for the case $m \ll M$.

21. Set up a variational principle for the frequency ω of a membrane with surface tension T, of mass σ per unit area, and with fixed edges; that is, find an integral over the area of the membrane, of which the extreme value is the frequency of the membrane.

22. If a watch is moved to a high altitude, does it run fast or slow?

23. A mass m is attached to a weightless string of length L, cross section S, and tensile strength T. The mass is suddenly released from a point near the fixed end of the string. How small should the Young's modulus, Y, of the string be, in order that it not break?

24. A train of mass M, moving with velocity v, is to be stopped with a coil-spring buffer of uncompressed length l_0 and spring constant k_0, which remains constant until the spring is fully compressed. At this point $l \ll l_0$ the spring constant k suddenly becomes very much greater than k_0. Assuming a free choice of k_0, what is the minimum value of l_0 if the absolute value of the maximum deceleration is not to exceed a_{max}?

25. (a) A cylinder of radius R, length h and density ρ floats upright in a fluid of density ρ_0. If it is given a small downward displacement of amplitude x, find the circular frequency ω of the resulting (undamped) harmonic motion.

(b) Show that for small oscillations, the motion of the fluid near the oscillating cylinder extends for a distance $\delta \sim \sqrt{\eta/\rho_0\omega}$ from the edge of the cylinder. The maximum gradient of velocity near the cylinder is thus $dV/dr \approx \omega x/\delta$. Neglecting the friction at the bottom of the cylinder, show that the maximum viscous retarding force on the cylinder is

$$ F \approx 2\pi R h\rho(\eta\omega^3/\rho_0)^{1/2}x. $$

26. A liquid film of surface tension τ is stretched between two circular loops of radius a as shown. Find the equation $r(z)$. For what ratio (d/a) is the configuration indicated in the figure stable?

27. A straight vertical strut, having length l and a square cross section with side a, is firmly fixed to the ground. Show that the maximum weight it can carry on the free end without bending is given by $W = \pi^2 a^4 Y/48l^2$, where Y is Young's modulus for the material of the strut.

28. A rectangular beam with cross section $(a \times a)$ and length L has one end anchored in a vertical brick wall. Calculate the deflection of its free end due to its own weight. The density is ρ and the Young's modulus is Y. Assume small bending.

29. A thin uniform chimney is pivoted at its low end. Show that a section through the chimney at any point undergoes a flexion stress, and calculate the most probable point of rupture as the chimney falls.

30. The free surface of a liquid is one of constant pressure. If an incompressible fluid is placed in a cylindrical vessel and the whole rotated with constant angular velocity ω, show that the free surface becomes a paraboloid of revolution.

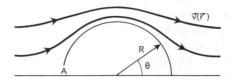

31. An aircraft hangar of semicylindrical shape (with length L and radius R) is exposed to wind directly perpendicular to its axis at infinity with a velocity v_∞. What force is exerted on this hangar if the door, located at A, is open? The velocity potential is given by

$$\phi = -v_\infty(r + R^2/r) \cos \theta,$$

$$L = 70 \text{ m}; \ R = 10 \text{ m}; \ v = 72 \text{ km/hr}; \ \text{air density} = 1.2 \text{ kg/m}^3.$$

32. An air mass of $T = 280°$ is separated by a horizontal plane from an air mass at $T = 300°K$, lying above it. Assume the presence of gravitational waves of wavelength λ and small amplitude, causing a sinusoidal wave on the interface. Find the velocity of the wave as a function of the wavelength, assuming the interface is far from other horizontal interfaces. Treat the oscillations of the air masses as incompressible.

33. Two perpendicular semi-infinite walls, OA and OB in the diagram, intersect at the origin O, and block the two-dimensional hydrodynamic flow of an incompressible fluid of density ρ from a point source of strength K situated at the coordinates (a, b). Calculate the pressure on the walls.

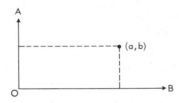

34. Let M and m be the masses of the sun and moon, and R and r be their respective distances from the earth. What is the ratio of the tides induced by these two bodies at the equator?

35. Find the fundamental period of oscillation of an isolated mass of incompressible water, having the radius of the earth (6300 km) and vibrating under its own gravitational attraction. Assume the velocity flow is irrotational.

36. The coordinate systems S_1 and S_2 move along the x-axis of a reference coordinate frame S, with velocities v_1 and v_2 respectively, referred to S. The time measured in S for the hand of a clock in S_1 to go around once, is t. What is the time interval t_2 measured in S_2 for the hand to go around?

37. A rocket is shot out from the earth into interstellar space. Except for a short time in the beginning, the acceleration of the rocket, as measured by the passengers, is constant. The rocket has been aimed at a star a fixed distance from the earth, and moves on a straight line. According to clocks inside the rocket, how long will it take to get to the star? Denote the constant distance and acceleration by D and a' respectively.

38. A particle of rest mass m moves on the x-axis of a Galilean frame of reference and is attracted to the origin O by a force (time rate of change of momentum) $m\omega^2 x$. It performs oscillations of amplitude a. Express the period of this relativistic oscillator in terms of a definite integral, and obtain an approximate value for this integral.

39. Antiprotons are captured at rest in deuterium, giving rise to the reaction $\bar{p} + D \longrightarrow n + \pi^0$ (In this problem we ignore other possibilities.). Determine the π^0 total energy. The rest masses are $M(\bar{p}) = M(p) = 938.2$ MeV, $M(D) = 1875.5$ MeV, $M(n) = 939.5$ MeV, $M(\pi^0) = 135.0$ MeV.

40. A positron (energy E_+, momentum $\mathbf{p}_+$) and an electron (energy E_-, momentum $\mathbf{p}_-$) are produced in a pair-creation process.
(a) What is the velocity of the frame in which the pair has zero momentum (barycentric frame)?
(b) Deduce the energy either particle has in this frame, and
(c) give an expression for the magnitude of the relative velocity between the particles, i.e. the velocity of one particle as seen by an observer attached to the other.

41. A fast (extremely relativistic) electron enters a condenser at an angle α as shown in the sketch. V is the voltage across the condenser and d is the distance between plates. Give an equation for the path of the electron in the condenser.

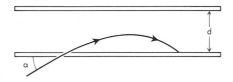

42. The neutral π^0 meson, of rest mass M, decays into two photons. The angular distribution of these γ-rays is isotropic in the rest system of the π^0. If in the laboratory the π^0 travels with velocity v in the z-direction, what is the probability $P(\theta)d\Omega$ that a photon is emitted in the solid angle $d\Omega$ about θ, when the meson decays in flight? Here θ is the angle as measured in the laboratory with respect to the z-axis, and v may be comparable to the speed of light.

43. (a) If neutrons from a cosmic-ray interaction one light-year from the earth were to reach here with a probability of $1/e$ or greater, what must their minimum energy be? (b) If they then decay, what is the maximum angle to the flight path at which their decay electrons could be produced? (c) What is the maximum angle for the decay neutrinos? (d) At the angle calculated in (c), what is the maximum energy of the neutrino?

44. A precession of the perihelion of planetary trajectories has been derived from the general theory of relativity. However, even the special theory of relativity predicts such an effect because of the dependence of inertial mass on velocity. Derive a formula for the *special-relativistic* precession for a planet of given angular momentum L, rest mass m, and energy E, moving in the gravitational potential of the sun. [*Hint:* Use polar coordinates $u = 1/r$ and θ, and find a differential equation involving u and θ, but not involving the time explicitly.]

45. A helium-filled balloon floats inside a closed container filled with air at STP, in interstellar space. The container accelerates in a given direction, with acceleration equal to that due to gravity at the surface of the earth. Which way does the balloon move relative to the acceleration?

3

ELECTROMAGNETISM

1. The edges of a cube consist of equal resistors of resistance R, which are joined at the corners. Let a battery be connected to two opposite corners of a face of the cube. What is the effective resistance?

2. A rectangular wire mesh of infinite extent in a plane has 1 A of current fed into it at a point A, as in the diagram, and 1 A taken from it at point C. Find the current in the wire AC.

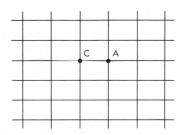

3. Given two iron bars, identical in appearance, one magnetized, the other not. Tell how to distinguish them without using external magnetic fields. (You are allowed to measure forces.)

4. A conductor is charged by repeated contacts with a metal plate which, after each contact, is recharged to a quantity of charge Q. If q is the charge of the conductor after the first operation, what is the ultimate charge on the conductor?

5. A variable capacitor is connected to a battery of *emf* E. The capacitor initially has a capacitance C_0 and charge q_0. The capacitance is caused to change with time so that the current I is constant. Calculate the power supplied by the battery, and compare it with the time rate-of-change of the energy stored in the capacitor. Account for any difference.

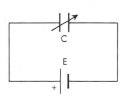

6. When a capacitor is immersed in a medium having conductivity g, a resistance R is measured between the terminals. Show that, regardless of the geometry of the plates, $RC = \epsilon/g$, where ϵ is the dielectric constant of the medium and C is the capacitance in the medium.

7. An eccentric hole of radius a is bored parallel to the axis of a right circular cylinder of radius b ($b > a$). The two axes are at a distance d apart. A currént of I amperes flows in the cylinder. What is the magnetic field at the center of the hole?

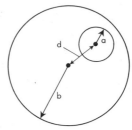

8. A conductor has the shape of an infinite conducting plane except for a hemispherical boss of radius a. A charge q is placed above the center of the boss at a distance p from the plane. Calculate the force on the charge.

9. Consider the thick hemispherical shell of inner radius a and outer radius b, shown in cross section in the accompanying figure. If the shell is uniformly magnetized along its axis of symmetry (z-axis of diagram), show that a small compass needle placed at the origin will swing freely.

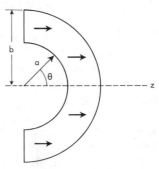

10. A thin uniform metal disk lies on an infinite conducting plane. A uniform gravitational field is oriented normal to the plane. Initially the disk and plane are uncharged; charge is slowly added. What value of charge density is required to cause the disk to leave the plate?

11. Calculate the capacity C of a spherical condenser of inner radius R_1 and outer radius R_2, which is filled with a dielectric varying as

$$\epsilon = \epsilon_0 + \epsilon_1 \cos^2 \theta,$$

where θ is the polar angle.

12. A long straight wire carrying current I is placed a distance a above a semi-infinite magnetic medium of permeability μ. Calculate the force per unit length acting on the wire; be sure to specify the direction of the force.

13. The electrical self-inductance of a circular loop of thin wire (such that the flux within the wire itself can be neglected) is measured in two different surroundings:

(A) The plane of the circle is placed in the XY-plane and all space below this plane is filled with a medium of permeability $\mu = 2$. All space above the XY-plane is evacuated;

(B) $\mu = 1$ everywhere.

What is the ratio of the self-inductance L in configuration (A) to that in configuration (B)?

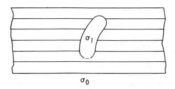

14. There is a small inclusion of conductivity σ_1 inside a metal of conductivity σ_0. The inclusion perturbs an otherwise constant electric field. Find the distance dependence of this perturbation far away from the inclusion. (Treat the problem only in the steady state.)

15. A long hollow cylindrical conductor of radius a is divided into two parts by a plane through the axis, and the parts are separated by a small interval. If the two parts are kept at potentials V_1 and V_2, show that the potential at any point within the cylinder is given by

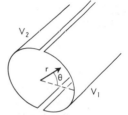

$$V = \frac{(V_1 + V_2)}{2} + 2\frac{(V_1 - V_2)}{\pi} \sum_{n=1}^{\infty} \frac{(-1)^{n-1}}{(2n-1)} \left(\frac{r}{a}\right)^{2n-1} \cos(2n-1)\theta,$$

where r is the distance from the axis of the cylinder, as shown in the figure.

16. Determine the manner in which an initial charge density at any point inside a conductor will decrease with time. Approximately how long would it take for an initial charge distribution inside a copper conductor to disappear? Comment on the validity of your solution. (Copper has a resistivity of 1.7×10^{-6} ohm-cm.) If the conductor is completely insulated, where does the charge go?

17. A small sphere of polarizability α and radius a is placed at a great distance from a conducting sphere of radius b, which is maintained at a potential V. Find an approximate expression for the force on the dielectric sphere valid for $a \ll r$.

18. Derive the Clausius-Mosotti relation connecting the dielectric constant ϵ with the polarizability α of a medium.

19. In a simple cubic lattice, the lattice constant is 2.00 Å and the refractive index (say for sodium light) is $n = 2.07$. Suppose that the medium is subjected to a stress leading to a 2% elongation along one cube edge and a 1% contraction along the other two cube edges.

Calculate the refractive index of the strained medium when the electric vector **E** is (a) parallel to and (b) normal to the principal strain axis. Consider the atomic polarizability α to be a constant scalar quantity.

[*Hints:* The local field acting on an atom in the medium described above can be found as follows. Imagine a spherical cavity about the atom, enclosing the six nearest neighbor atoms. Outside the cavity, the medium may be regarded as continuous and isotropic. The local field at the center of the cavity may be expressed as

$$\mathbf{E}_l = \mathbf{E} + \mathbf{E}' + \sum_{j=1}^{6} \mathbf{E}_j'',$$

where **E** is the applied field, $\mathbf{E}'$ is the contribution from the polarized continuum outside the spherical cavity, and $\mathbf{E}_j''$ is the contribution from the dipole induced in the jth atom within the cavity. In an anisotropic medium, $\sum_{j=1}^{6} \mathbf{E}_j''$ does not vanish; moreover it depends upon the direction of the applied **E**.]

20. What is the critical angle for total external reflection for high-energy x-rays of wave length λ, falling on a metal plate in which all N electrons per unit volume are essentially "free"?

21. The ionosphere can be considered as an ionized medium containing N essentially free electrons per unit volume. Show that if a linearly polarized wave propagates in the ionosphere in a direction parallel to that of the small uniform magnetic field **H** produced by the earth, its plane of polarization will be rotated through an angle proportional to the distance traveled by the wave. Calculate the constant of proportionality.

22. Show that it is possible for electromagnetic waves to be propagated in a hollow metal pipe of rectangular cross section with perfectly conducting walls. What are the phase and group velocities? Show that there is a cutoff frequency below which no waves are propagated.

23. Inside a superconductor, instead of Ohm's Law ($\mathbf{J} = \sigma\mathbf{E}$), we assume London's equations to be valid for the current density **J**:

$$c \, \mathbf{curl} \, (\lambda\mathbf{J}) = -\mathbf{B}, \qquad \frac{\partial}{\partial t}(\lambda\mathbf{J}) = \mathbf{E}$$

(in Gaussian units), and regard λ as a constant. Otherwise, Maxwell's equations (with $\epsilon = 1$, $\mu = 1$) and the corresponding boundary conditions are unchanged.

Consider an infinite superconducting slab of thickness $2d$ $(-d \leq z \leq d)$, outside of which there is a given constant magnetic field parallel to the surface:

$$H_x = H_z = 0, H_y = H_0 \quad (same \text{ value for } z > d \text{ and } z < -d),$$

with $\mathbf{E} = \mathbf{D} = 0$ everywhere. If surface currents and charges are absent, compute $\mathbf{H}$ and $\mathbf{J}$ inside the slab.

24. Polarized light is passed parallel to the axis through a solid glass cylinder of length L which is rotating with an angular velocity Ω around its axis. Find the rotation of the plane of polarization. (Assume constant index of refraction n and permeability 1.)

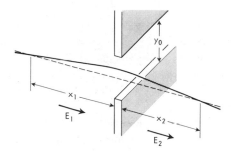

25. The slit lens in the figure above, containing an aperture long in comparison with its width y_0, separates a region in which the electric field is E_1 from a region in which the electric field is E_2. A beam of charged particles focusing at a distance x_1 to the left of the aperture is refocused at a distance x_2 to the right of the aperture. If V_0 is the voltage through which the particles were accelerated before reaching the lens, show that

$$\frac{1}{x_2} + \frac{1}{x_1} \simeq \frac{(E_2 - E_1)}{2V_0}.$$

Use the approximations $V_0 \gg E_1 x_1$ and $E_2 x_2$, and x_1 and $x_2 \gg y_0$.

26. An ion moves in a helical path around the axis of a long solenoid wound so that the ion encounters a region in which the field intensity increases gradually from B_1 to B_2. Under what circumstances will the ion be reflected?

27. The accompanying figure shows a section through the cylindrical plate (radius b) and filament (radius a) of a magnetron. The filament is grounded, the plate is at V volts positive, and a uniform magnetic field H is directed along the axis of the cylinder. Electrons leave the filament with zero velocity and travel in curved paths toward the plate. Below what level of V will the current be suppressed by the field H?

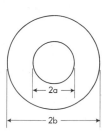

28. The magnetic quadrupole shown in the figure has the property of focusing a beam of charged particles traveling nearly parallel to the z-axis. The focusing is in the $x = 0$ or $y = 0$ planes, depending on the sign of the charge of the particles. Find the simplest distribution of magnetic "poles" which focuses, in a similar way, *uncharged* particles with a magnetic moment μ polarized parallel (or antiparallel) to the x-axis.

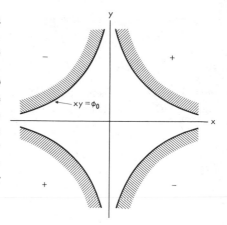

29. A well-collimated beam of protons moves through space in the form of a cylinder of radius R. The velocity of the protons is v and the number per unit volume is ρ. Find the forces on a proton at a radius r from the beam axis. Discuss qualitatively the stability of the beam.

30. Set up the nonrelativistic equation of motion of an electrically charged particle about a fixed magnetic monopole of strength Γ. Find the constants of motion.

31. A standard method for calibrating the orbits of charged particles of momentum p in static magnetic fields is to empirically determine the configuration in those fields assumed by a perfectly flexible wire carrying current I and under tension T. Derive the physical basis of this method. [*Hint:* Derive the general differential equations for: (a) a particle orbit, $d^2\mathbf{r}/ds^2 = ?$, and (b) the equilibrium configuration of a current carrying wire.]

32. Given an atom with spherically symmetrical charge distribution in an external magnetic field H, show that the field at the nucleus caused by the diamagnetic current is $\Delta\mathbf{H} = -(e\mathbf{H}/3mc^2)\,\phi(0)$ where $\phi(0)$ is the electrostatic potential produced at the nucleus by the atomic electrons. Estimate very roughly the magnitude of $\Delta H/H$ for an atom with atomic number $Z = 50$.

33. A spherically symmetric charge distribution of finite extent pulsates radially with some frequency ω. How could one detect these radial pulsations? Explain your answer.

34. A flywheel of radius R, with charge Q uniformly distributed along the rim, rotates with angular velocity ω. What is the rate at which energy is radiated by the system?

35. Show that in the collision of nonrelativistic, spinless, identical particles, the emission of electric and magnetic dipole radiation does not occur, according to classical radiation theory.

36. A linearly polarized plane wave of electromagnetic radiation is incident upon an atom of polarizability α. According to classical electromagnetic theory, what is the electric field of the scattered wave at large distances? What is the total scattering cross section?

37. A thin copper ring rotates about an axis perpendicular to a uniform magnetic field $\mathbf{H}_0$. (See figure.) Its initial frequency of rotation is ω_0. Calculate the time it takes the frequency to decrease to $1/e$ of its original value under the assumption that the energy goes into Joule heat. (Copper has conductivity $\sigma = 5 \times 10^{17}$ cgs, and density $\rho = 8.9$ gm/cm³. $H_0 = 200$ G.)

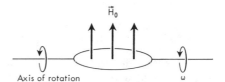

38. A particle of mass m and charge e is suspended on a string of length L. At a distance d under the point of suspension there is an infinite plane conductor. Compute the frequency of the pendulum if the amplitude is sufficiently small that Hooke's Law is valid. Compute how much energy the mass point will lose by radiation per second if it oscillates with small amplitude a.

39. Seven antennas, radiating as electric dipoles polarized along the z-direction are placed along the x-axis, in the xy- plane, at coordinates $x = 0$, $\pm\lambda/2$, $\pm\lambda$, $\pm3\lambda/2$. The antennas all radiate at wave length λ and are all excited in phase.

(a) Calculate the angular distribution of the radiated power as a function of the polar angle θ and the azimuthal angle ϕ (neglect constant multiplying factors).

(b) Make a rough diagram of radiated power vs. angle ϕ in the xy-plane.

(c) Consider the direction in which the radiated intensity is a maximum for (1) this array, and (2) for a single antenna by itself. How do these intensities compare?

40. Show that the energy loss per revolution of a singly charged particle due to radiation is proportional to:

$$L = \frac{\beta^3}{(1 - \beta^2)^2} \frac{r_0}{R} \quad \left(\beta = \frac{v}{c}\right),$$

where R is the radius of the orbit and $r_0 = e^2/m_0c^2$.

41. A light beam of intensity I_0 and frequency ν_0 directed along the positive z-axis is reflected normally by a perfect mirror moving along the positive z-axis with a velocity v_0. What is the intensity I and frequency ν of the reflected light in terms of I_0 and ν_0?

42. Two thin, parallel, infinitely long, nonconducting rods, a distance a apart, with identical constant charge density λ per unit length in their rest frame, move with a velocity v, not necessarily small compared to the speed of light. Calculate the force per unit length between them in a frame of reference that is at rest, and in a frame of reference moving with the rods, and compare the results.

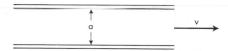

43. Consider an electron moving in a time-dependent axially symmetric magnetic field with $B_\theta = 0$. Given the Lagrangian

$$L = -mc^2\left(1 - \frac{v^2}{c^2}\right)^{1/2} + \frac{e\mathbf{v}}{c}\cdot\mathbf{A} \qquad (\mathbf{B} = \nabla \times \mathbf{A})$$

what conditions must be satisfied in order to have a circular orbit whose position and radius are constant in time? What is the angular frequency and energy of the electron in this orbit? Investigate the stability of the circular orbits. Assume that the shape of the field near the orbit may be represented by $B_z = B_0(r_0/r)^n$, where B_z is the instantaneous value of the field at the equilibrium orbit $r = r_0$; the z-axis is the axis of symmetry, n is positive, and $B(r, z, t) = B(r, z)T(t)$. Assume that the time variations of the external field are small in the time required for a single revolution.
(a) Show that if $n > 1$, the orbit is unstable against radial oscillations.
(b) Show that the sum of the squares of the radial and vertical oscillation frequencies is equal to the square of the equilibrium orbit frequency.

44. Find a covariant generalization for (a) the Lorentz force equation $\mathbf{F} = e[\mathbf{E} + \mathbf{v} \times \mathbf{B}]$; and (b) the equation of motion for a particle with spin $\mathbf{S}$,

$$\frac{d\mathbf{S}}{dt} = \frac{ge}{2m}\mathbf{S} \times \mathbf{B}$$

where g is the gyromagnetic ratio.

ELECTRONICS

1. A vacuum tube is used in a conventional manner in combination with an LC circuit as an rf generator. If the distances between the electrodes are of the order of magnitude of 1 mm and the plate voltage is 200 V, what is the order of magnitude of the frequency above which the circuit will fail to operate as a generator?

2. Show that the infinite chain of inductances L and capacitors C indicated in the figure below acts as a low-pass filter. Calculate the cutoff frequency ω_c in terms of $\omega_0 = 1/\sqrt{LC}$.

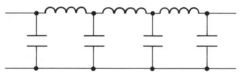

3. A waveform produced by an electronic circuit is to be attenuated with minimum waveform distortion (maximum band width). Using the high-impedance attenuation circuit shown below, draw in any additional components required to fulfill the above requirements, and determine the values of the required components.

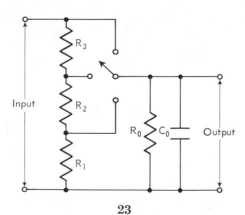

4. The triode in the tuned plate oscillator shown below has plate amplification factor μ and plate resistance R_p. At what frequency is oscillation expected? Under what conditions will the circuit fail to oscillate? [*Hint:* Assume that the grid current is zero.]

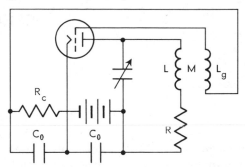

5. The diode circuit shown in part (a) of the accompanying figure is subjected to an input wave form at A shown in part (b) of the figure.

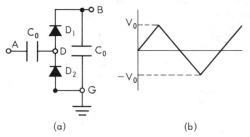

(a)   (b)

Assuming that the capacitors are initially uncharged, plot the voltage at points B and D over three cycles, assuming the diodes are perfect switches. What is the limiting voltage at point B?

OPTICS

1. A thin lens with index of refraction n and radii of curvature R_1 and R_2 is located at the interface between two media with indices of refraction n_1 and n_2, as shown in the figure below. If S_1 and S_2 are the object and image distance respectively, and f_1 and f_2 the respective focal lengths, show that $(f_1/S_1) + (f_2/S_2) = 1$.

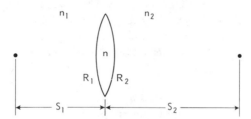

2. Light is normally incident on the interface between two media labeled 1 and 2, respectively. A wave of unit amplitude in medium 1, incident on the interface, is found to have reflected and transmitted amplitudes r and t respectively. Similarly when the wave is incident in medium 2, the reflected and transmitted amplitudes are found to be r' and t'. Using the superposition principle and time-reversal invariance, derive the Stokes' relations $r^2 + tt' = 1$ and $r = -r'$.

3. A plane electromagnetic wave is normally incident from vacuum on a plane surface film of uniform thickness d covering a semi-infinite dielectric. Assume that $\mu = 1$ for both media, and that the film and substrate have indices of refraction n_1 and n_2. Find an expression for the wave reflected into the vacuum in terms of n_1, n_2 and the vacuum wavelength λ. Under what conditions will the reflected wave vanish?

4. A method of color photography was invented by Lippmann in 1881. The photographic plate consists of an extremely fine-grain emulsion on a glass plate which is covered on the emulsion side with a layer of mercury to form a reflecting surface. It is exposed in a camera with the glass side toward the light. After development and with the layer of mercury in place, the plate is now illuminated with white light and viewed by reflection. A faithfully colored picture of the image is seen. Explain what is happening.

Light

$\downarrow$

Glass
Emulsion
Hg

5. In a pin-hole camera the distance of the pin hole from the photographic plate is 10 cm. You want to take a picture of the sun in the visible spectrum ($\lambda \approx 5000$ Å). What diameter of the pin hole should you use in order to obtain the sharpest resolution?

6. A transmission grating 5 cm wide is used to analyze, at normal incidence, the spectrum of sodium. Determine the minimum number of lines needed to resolve the sodium D doublet at 5890 Å and 5896 Å respectively, in the first order. In these conditions, what will be the angular separation of the two members of the doublet?

7. A microwave detector is located at the shore of a lake 0.5 m above the water level. As a radio star emitting monochromatic microwaves of 21 cm wavelength rises slowly above the horizon, the detector indicates successive maxima and minima of signal intensity. At what angle θ above the horizon is the radio star when the first maximum is received?

8. An opaque screen has two very narrow parallel slits a distance d apart. The screen is illuminated by a long, straight, incandescent metal ribbon of small width w and located a distance L in front of the diagram screen. A colored-glass filter placed in front of the screen transmits light of wavelength λ. The transmitted light is made to interfere on an observing screen placed a large distance behind the opaque screen. As the distance d between

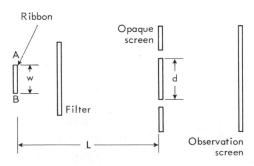

the slits is increased, it is found that for a distance $d = d_0$, the interference fringes on the observing screen disappear. What is the width w of the ribbon?

9. A diffraction grating is made up of N slits each half as long as the previous one. The slits are separated by a distance d. What is the angular distribution of the intensity for light of wavelength λ?

10. Consider a reflection grating whose grooves, though equidistant, have alternating reflecting powers, like $(1 + \alpha)$, $(1 - \alpha)$, $(1 + \alpha)$, $(1 - \alpha)$, and so on. Discuss the effect on the diffraction pattern of increasing α from 0 to some small value much less than 1.

11. A black screen with a circular opening of radius a is located in the xy-plane, with center at the origin. It is irradiated by a plane wave

$$\psi = \exp{(ikz)}, \qquad k = 2\pi/\lambda.$$

Determine the approximate zeros of intensity on the positive z-axis for $z \gg a$.

12. Light is passed through an array of perfect polarizers. The polarization planes are approximately lined up in a fixed direction, but there is a random error between two neighboring planes with a Gaussian distribution $B \exp{(-a\theta^2)}$ where θ is the relative angular deviation. Find the average attenuation coefficient of the system per polarizer for a beam of light after passing the first polarizer. Assume $a \gg 1$.

13. A plane monochromatic wave is incident upon an imperfect linear diffraction grating consisting of N identical apertures. The imperfection is due to the fact that the apertures oscillate, independently of one another, parallel to the grating. The equilibrium positions of the apertures correspond to a perfect linear grating with grating spacing d.

The time needed to photograph the diffraction pattern is very long compared to the periods of oscillation involved. The probability distribution for the displacement of an aperture from its equilibrium position is Gaussian, the root-mean-square displacement being the same for all apertures.

Show that the intensity distribution (i.e. intensity as a function of angle between direction of incidence and direction of observation) in the Fraunhofer diffraction pattern of such an imperfect grating can be expressed as

$$I = \phi I_0 + N(1 - \phi)i_0.$$

I_0 is the intensity distribution for the corresponding perfect grating formed by the apertures when they are in their equilibrium positions; i_0 is the intensity distribution in the diffraction pattern for a single aperture. Express ϕ in terms of the rms displacement.

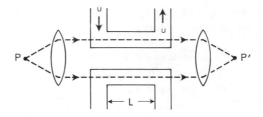

14. Light from a source of frequency f is led through the system shown. If the upper conduit carries a liquid having index of refraction n moving with velocity u, and the lower conduit contains the same liquid at rest, what is the minimum value of u that will cause destructive interference at P'?

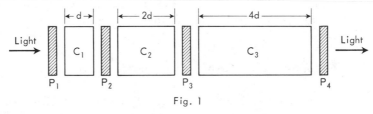

Fig. 1

15. In order to observe the sun in monochromatic light, the French astronomer B. Lyot invented the birefringent filter, consisting of a series of birefringent crystals (C). Each element, after the first, is twice as thick as the previous one. Polarizing films (P) are mounted between the crystals and at each end. (See diagram 1.) All the crystals are mounted

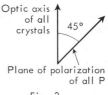

Fig. 2

with their optical axes parallel and at right angles to the direction of propagation of the light. The polarization axes of the polarizing films are also all parallel and at 45° to the direction of the optic axes. (See diagram 2.) Only certain bands of light are able to penetrate the filter. For a filter containing s elements, calculate the transmission as a function of wavelength λ. Also find the width $\Delta\lambda$ of the bands that can pass the filter, and the wavelength separation between such bands.

6

QUANTUM MECHANICS

1. Let B and C be two anticommuting quantum-mechanical operators, i.e.

$$\{B, C\}_+ \equiv BC + CB = 0.$$

Let ψ be an eigenstate of both B and C. What can be said about the corresponding eigenvalues? For $B =$ baryon number and $C =$ charge conjugation, the relations $\{B, C\}_+ = 0$ and $C^2 = 1$ hold. What does your result imply in this case?

2. Three matrices M_x, M_y, M_z, each with 256 rows and columns, are known to obey the commutation rules $[M_x, M_y] = iM_z$ (with cyclic permutations of x, y, z). The eigenvalues of one matrix, say M_x, are ± 2, each once; $\pm\frac{3}{2}$ each 8 times; ± 1, each 28 times; $\pm\frac{1}{2}$, each 56 times; 0,70 times. State the 256 eigenvalues of the matrix $M^2 = M_x^2 + M_y^2 + M_z^2$.

3. Find the eigenvalues of the matrix

$$\lambda_{ik} = [x_i, [L^2, x_k]], \qquad i, k = 1, 2, 3, \qquad L^2 = (\mathbf{r} \times \mathbf{p})^2.$$

4. If one considers systems capable of emitting particles of half-integral spin, one encounters operators U obeying the commutation relations

$$(1) \quad [U, J_z] = \frac{1}{2} U$$

$$(2) \quad [[U, \mathbf{J}^2], \mathbf{J}^2] = \frac{1}{2}(U\mathbf{J}^2 + \mathbf{J}^2 U) + \frac{3U}{16},$$

where $\mathbf{J}$ is the angular momentum of the emitting system. Find selection rules following from (1) and (2), in a matrix representation which makes J_z and $\mathbf{J}^2$ diagonal (eigenvalues m and $j(j+1)$, respectively). In other words, what matrix elements $(m'j'|U|mj)$ can be nonzero? [*Hint:* Let $X_j = j(j+1)$.]

5. Prove that all the wave functions belonging to the maximum eigenvalue of the square of the total spin operator of a system of N electrons are symmetric in the spin coordinates of the individual electrons.

6. Prove the Thomas-Reiche-Kuhn sum rule

$$\sum_n \frac{2m\,|x_{n0}|^2}{\hbar^2}(E_n - E_0) = 1.$$

The sum is over the complete set of eigenstates ψ_n of energy E_n of a particle of mass m, which moves in a potential; ψ_0 represents a bound state.

7. Show that the source-free Maxwell equations may be put into the "Dirac form"

$$p_\beta S_\beta \mathbf{F} = \frac{i}{c}\frac{\partial}{\partial t}\mathbf{F}$$

through the introduction of the Kramers vector $\mathbf{F} = \mathbf{E} + i\mathbf{H}$, $\mathbf{F}^* = \mathbf{E} - i\mathbf{H}$, and the definition of suitable matrices S. Use this representation of the electromagnetic field to show that the photon has spin 1. In the equation above, $\mathbf{p}$ is a momentum operator.

8. Use the Bohr-Sommerfeld quantization rule to calculate the allowed energy levels of a ball which is bouncing elastically in a vertical direction.

9. A three-dimensional isotropic harmonic oscillator has the energy eigenvalues $\hbar\omega(n + \frac{3}{2})$ where $n = 0, 1, 2 \ldots$ What is the degree of degeneracy of the quantum state n?

10. Three mass points of mass m are confined to move on a circle of radius r. Their mutual distances are fixed and equal so that they form an equilateral triangle. The three mass points obey Bose statistics and have no spin. Discuss the rotational energy levels of the system.

11. Derive the dipole-dipole magnetic interaction energy of a proton and an antiproton at a fixed distance a, in eigenstates of total spin, in terms of the proton magnetic moment μ_0. Two magnetic dipoles have the interaction energy

$$V = \frac{1}{r^3}\left\{(\boldsymbol{\mu}_1 \cdot \boldsymbol{\mu}_2) - 3\frac{(\boldsymbol{\mu}_1 \cdot \mathbf{r})(\boldsymbol{\mu}_2 \cdot \mathbf{r})}{r^2}\right\}.$$

12. Solve and classify the eigenvalues of the Hamiltonian

$$H = A(\sigma_Z^{(1)} + \sigma_Z^{(2)}) + B\boldsymbol{\sigma}^{(1)} \cdot \boldsymbol{\sigma}^{(2)},$$

where $\boldsymbol{\sigma}_1$, $\boldsymbol{\sigma}_2$ are the Pauli spin matrices for the particles (1) and (2) respectively. (The Pauli principle is not considered.)

13. A system consists of two distinguishable particles, each with intrinsic spin $\frac{1}{2}$. The spin-spin interaction of the particles is $J\boldsymbol{\sigma}_1 \cdot \boldsymbol{\sigma}_2$, where J is a constant. An external magnetic field H is applied. The magnetic moments of the two particles are $\alpha\boldsymbol{\sigma}_1$ and $\beta\boldsymbol{\sigma}_2$. Find the exact energy eigenvalues of this system.

14. An electron is contained inside a sphere of radius R. What is the pressure P exerted on the surface of the sphere, if the electron is in (a) the lowest S state? (b) the lowest P state?

15. A particle of mass m moves in a potential $V(r) = -V_0$ when $r < a$, and $V(r) = 0$ when $r > a$. Find the least value of V_0 such that there is a bound state of zero energy and zero angular momentum.

16. Consider the one-dimensional Schrödinger equation with

$$V(x) = \begin{cases} \dfrac{m}{2}\,\omega^2 x^2 & \text{for } x > 0, \\ +\infty & \text{for } x < 0. \end{cases}$$

Find the energy eigenvalues.

17. An electron in free space moves under the influence of a uniform magnetic field $\mathbf{B}$. Find the energy levels. If the orbit is large, show that the magnetic flux through the electron orbit is quantized. Neglect electron spin. Show also how knowledge of the energy levels found here nonrelativistically may be used to determine the relativistic corrections to the energies.

18. A quantum-mechanical system in the absence of perturbations can exist in either of two states 1 or 2 with energies E_1 or E_2. Suppose that it is acted upon by a time-independent perturbation

$$V = \begin{pmatrix} 0 & V_{12} \\ V_{21} & 0 \end{pmatrix},$$

where $V_{21} = V_{12}^*$. If at time $t = 0$, the system is in state 1, determine the amplitudes for finding the system in either state at any later time.

19. Use the variation principle to estimate the ground-state energy of a particle in the potential

$$V = \infty \qquad \text{for } x < 0,$$
$$V = cx \qquad \text{for } x > 0.$$

Take xe^{-ax} as the trial function.

20. An electron with charge e and mass m is confined to move on a circle of radius r. It is perturbed by a uniform electric field F parallel to one of the diameters of the circle. Find the perturbation of the energy levels up to terms of the order of F^2. Notice in particular the anomalous behavior of the first excited state.

21. In the Stark effect on the ground state of an atom, if the applied electric field F is small, the energy of the state is perturbed by an amount proportional to the square of the applied field F. Thus $\Delta E = -\frac{1}{2}\,\alpha\,F^2$ where α is the polarizability of the atom. Obtain an expression for the polarizability of the hydrogen atom in its ground state by using perturbation theory. Evaluate the expression in an approximate way which gives an upper and lower limit to the polarizability, thus showing that $4a^3 < \alpha < (16/3)\,a^3$, where a is the Bohr radius.

22. Two identical particles of spin $\frac{1}{2}$ obey Fermi statistics. They are confined in a cubical box whose sides are 10^{-8} cm in length. There exists an attractive potential between pairs of particles of strength 10^{-3} eV, acting whenever the distance between the two particles is less than 10^{-10} cm. Using nonrelativistic perturbation theory, calculate the ground-state energy and wave function. (Take the mass of the individual particles to be the mass of the electron.)

23. Consider an atom with one $2p$ electron placed in an electric field with orthorhombic symmetry. The potential V of the field is $V = Ax^2 + By^2 - (A + B)z^2$. Show that the expectation value of L_z is zero. Neglect electron spin. You may assume that V is small compared to the strength of the atomic potential.

24. Two identical plane rotators with coordinates θ_1, θ_2 are coupled according to the Hamiltonian

$$H = A(p_{\theta_1}^2 + p_{\theta_2}^2) - B\cos(\theta_1 - \theta_2),$$

where A and B are positive constants. (*Note:* $\theta_i + 2\pi$ is equivalent to θ_i.) From the Schrödinger equation determine the energy eigenvalues and eigenfunctions when the following conditions hold:
(a) In the case $B \ll A\hbar^2$, discussing only terms linear in B. Watch out for degeneracies.
(b) In the case $B \gg A\hbar^2$, by reducing the problem to an oscillator problem (small oscillations).

25. A particle of mass m moves in a two-dimensional potential well

$$V(x, y) = 0 \qquad \text{for } |x|,\, |y| \text{ both} < a$$
$$= \infty \qquad \text{otherwise.}$$

Determine the expectation values of the coordinate operators x, y for the ground state when a small perturbation $V = F_1 x + F_2 y$ (F_1, F_2 are constants) is applied. Consider terms only to first order in F_1 and F_2. You need not compute the final matrix elements. It is sufficient to express them in integral form and state explicitly which ones are not zero.

26. Consider two identical linear oscillators with spring constant k. The interaction potential is given by $H' = cx_1x_2$ where x_1 and x_2 are the oscillator variables.

(a) Find the exact energy levels.

(b) Assume $c \ll k$ and compute the lowest pair of excited states in first-order perturbation theory. (Give energy levels in first order and eigenfunctions in zeroth order.)

27. A two-dimensional oscillator has the Hamiltonian

$$H = \frac{1}{2}(p_x^2 + p_y^2) + \frac{1}{2}(1 + \delta xy)(x^2 + y^2),$$

where $\hbar = 1$ and $\delta \ll 1$. Give the wave functions for the three lowest energy levels for $\delta = 0$; evaluate the first-order perturbation of these levels for $\delta \neq 0$.

28. Consider an electron in a uniform magnetic field in the positive z-direction. The result of a measurement has shown that the electron spin is along the positive x-direction at $t = 0$. For $t > 0$ compute quantum-mechanically the probability for finding the electron in the state (a) $S_x = \frac{1}{2}$, (b) $S_x = -\frac{1}{2}$, and (c) $S_z = \frac{1}{2}$.

29. Tritium (the isotope H^3) undergoes spontaneous beta decay, emitting an electron of maximum energy about 17 keV. The nucleus remaining is He^3. Calculate the probability that an electron of this ion is left in a quantum state of principal quantum number 2. Neglect nuclear recoil, and assume that the tritium atom was initially in its ground state.

30. An atom with $J = \frac{1}{2}$, $m_J = \frac{1}{2}$ is in a uniform magnetic field. Suddenly the field is rotated by $\phi = 60°$. Find the probability that the atom is in the sublevels $m_J = +\frac{1}{2}$ or $m_J = -\frac{1}{2}$, relative to the new field, immediately after the change in field.

31. What is the physical basis for the selection rule that a transition from one state of angular momentum zero to another state of angular momentum zero by emitting a photon, is forbidden? Is there any way in which a zero-zero transition could be made by the emission of light? What is the physical basis for the fact that a radiative transition requiring a large spin change proceeds slowly?

32. Show that the photoelectric absorption cross section of an atom, for the ejection of an electron from the K-shell, goes like Z^5 for photon energies large compared with the binding energy of a K electron. Use simple perturbation theory, and neglect recoil.

33. A spherical square well has depth V_0 and radius a. A particle of positive energy E and mass m is caught in a state of angular momentum $L \neq 0$. Estimate the lifetime τ of the particle, ignoring the angular momentum inside the well.

34. A quantum-mechanical system is initially in a state of angular momentum $L_1 = 0$. It decays by electric dipole emission of light, to a lower state of angular momentum $L_2 = 1$. This state in turn decays by an electric dipole transition, after a short time, to the ground state of the system, of angular momentum $L_3 = 0$. Both quanta are observed with suitable detectors. Calculate the probability, $W(\phi)$, that the directions of propagation of the two quanta form an angle ϕ. Does the result depend on whether the system is an atom or nucleus? [*Hint:* Use second-order perturbation theory.]

35. Consider the scattering of a particle by a simple cubic structure with lattice spacing d. The interaction with the lattice points is

$$V = \frac{-2\pi a \hbar^2}{m} \sum_i \delta(\mathbf{r} - \mathbf{r}_i).$$

Treat the scattering in Born approximation. Show from your result that the condition for nonvanishing scattering is that the Bragg law be satisfied.

36. The expression

$$\mathbf{J} = \frac{\hbar}{2m_i}[\psi^*\nabla\psi - \psi\nabla\psi^*]$$

gives the probability that one particle per second will pass through a unit area normal to the direction of $\mathbf{J}$. A beam of particles with uniform velocity v enters a region where some of them are absorbed. This absorption may be represented by the introduction of a constant complex potential $V_r - iV_i$ into the wave equation. Show that the cross section per atom for absorption is $\sigma = 2V_i/\hbar Nv$ where N is the number of absorbing atoms per unit volume.

37. A particle is scattered by a completely absorptive ("black") sphere of radius a which is large compared to the de Broglie wavelength $\lambda/2\pi = 1/k$. How do the η_l and δ_l in the scattering amplitude

$$f(\vartheta) = \frac{1}{2ik} \sum_{l=0}^{\infty} (2l + 1)(\eta_l e^{2i\delta_l} - 1)P_l(\cos\theta)$$

behave as a function of l? Evaluate the elastic scattering cross section, the reaction cross section, and the total cross section.

38. Calculate the differential and total cross sections for scattering of a spinless particle of mass m incident on an infinitely massive, infinitely hard sphere of radius a. Treat the case for which the particle moves sufficiently slowly that the D-wave phase shift is negligible. Give the answer in the form of a polynomial in (ka) and retain only terms in the cross sections which are of lower order than $a^2 (ka)^4$. You may use the following recursion relation, valid for both regular and irregular solutions: If F_l satisfies

$$F_l''(x) + \frac{2}{x} F_l'(x) + F_l(x)\left[1 - \frac{l(l+1)}{x^2}\right] = 0,$$

then

$$F_{l+1}(x) = -x^l \frac{d}{dx}\{F_l(x)x^{-l}\}.$$

39. Consider the scattering of a beam of spinless particles by a hard sphere of radius a:

$$V = \infty \qquad \text{for } r < a,$$
$$V = 0 \qquad \text{for } r > a.$$

(a) Find the total cross section where $a \ll \lambda$, if $\lambda = 1/k$.

(b) Consider the case $a \gg \lambda$. Show that in the forward direction the various partial wave contributions to the scattering amplitude $f(\theta)$ add up coherently to produce a diffraction pattern of the Fraunhofer type.

Useful formulas:

$$P_n(\cos \theta) \approx J_0(n \theta) \qquad \text{for large } n, \text{ small } \theta$$

$$\frac{d}{dz}\{z^{n+1} J_{n+1}(z)\} = z^{n+1} J_n(z)$$

$$f(\theta) = \frac{1}{2ik} \sum_{n=0}^{\infty} (2n + 1)(e^{2i\delta_n} - 1) P_n(\cos \theta)$$

40. Find the bound states of a one-dimensional attractive delta-function potential. Suppose a stream of particles is incident from the left. Find the relative intensities of reflected and transmitted beams.

7

THERMODYNAMICS

1. Derive the following relations (*the Maxwell equations*):

$$\left(\frac{\partial T}{\partial V}\right)_S = -\left(\frac{\partial P}{\partial S}\right)_V, \qquad \left(\frac{\partial T}{\partial P}\right)_S = \left(\frac{\partial V}{\partial S}\right)_P, \qquad \left(\frac{\partial P}{\partial T}\right)_V = \left(\frac{\partial S}{\partial V}\right)_T, \qquad (1)$$

and

$$\left(\frac{\partial V}{\partial T}\right)_P = -\left(\frac{\partial S}{\partial P}\right)_T;$$

$$T\,dS = C_V\,dT + T\left(\frac{\partial P}{\partial T}\right)_V dV; \qquad (2)$$

and

$$\left(\frac{\partial U}{\partial V}\right)_T = T\left(\frac{\partial P}{\partial V}\right)_V - P.$$

2. Calculate the Joule-Thomson coefficient $(\partial U/\partial V)_T$ for a van der Waals gas, where

$$P = \frac{RT}{V-B} - \frac{a}{V^2}.$$

3. In certain phase transitions it is noted that there are no discontinuous changes in enthalpy H or volume V, but rather in their first derivatives with respect to temperature. Derive the two thermodynamic relations which supplant the Clapeyron equation. Let the subscripts 1 and 2 refer to the two phases.

4. For water at 27°C, $(1/V)(\partial V/\partial T)_P = 0.00013$ deg^{-1}. Find the change of temperature in a large body of water carried by a current to a depth of 1 km.

5. Consider a pressure-volume diagram for a given mass of a substance, on which there is a family of adiabatic curves. Prove that no two of these adiabatic curves can intersect.

6. One mole of H_2O is cooled from 25°C to 0°C and frozen. All the heat taken up by the refrigerating machine, operating at maximum theoretical efficiency (no entropy created), is delivered to a second mole of H_2O at 25°C, heating it to 100°C. (a) How many moles of H_2O are converted to steam at 100°C? The heat of vaporization λ' at 100°C is 9730 cal/mole. The heat of fusion λ at 0°C is 1438 cal/mole. How much work must be done by the refrigerator?

7. Two identical perfect gases with the same pressure P and the same number of particles N, but with different temperatures T_1 and T_2, are confined in two vessels, of volume V_1 and V_2, which are then connected. Find the change in entropy after the system has reached equilibrium.

8. A gas-tight, frictionless piston of small thermal conductivity slides in a thermally insulated cylinder. Both compartments A and B contain equal amounts of an ideal monatomic gas. Assume that initially the temperature of the gas is T_0 in volume A, and $3T_0$ in B. The system is to be considered as in mechanical equilibrium at all times, and eventually it will be in thermal equilibrium as well.

(a) What will be the ratio of the volume of A to that of B initially, and at $t = \infty$?

(b) What will be the total change in the entropy of the entire system per mole of the gas in it between $t = 0$ and $t = \infty$?

(c) How much useful work could have been done by the system (via a suitable transfer mechanism) per mole of gas if the transfer of heat from one compartment to the other had been accomplished reversibly?

9. If one mole of a monatomic gas undergoes a free expansion, what is the change in temperature in terms of the initial and final volumes and the constants of the van der Waals equation of the gas? What is the approximate change in entropy? What is the approximate change in the enthalpy?

10. One gram of water at 20°C is forced through an insulated porous plug under a pressure of 10^4 atm into a lab where the pressure is 1 atm. Find the state of the water emerging from the plug. Assume that the density at 10^4 atm is the same as that at 1 atm. The latent heat of evaporation is 540 cal/gm.

11. A vessel contains helium gas at 10°K (above critical point). The vessel is thermally isolated and the gas is allowed to escape slowly through a capillary tube until the pressure within the vessel is 1 atm and the temperature is 4.2°K (normal boiling point of helium). If the gas is perfect, find the initial pressure P_i for the vessel to be entirely filled with liquid at the end of the process. *Data:* Latent heat of He at 4.2°K = 20 cal/mole; C_v of helium gas = 3 cal/mole deg.

12.   A long thin metal bar vibrates in its fundamental mode of longitudinal oscillation. In what range of frequency does the oscillation tend to be isothermal? Take Young's modulus Y as 10^{12} dyn/cm^2, the density ρ as 10 gm/cc, the heat conductivity k as 1 cal/cm·deg·sec, and the specific heat C as 0.1 cal/gm·deg.

13.   Two vessels, each of volume V, are connected with a tube of length L and small cross section A ($LA \ll V$). Initially, one vessel contains a mixture of carbon monoxide at a partial pressure P_0 and nitrogen at a partial pressure $P - P_0$, while the other vessel contains nitrogen at a pressure P. The coefficient of diffusion of CO into N_2 or N_2 into CO is D. Calculate the partial pressure of CO in the first vessel as a function of time.

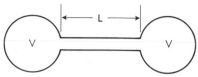

14.   An iron sphere of radius 20 cm is heated to a temperature of 100°C throughout; its surface is then kept at constant temperature 0°C. Find the temperature of the center 15 min after the cooling has begun. The ratio of thermal conductivity to specific heat per unit volume is $K/C\rho = 0.185$ in cgs units.

15.   A sphere of radius R is immersed in an infinite liquid at temperature T_0. At time $t = 0$, the sphere is brought to a temperature $T_1 > T_0$ and is maintained at that temperature thereafter. The conductivity of the liquid is K, its specific heat C, and its density ρ.
(a)   Express the temperature at any point outside the sphere, and at any time $t > 0$ as a definite integral.
(b)   Find explicitly the limiting temperature distribution as $t \longrightarrow \infty$.

16.   Find the temperature dependence of the electromagnetic energy density E within a cavity with perfectly reflecting walls, using thermodynamic arguments.

17.   A spherical satellite of radius r, painted black, travels around the sun in a circular orbit at a distance D from its center ($r \ll D$). The sun, a sphere of radius R, radiates as a blackbody at a temperature $T_0 = 6000°$K and subtends an arc $2\alpha = 32'$ as seen from the satellite. What is the equilibrium temperature T of the satellite?

18.   The Helmholtz free energy A of a ferromagnet can be assumed to depend on the magnetization for no external field as follows:

$$A = A_0 + \alpha(T - T_c)M^2 + \beta M^4 \qquad (1)$$

for temperature T near the Curie temperature T_c; α and β are positive and approximately temperature-independent. A_0 is the free energy of the un-magnetized state, and may be regarded as approximately temperature independent. Derive the temperature dependence of the average magnetization $M(T)$ which follows from Eq. (1) if fluctuations are neglected. What is the influence of fluctuations on $M(T)$? Obtain also the magnetic suscepti-bility above the Curie temperature.

19. The velocity of sound v in a paramagnetic gas is altered by a magnetic field H. Calculate $(v(H) - v(0))/v(0)$, assuming magnetization per mole $M = \gamma H/T$, to lowest order in γ. The specific heat (for $H = 0$) may be regarded as temperature independent.

20. According to Meissner, $B = 0$ in a superconductor. In the normal state, M is negligible. At a fixed temperature $T < T_c$, as the external field H is lowered below the critical field $H_c(T) = H_0 [1 - (T/T_c)^2]$, the normal state undergoes a phase transition to the super-conducting state. Show that this viewpoint is correct by finding the difference in Gibbs free energies of the normal and supercon-ducting metal at temperature $T \leq T_c$, i.e. $G_n(T, H) - G_s(T, H)$. At $H = 0$, compute the latent heat of transition L from the normal to the superconducting state, and the discontinuity in the specific heat. Is the phase transition first or second order? The Gibbs function in a magnetic field is given by $G = U - TS - HM$.

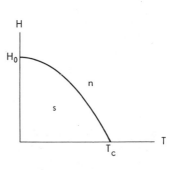

21. From thermodynamic principles, determine the vapor pressure P_r for a very small droplet of liquid of radius r, in terms of the vapor pressure P_∞ of a large body of the same liquid having a negligible surface-to-volume ratio. [*Hint:* For finely divided matter the ratio of area to volume increases greatly, and surface phenomena become dominant.]

22. A rocket weighing 1000 kg is shot into space. Under the assumption that all stellar bodies have an average mass of 10^{30} kg and move about with a random velocity averaging 10 kg/sec, what is the average velocity the rocket will tend to assume after a very long time? (Neglect the possibility of the rocket falling into a star.)

23. Consider hydrogen gas at temperature T (around liquid nitrogen tem-perature) and pressure P (about 1 cm Hg), with a concentration of ortho-hydrogen equal to x. Derive an expression for (a) the molar specific heat of this gas, and (b) the thermal conductivity K of this gas.

Besides T, P, x and fundamental constants, your expressions should contain only three parameters characterizing the H_2 molecule, the molecular mass M, the internuclear separation R, and the collision cross section σ. You may assume that at liquid nitrogen temperature, the number of molecules in rotational states ($J > 2$) is negligibly small.

24. Compute the ratio of thermal conductivity of helium gas at $P = 0.1$ atm and 300°K to that at $P = 0.5$ atm and 300°K. Compute also the ratio of viscosities at the two pressures.

25. Obtain the mean vector velocity of gas molecules escaping through a small hole in an enclosure kept at temperature T. Let N be the number of particles per unit volume of the gas.

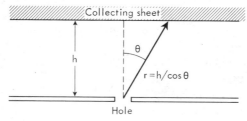

26. A small circular opening of radius a, which is small compared to the mean free path of mercury, is cut in the wall of a very thin-walled rectangular tank containing mercury vapor at a temperature T and very low pressure P. At a distance h above the hole and parallel to the wall of the tank, a collecting sheet of metal is placed and cooled so that when any of the mercury atoms strike it, the vapor condenses at once. Derive an expression for the distribution of mercury in gm/cm² on the collecting sheet at time t in terms of polar angle θ between the normal to the hole and the point of collection. See the figure above.

8

STATISTICAL PHYSICS

1. A young man, who lives at location A of the city street plan shown in the figure, walks daily to the home of his fiancee, who lives m blocks east and n blocks north of A, at location B. Because he is always anxious to see his fiancee, his route always approaches B, i.e., he never doubles back. In how many different ways can he go from A to B?

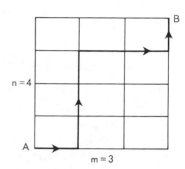

2. According to the Stefan-Boltzmann law, the energy of Hohlraum (blackbody) radiation depends on the temperature T as T^K with $K = 4$. Replace the ordinary three-dimensional Hohlraum by one of N dimensions (N integral). Derive the exponent K in the energy-temperature relation for a photon gas enclosed in an N-dimensional Hohlraum.

3. If the sun behaves as a blackbody at 6000°K with diameter 10^6 km, what is its total microwave-emitted power per megacycle bandwidth at 3 cm?

4. Three particles at the corners of an equilateral triangle each carry a quantum-mechanical spin $\frac{1}{2}$, their mutual spin Hamiltonian being given by

$$H = \frac{\lambda}{3}(\boldsymbol{\sigma}_1 \cdot \boldsymbol{\sigma}_2 + \boldsymbol{\sigma}_1 \cdot \boldsymbol{\sigma}_3 + \boldsymbol{\sigma}_2 \cdot \boldsymbol{\sigma}_3).$$

List the energy levels of this spin system, giving their total spin values and degeneracies. Deduce the partition function, Z.

5. Consider a gas contained in volume V at temperature T. The gas is composed of N distinguishable particles of zero rest mass, so that energy E and momentum p of the particle are related by $E = pc$. The number of single-particle energy states in the range p to $(p + dp)$ is $4\pi V p^2 dp/h^3$. Find the equation of state and the internal energy of the gas and compare with an ordinary gas.

6. 10^{10} weakly interacting spinless particles, each with the mass of the electron, are identical in appearance but obey classical statistics. They are confined in a cubical box which is 10^{-6} cm on an edge. Each particle undergoes a potential interaction with the box which is of two sorts. One is attractive and leads to a bound state well localized near the center of the box and having an energy of -1 eV. The other interaction is a strong repulsion which prevents the particle from escaping through the walls of the box. Find at what temperature the pressure in the box is one atm.

7. A surface has a temperature of 800°C. Atoms of sodium which strike the surface are found to be 99% ionized when they evaporate. Atoms of chlorine are found to be ionized negatively by one part in 10^6 when they evaporate from the same surface. What is the electron affinity of chlorine? The ionization potential of Na is $\phi = 5.1$ V.

8. Helium atoms can be adsorbed on the surface of a metal, an amount of work ϕ then being necessary to remove a helium atom from the metal surface to infinity. The helium atoms are completely free to move, without mutual interaction, on the two-dimensional metal surface. If such a metal surface is in contact with helium gas at a pressure P, and the whole system is in equilibrium at temperature T, what is the mean number of atoms adsorbed per unit area of the metal surface? Express your answer in terms of quantities given in the problem, and fundamental constants.

9. An LC circuit is used as a thermometer by measuring the noise voltage across an inductor and capacitor in parallel. Find the relation between the rms noise voltage and the absolute temperature T.

10. A solid contains N mutually noninteracting nuclei of spin 1. Each nucleus can therefore be in any of three quantum states labeled by the quantum number m, where $m = 0, \pm1$. Because of electric interactions with internal fields in the solid, a nucleus in the state $m = 1$ or in the state $m = -1$ has the same energy $\epsilon > 0$, while its energy in the state $m = 0$ is zero.

Derive an expression for the entropy of the N nuclei as a function of the temperature T, and an expression for the heat capacity in the limit $\epsilon/kT \ll 1$.

11. Consider a system of three-dimensional rotators (with two degrees of freedom and no translational motion) in thermal equilibrium according to Boltzmann statistics; take account of the quantization of energy. Calculate the free energy, entropy, energy and heat capacity (per rotator) in the case of high temperatures, making use of Euler's approximation formula:

$$\sum_{J=0}^{\infty} f\left(J + \frac{1}{2}\right) = \int_0^{\infty} f(x)dx + \frac{1}{24}[f'(0) - f'(\infty)] + \cdots$$

12. The average energy of a system in thermal equilibrium is $\langle E \rangle$. Prove that the mean square deviation of the energy from $\langle E \rangle$, $\langle (E - \langle E \rangle)^2 \rangle$ is given by

$$\langle (E - \langle E \rangle)^2 \rangle = kT^2 C_v,$$

where C_v is the heat capacity of the entire system at constant volume. Use this result to show that the energy of a macroscopic system may ordinarily be considered constant when the system is in thermal equilibrium.

13. An assembly of N particles of spin $\frac{1}{2}$ are lined up on a straight line. Only nearest neighbors interact. When the spins of the neighbors are both up or both down, their interaction is J. When one is up and one is down, the interaction energy is $-J$. (In quantum-mechanical language, the energy is $J\sigma_z^i \sigma_z^j$ between a neighboring pair i and j.) What is the partition function Z of the assembly at temperature T?

14. A highly simplified theory for the temperature dependence of the molar specific heat c associated with the transition from paramagnetism to ferromagnetism of ions with spin $s = \frac{1}{2}$, gives a curve of the type shown in the accompanying diagram; i.e., $c = c_{max} (2T/T_0 - 1)$ for $T_0/2 < T < T_0$, and $c = 0$ otherwise. What is c_{max} in terms of fundamental constants?

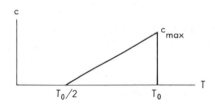

15. Show how a consideration of the appropriate partition functions leads to Fermi-Dirac and Bose-Einstein statistics. Find also the distribution resulting from "parastatistics," in which no more than two particles may be characterized by a given set of quantum numbers.

16. How would Debye's theory of the specific heat of solids be altered if the phonons (sound quanta) obeyed Fermi-Dirac (instead of Bose-Einstein) statistics? Find, under this assumption, the temperature dependence of the specific heat for temperatures very low and very high compared with the Debye temperature. (Constant factors need not be evaluated.)

17. Assume that the inside of a metal represents a square potential well for electrons. Starting from Pauli's principle, derive the velocity distribution of the electrons as a function of temperature. Derive the formula for the current in a vacuum diode with planar electrodes for negative plate voltage.

18. Show that a neutrino star, if sufficiently dense, may be treated as a degenerate gas of relativistic fermions. Derive a condition relating the mass and radius for the equilibrium of such a star.

19. Derive an expression for the magnetic susceptibility of a dilute solution of permanent dipoles, each of magnetic moment M, if the dipole can assume (a) any position, and (b) only two positions, with respect to a weak magnetic field.

20. If a magnetic field H, is applied to a gas of uncharged particles having spin $\frac{1}{2}$ and magnetic moment μ, and obeying Fermi-Dirac statistics, the lining up of the spins produces a magnetic moment/volume. Set up general expressions for the magnetic moment/volume at arbitrary T and H.

Then for low enough temperature, determine the magnetic susceptibility of the gas in the limit of zero magnetic field, correct to terms of order T^2. Note the integral

$$\int_0^\infty \frac{\sqrt{E}\, dE}{\exp\left[(E - \xi)/kT\right] + 1} = \frac{2}{3}\xi^{3/2}\left[1 + \frac{\pi^2}{8}\left(\frac{kT}{\xi}\right)^2 + \cdots\right].$$

21. Consider an oil drop of radius 0.0001 cm, in a gas of viscosity 180 micropoise and temperature 27°C. What is the rms displacement of this drop after 10 sec? Neglect gravitational effects.

22. In a hot plasma, all the atoms may be regarded as completely ionized. Although the ions have long-range forces due to Coulomb interactions, macroscopically the plasma is electrically neutral. This suggests that the Coulomb interactions are screened, and so become short-range. Estimate this range, making suitable approximations.

9

ATOMIC PHYSICS

1. Give a labeled energy-level diagram of the hydrogen atom for principal quantum numbers 1, 2, 3, neglecting relativistic effects and assuming the proton to be a point charge. Give a similar energy-level diagram for those levels of the helium atom which correspond to the excitation of one electron only out of the ground state. Point out and explain the similarities and differences between the two diagrams.

2. Find the three lowest terms of the carbon atom. Neglect spin-orbit coupling. Write wave functions for the three terms.

3. Find the three lowest terms of the nitrogen atom.

4. Give the electronic configurations for zirconium and for hafnium, and explain why the chemical separation of these two elements is difficult. The atomic numbers are 40 and 72, respectively.

5. Assume the lowest-term values in sodium (expressed in cm^{-1}) to be

$3s$	$^2S_{1/2}$	41,448;	$3p$	$^2P_{1/2,3/2}$	24,484;
			$3d$	$^2D_{3/2,5/2}$	12,274;
			$4f$	$^2F_{5/2,7/2}$	6,858;
$4s$	$^2S_{1/2}$	15,705;	$4p$	$^2P_{1/2,3/2}$	11,180;
			$4d$	$^2D_{3/2,5/2}$	6,897;
$5s$	$^2S_{1/2}$	8,246;	$5p$	$^2P_{1/2,3/2}$	6,407;

Indicate what transitions would be observed when excited by
(a) illumination with light at 4123 Å,
(b) bombardment by 3.3 eV electrons, if the sodium is initially in 3S.

6. Obtain an approximate expression for the energy shift of the ground state of the hydrogen atom due to the finite size of the proton, assuming that the proton is a uniformly charged sphere of radius $R = 10^{-13}$ cm.

7. An electron is placed in a potential

$$V = -\frac{e^2}{r} + \alpha(x^2 + y^2) + \beta z^2,$$

where $0 < \alpha < -\beta \ll e^2/a_0^3$. Neglect spin. What are the five lowest orbital states? Sketch their relative positions. Calculate the linear Zeeman effect for B parallel to z and for B parallel to x.

8. In a hydrogen-like atom, the 2S and 2P levels are separated by a small energy difference Δ, due to a small effect which has a negligible influence on the wave functions of these states. The atom is placed in an electric field E. Neglecting the influence of more distant levels, obtain a general expression for the energy shifts of the $n = 2$ levels as a function of the field strength E. [*Note:* Neglect electron spin in this problem. Do not evaluate explicitly any nonzero integrals which may occur in your discussion.]

9. The spectral line of mercury (Hg) at 1849 A splits under the influence of a magnetic field of 1000 G, into three components separated by 0.0016 Å intervals. Determine whether the Zeeman effect is normal or anomalous.

10. The electron and positron have the same (absolute) magnetic moment, but opposite g-factors. Show that the "ground state" of the e^+e^- atom (positronium), a $^1S_0$, $^3S_1$ doublet, cannot have a linear Zeeman effect if this is true. Argue in terms of the total magnetic-moment operator.

11. Solve exactly for the energy eigenvalues when a magnetic field $\mathbf{B}$ is applied to a doublet P level (say of sodium), ignoring hyperfine structure. The Hamiltonian is

$$H = \frac{2\epsilon}{3}\mathbf{S}\cdot\mathbf{L} + \mu_0(\mathbf{L} + 2\mathbf{S})\cdot\mathbf{B},$$

where ϵ is the fine-structure splitting when $\mathbf{B} = 0$.

12. Calculate the hyperfine splitting of energy levels of the hydrogen atom. Assume the electron is in an S state, and give the answer in electron volts.

13. The ground state of the hydrogen atom is split into two hyperfine states separated by an interval $\Delta E = 1.42 \times 10^9$ cps. What is the hyperfine splitting in the deuterium atom? The respective magnetic moments are $\mu_p = 2.8\mu_n$ and $\mu_d = 0.86\mu_n$, where μ_n is the nuclear magneton.

14. A $D_{5/2}$ term in the optical spectrum of $_{19}K^{39}$ has a hyperfine structure with four components.
(a) What is the spin of the nucleus?
(b) What interval ratios in the hyperfine quadruplet are expected?

15. An atom with no permanent magnetic moment is said to be diamagnetic. If one neglects the spin of the electron, what is the induced diamagnetic moment for a hydrogen atom in its ground state when a weak magnetic field B is applied?

16. (a) Find the magnitude of the Doppler broadening for an argon glow-tube of temperature 300°K. Assume a wavelength of 0.5 μ for the radiation. (b) At what pressure should collision broadening and Doppler broadening become of equal magnitude for the above-mentioned argon source? Treat the argon atoms as solid spheres of radius 1 Å.

17. Find the alteration of intensity (ratio of intensities of succeeding lines in an electronic band spectrum) of the following molecules, which are assumed to be in perfect thermal equilibrium:

(a) $(H^1)_2$ (b) $(H^2)_2$ (c) $(He^3)_2$ (d) $(He^4)_2$

Assume kT so large that one may ignore the variation of intensity due to the Boltzmann factor.

18. The potential of a diatomic molecule with nuclei of masses M_1 and M_2, can be approximated by

$$V(r) = -2V_0\left(\frac{1}{\rho} - \frac{1}{2\rho^2}\right),$$

where $\rho = r/a$ (a being a characteristic length). Find the rotational, vibrational and rotational-vibrational energy levels for small oscillations by a power-series development of the effective potential.

19. The far-infrared spectrum of HBr consists of a series of lines spaced 17 cm^{-1} apart. Find the internuclear distance of HBr.

20. The dissociation energy of the H_2 molecule is 4.46 eV, while that of the D_2 molecule is 4.54 eV. Find the zero-point energy of the H_2 molecule.

21. The muon has electric charge equal to that of the electron and a mass roughly two hundred times that of the electron. It is known that a μ-mesic hydrogen atom may exist. This atom is known to combine with a proton to form a "mulecule" H_2^+, consisting of $p^+p^+\mu^-$. Assuming that the muon acts in a fashion analogous to the electron in H_2^+, make a plausible estimate for the internuclear equilibrium distance r, the zero-point energy of the molecule, and the binding energy of the mulecule. The following data are given: internuclear distance in H_2^+ is 1 Å, the zero-point energy is 0.14 eV, and the binding energy is 2.7 eV.

10

SOLID STATE

1. Many metals can occur with both the body-centered cubic and the face-centered cubic structure, and it is observed that the transition from one structure to the other involves only an insignificant volume change. Assuming *no* volume change, find the ratio D_{fc}/D_{bc} where D_{fc} and D_{bc} are the closest distances of approach of metal atoms in the two structures.

2. Tolman obtained a value for e/m for electrons by the mechanical acceleration of a sample of metal. How can this be done? Assume the electrons free.

3. Given the following data concerning lithium chloride:
  (a) lattice energy = 192 kcal/mole = A;
  (b) heat of formation = 97 kcal/mole = B;
  (c) ionization potential of lithium atom = 5.29 V = C;
  (d) heat of sublimation of lithium = 38 kcal/mole = D;
  (e) heat of dissociation of chlorine molecule = 58 kcal/mole = E.
Compute the electron affinity F of the chlorine atom.

4. A sample of germanium shows no Hall effect. If the mobility of electrons in germanium is 3500 cm²/V-sec and that of the holes is 1400 cm²/V-sec, what fraction of the current in the sample is carried by electrons?

5. The room-temperature modification of iron is body-centered cubic (bcc), and this transforms at 910°C into a face-centered cubic (fcc) modification; the heat of transformation L is 253 cal/mole. The addition of a small amount of carbon lowers the transition temperature. Estimate the change in the transition temperature induced by the addition of 0.1% atomic carbon. (The solubility of carbon is very much greater in the fcc than in the bcc modification.)

6. A crystalline body in a state of thermally excited elastic vibrations may be treated as a system of N *distinguishable* independent quantum-harmonic oscillators of the same angular frequency ω (Einstein's model). Give the expression for the distribution law of the system. Compute the average energy of the system at high and low temperatures; find the molar specific heat C in those limiting cases, and discuss the validity of the model in those cases.

7. The lattice specific heat of a certain form of carbon has a temperature dependence T^2, instead of the more common T^3 dependence for solids. What can you infer about the structure of this particular phase of carbon?

8. A pure semiconductor is found to have a conductivity of 0.01 mho/meter at $T_1 = 273°$K. From optical data, the valence band is found to be 0.1 eV below the conduction band. Calculate the conductivity at $T_2 = 500°$K.

9. A parallel beam of 25 eV electrons is incident on a thin polycrystalline sheet of a metal which is known to have a cubic lattice with lattice constant 5 Å. When the diffraction pattern formed by the electron passing through the sheet is photographed, it is found that the smallest ring has an angular diameter of 120°. What is the depth of the potential well for the metal?

10. A crystalline insulator is placed in an electric field **E**. Show that an electron in the crystal will oscillate according to

$$e(\mathbf{x} - \mathbf{x}_0) \cdot \mathbf{E} = \epsilon\left(\mathbf{k}_0 + \frac{e\mathbf{E}t}{h}\right) - \epsilon(\mathbf{k}_0),$$

where ϵ and **k** are the energy and momentum of the electron. Estimate a typical amplitude for the oscillation, and a typical period.

11. Positrons annihilate electrons in condensed media with the cross section $\sigma(v) = \sigma_0 c/v$, where v is the relative velocity between the annihilating particles. Assume that most positrons annihilate from rest, and that the two annihilation photons are emitted isotropically in the center-of-mass frame of the electron and positron.

(a) If only the outermost electrons are annihilated, show that the two annihilation photons are emitted in opposite directions within a cone of half-angle, δ radians, where $\delta \approx 1/137$.

(b) Predict the distribution $W(\delta)$ for a metal, knowing the number of conduction electrons per unit volume. In this case, will there be a temperature effect?

12. Consider the propagation of a sound wave with frequency ω and wave number k, in a metal of N atoms per unit volume in the free-electron model. Derive an expression for the velocity of sound in the metal. Estimate numerically for aluminum, which has atomic number 27 and $Z = 3$ valence electrons. [*Hint:* Consider the lattice as a "heavy" plasma with screening by the electrons.]

11

NUCLEAR PHYSICS

1. Pions and muons each of 140 MeV/c momentum pass through a transparent material. Find the range of the index of refraction of this material over which the μ's alone give Cerenkov light. Assume $m_\pi c^2 = 140$ MeV; $m_\mu c^2 = 106$ MeV.

2. The existence of a massive meson was suggested by Yukawa to explain the nuclear forces. From the uncertainty principle, derive a relation between the range of the force and the mass of the meson. Estimate the mass of the meson.

3. A high-energy beam of antiprotons enters a liquid-hydrogen bubble chamber of length l. Let σ_e be the cross section for elastic scattering and σ be the total cross section; assume both are independent of energy. Derive an expression $P_2(l)$ for the probability that an incident $\bar{p}$ undergoes double elastic scattering, i.e. is elastically scattered twice and leaves the chamber in the distance l.

4. The lifetime of an unstable nucleus is determined by the interval between two events representing its creation and decay. The mean life of such nuclei may be determined in the following way: Each pulse from detection of creation events is put into a coincidence circuit after passing through a known time delay. Each pulse from decay events is put into the same coincidence circuit with *no* delay. The coincidence rate is observed for two different delays, t_1 and t_2. Assume that, in a certain measurement, the decay rate λ is approximately known, and that $1/\lambda$ is much larger than the resolving time of the coincidence circuit. Assume no background or accidental coincidence problems.

How would you find λ from the observed coincidence rates C_1 and C_2 corresponding to delay times t_1 and t_2? If you have a total time T for the experiments, how would you divide your time between observing C_1 and C_2? (Assume you cannot observe the two rates simultaneously.) What delays t_1 and t_2 would you use?

5. Experiments (deuteron "stripping") show that the ground state of O^{17} is formed from that of O^{16} only by acceptance of a neutron of orbital angular momentum $l = 2$. The first excited state is formed by the acceptance of a neutron with angular momentum $l = 0$. What can you conclude about spin and parity of (a) the ground state, (b) the first excited state of O^{17}?

6. Which of the following particles may undergo two-pion decay? Give your reasoning for each of the three cases.

$$f^0 \qquad (J^P, I) = (2^+, 0)$$
$$\omega^0 \qquad (J^P, I) = (1^-, 0)$$
$$\eta^0 \qquad (J^P, I) = (0^-, 0).$$

Here, J and P are the intrinsic spin and parity, I is the isospin. Assume strict isospin and parity conservation.

7. Particle A decays by strong or electromagnetic interactions to particle B and particle C. If A has spin $\frac{1}{2}$, prove that the decay products must come out isotropically even if A is polarized.

8. Knowing that the spin of the π^- is zero and the parity of the deuteron is even, show how the existence of the reaction

$$\pi^- \text{ (at rest)} + d \longrightarrow n + n$$

determines the parity of the π^-.

9. The elementary particles Λ, p, n, π^+, π^0, π^- have the following (internal) quantum numbers ($I =$ isospin, $I_3 =$ third component of isospin, $S =$ strangeness).

Multiplet	Particle	S	I	I_3
Λ	Λ	-1	0	0
N	p	0	$\frac{1}{2}$	$+\frac{1}{2}$
	n			$-\frac{1}{2}$
π	π^+	0	1	$+1$
	π^0			0
	π^-			-1

The weak nonleptonic decays

$$\Lambda \longrightarrow N + \pi \qquad\qquad (1)$$

obey the selection rules

$$|S_{\text{initial state}} - S_{\text{final state}}| = 1, \qquad |I_{\text{initial state}} - I_{\text{final state}}| = \tfrac{1}{2}$$

(the so-called "$\Delta I = \frac{1}{2}$ rule"), and of course charge conservation. Calculate the branching ratio

$$A = \frac{\text{Rate}(\Lambda \rightarrow p\pi^-)}{\text{Rate}(\Lambda \rightarrow n\pi^0)}.$$

The strong interactions conserve I, I_3, S, and the electric charge. In particular, the process

$$\pi + N \longrightarrow K + \Lambda \tag{2}$$

proceeds via strong interactions. The ratio of the cross sections is

$$\frac{(\pi^- + p \rightarrow K^0 + \Lambda)}{(\pi^0 + n \rightarrow K^0 + \Lambda)} = B.$$

Show how this determines the isospin of the K^0.

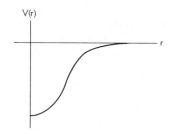

10. The nuclear shell model describes the nucleons in the nucleus as moving in a common nuclear potential as represented in the diagram above, with spin and angular momentum coupled by an interaction $-2a\mathbf{S}\cdot\mathbf{L}$, where a is a positive constant. Use this model to predict the spins and parity of the following nuclei:

(a) $_1H^3$ (b) $_3Li^7$ (c) $_5B^{11}$ (d) $_7N^{15}$.

11. A radioactive element is believed to have the following decay scheme:

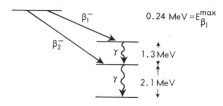

Suggest a series of experiments to measure the energies of the radiations and to establish the correctness of this scheme.

12. The radiations emitted by a certain radioactive species were studied in a β-ray spectrometer. The β-spectrum was resolved into two components of 0.61 MeV and 1.436 MeV maximum energies. The higher-energy component was about four times as abundant as the low-energy one. When the γ-rays accompanying the β-spectrum were allowed to strike a thin silver foil placed in the source position of the spectrometer, the following five photoelectron energies were measured:

Photoelectron	E (MeV)	Intensity
A	0.216	strong
B	0.237	weak
C	0.801	weak
D	0.822	very weak
E	1.042	very weak

The K and L binding energies in silver are 25 and 4 KeV, respectively. Draw a plausible decay scheme for the radioactive species under investigation.

13. Show that the conversion of a high-energy photon into an electron-positron pair can occur only in the presence of matter.

14. Pions are produced in nuclear explosions (stars) and registered in photographic emulsions. It is observed that up to a kinetic energy of approximately five MeV only negative mesons emerge from silver nuclei which are in the emulsion. Why are positive pions not observed below this energy?

15. $_{14}Si^{27}$ decays to its "mirror" nucleus A^{27} by positron emission. The maximum positron energy is 3.48 MeV. Assuming that the nuclear radii may be given in the form $r_0 A^{1/3}$, where A is the mass number, estimate r_0 from the above data.

16. N^{17} decays to an excited state of O^{17} with β^- emission of maximum energy 3.72 MeV. This excited O^{17}-state decays by neutron emission. F^{17} decays by positron emission of maximum energy 1.72 MeV; no radiation occurs following this decay. The mass differences are:

$$n - H^1 = 0.78 \text{ MeV}, \tag{1}$$

$$N^{17} - O^{17} = 8.80 \text{ MeV}, \tag{2}$$

$$O^{16} + H^1 - F^{17} = 0.59 \text{ MeV}. \tag{3}$$

O^{16} has excited levels at 6.05, 6.13, 6.9, 7.1 MeV and higher.

(a) Using only the above data, calculate the energy of the emitted neutrons in the laboratory system.

(b) Show all energies involved in a level diagram for the nuclei with $A = 17$.

(c) Which qualitative features of the level diagram follow from charge independence? Add the further levels which can be predicted from charge independence.

17. The threshold for the reaction $N^{14}(n, 2n)N^{13}$ is 10.6 MeV. Suppose the nitrogen in the air is exposed to the radiation from a radioactive series containing the following alpha emitters:

Element	Half-life	Observed alpha energy
X	10^{10} sec	5.0 MeV
Y	1 sec	not measured
Z	10^3 sec	10.0 MeV

Would you expect any N^{13} to be formed by the reaction N^{14} $(\alpha, \alpha n)N^{13}$?

18. A certain nucleus may be considered to be a sphere of charge Z and radius a, which will undergo a nuclear reaction when hit by either a neutron or proton. Compute the ratio of the reaction cross sections (σ_p/σ_n) according to classical mechanics.

19. A charged particle is slowed down in a photographic emulsion from a velocity of 10^9 cm/sec to thermal velocities. Does the grain density increase or decrease if the particle is
(a) an electron?   (b) a nucleus of charge $Z = 11$?

20. Calculate classically and nonrelativistically the differential scattering cross section at small angles, for fast magnetic monopoles of strength g scattering from fixed nuclei of charge Ze. Estimate the energy loss dE/dx of a monopole traversing a nonmagnetic sample of these nuclei, having N nuclei per unit volume.

21. Derive a formula for the low-energy differential scattering cross section from a potential which is capable of producing one weakly bound state with the scattered particle.

22. A neutron is bound to a force center by an attractive force of range $r_0 = 10^{-13}$ cm. The ground state of the system has a binding energy of 1 KeV. What is the neutron scattering cross section at zero energy for this force center? Assume the potential function is well behaved.

23. A beam of 100-KeV neutrons is attenuated to 50% of its initial intensity in passing through 10 gm/cm² of carbon. What can you say about the s-wave phase shift for the scattering of neutrons from carbon nuclei?

24. It is required to make a quick estimate (in the absence of tables of cross section) of the fraction of 14-MeV neutrons which will be elastically scattered (over all angles) from a sheet of lead 2 cm thick (of density 11 gm/cm³) placed in a 14-MeV neutron beam. Make your best guess, listing your assumptions.

25. A thick target of Mn^{55} is bombarded during a time t with a deuteron beam (current i) in order to produce Mn^{56}, which decays with a half-life $T_{1/2}$. Calculate the number of active nuclei present at the end of the irradiation, assuming a range R of the deuterons and an average cross section over the range of the deuterons.

Numerical example:

$$i = 4.8 \times 10^{-6} \, A \qquad T_{1/2} = 2.6 \, hr \qquad t = 5.2 \, hr$$

$$R = 110 \, mg/cm^2 \qquad \sigma = 10^{-25} \, cm^2.$$

26. Show, by explicit calculation of the scattering cross section of ortho- and para-hydrogen for thermal neutrons, how the relative sign of the triplet and singlet scattering length may be determined.

27. Suppose that the differential cross section $d\sigma/d\Omega$ in the center-of-mass frame for the reaction $p + p \rightarrow \pi^+ + D$ is $A + B \cos^2 \theta$ at energy E. What is the $d\sigma/d\Omega$ for the inverse reaction $\pi^+ + D \rightarrow p + p$ at the same center-of-mass energy? The experiment is done with unpolarized beams.

28. Neutrons can be scattered in the Coulomb field of a nucleus because of their magnetic moment. Write down the Hamiltonian of the interaction, and calculate the spin-averaged differential cross section in the Born approximation. (Treat nonrelativistically.)

29. Assume that an iron atom has an average magnetic-dipole moment per unit volume given by $\boldsymbol{\mu}_I g(r)$, where $\boldsymbol{\mu}_I$ is the magnetic moment of the iron atom. Thermal neutrons of momentum $\mathbf{k}_0$, polarized along the direction of $\boldsymbol{\mu}_I$, with $\mathbf{k}_0$ perpendicular to $\boldsymbol{\mu}_I$, are scattered in iron. Note that the scattering has two sources; that due to the nuclear force between neutrons and iron nuclei, and that due to magnetic forces between neutrons and iron atoms. Compare these contributions to the scattering: Also calculate the total spin and nonspin flip cross sections. Treat the scattering as though due to a single iron atom. [The vector potential $\mathbf{A(r)}$ due to a magnetic dipole $\boldsymbol{\mu}$ is given by $\mathbf{A} = (\boldsymbol{\mu} \times \mathbf{r})/r^3.$]

30. The dominant decay mode of the neutral sigma hyperon (spin $\frac{1}{2}$) is radiative, i.e. $\Sigma^0 \rightarrow \Lambda^0 + $ photon. Although electric-dipole radiation is forbidden because Σ^0 has no charge, the decay may proceed via magnetic-dipole radiation, through an effective interaction

$$H = \frac{ge\hbar}{(M_\Sigma + M_\Lambda)c} \tau_{\Sigma\Lambda} \boldsymbol{\sigma} \cdot \nabla \times \mathbf{A}.$$

Here $\tau_{\Sigma\Lambda}$ denotes the operator that converts Σ into Λ, and $ge\hbar \, \boldsymbol{\sigma}/(M_\Sigma + M_\Lambda)c$ may be interpreted as a transition magnetic moment, which interacts with the magnetic field $\mathbf{B} = \nabla \times \mathbf{A}$ of the radiation field. Taking the amplitude for emitting a photon to be the plane wave*

$$\mathbf{A} = c\left(\frac{\hbar}{2\omega V}\right)^{1/2} \boldsymbol{\epsilon} \exp(i \cdot \mathbf{k} \cdot \mathbf{x} - i\omega t),$$

estimate the Σ° lifetime, for $g = 1$. The masses of the particles are

$$M(\Sigma) = 1190 \text{ MeV} \quad \text{and} \quad M(\Lambda) = 1115 \text{ MeV}.$$

31. An element of low atomic number Z decays with positron and neutrino emission; the maximum kinetic energy of the positron is $W = 50$ KeV, and the probability per second of the decay process is Γ_1. Using Fermi's β-decay (for an allowed transition), calculate the probability Γ_2 for K-capture by the same nucleus.

32. A system which may be regarded as having infinite mass emits spontaneously two nonidentical particles with relativistic energies. Assuming the partition of energy among the two particles to be governed entirely by phase-space considerations (energy-independent matrix element), obtain an expression for the energy spectrum of the particles in terms of E, the energy liberated in the process, and m_1 and m_2, the rest masses of the emitted particles. Discuss the relation of the result to the process of β-decay.

33. (a) Assume that the muon decays into an electron and two distinct neutrinos with a transition rate

$$\Gamma = \frac{2\pi}{\hbar}(g/V)^2 \frac{V p^2 \, dp}{2\pi^2 \hbar^3} \frac{dn}{dw},$$

where g is a coupling constant, V a normalization volume, w the energy of the three emitted particles, and $\mathbf{p}$ the momentum of the electron. By calculating the number of neutrino states per unit energy interval, dn/dw, obtain an expression for the momentum spectrum of the electron. (Assume electron energy $\approx pc$.)

b) Calculate g^2 from the mean life of the muon,

$$\tau_\mu = 2.2 \times 10^{-6} \text{ sec} \quad (m_\mu = 207 \, m_e).$$

34. Assume the existence of a particle of electronic charge e, spin $\frac{1}{2}$, and rest mass M, which disintegrates spontaneously into an electron (mass m) and a photon. The mean lifetime of the particle is T (in its rest system). Conversely, these particles can be created by irradiated electrons with light of appropriate frequency. What is the frequency ω_0 required if the electrons are (practically) at rest? And what is the probability of the process per

---

* In these units, $e^2/\hbar c = 4\pi/137$. In units where $e^2/\hbar c = 1/137$, we have $\mathbf{A} = c(2\pi\hbar/\omega V)^{1/2}\boldsymbol{\epsilon} \exp(i \, \mathbf{k} \cdot \mathbf{x} - i\omega t)$.

second (per electron) if the incident photons have an energy $U(\omega) \, d\omega$ per unit volume and per frequency interval $d\omega$? What is the answer to the second question if the heavy particle has spin $\frac{3}{2}$ instead of $\frac{1}{2}$?

35. In a chain-reacting device consisting of a homogeneous mixture of uranium and carbon, neutrons emitted by the uranium move about with an average velocity v and mean free path $\lambda = 10$ cm. They will, on the average, be captured after $N = 100$ collisions. For each neutron captured, $k = 1.04$ new neutrons are emitted on the average. Give the differential equation for the time rate of increase in the neutron density. If the device is built in the form of a cube, find the critical size.

36. Neutrons are slowed down in hydrogen.
(a) After n collisions what is their average energy and the distribution function $\rho_n(E)$ for the number of neutrons per unit energy? Assume that the scattering of a neutron by a proton is spherically symmetric in the center-of-mass frame of reference.
(b) If neutrons of energy E_0 are produced at the rate of $q/\text{cm}^3 \ \text{sec}^{-1}$ and the total scattering and absorption cross sections are known functions of the energy, find the steady-state neutron flux (product of neutron density and velocity) as a function of energy and the known cross sections.

37. Slow positive and negative μ-mesons (μ^+, μ^-) have, in a vacuum, a mean life τ_0 for decaying into an electron and two neutrinos. Negative μ's can also be captured into atomic orbits and very rapidly fall into the K-shell, where they are close enough to the nucleus to be absorbed by it.
(a) Making the plausible assumption that the probability of this nuclear absorption is proportional to the fraction of time the μ-meson spends inside the nucleus, how does the probability depend on the atomic number of the material in which the meson is stopped?
(b) If the lifetime in the K-shell of hydrogen is τ_H, what are the mean lives τ_+, τ_- of positive and negative muons in zinc ($Z = 30$)?

$$\tau_0 = 2.10 \times 10^{-6} \text{ sec} \qquad \tau_H = 2.075 \times 10^{-6} \text{ sec}$$

Note: Neglect time required for transition from outer shells to K shell.

38. A cyclotron produces a beam of deuterons of 200 MeV energy. When incident on a Be target, a narrow neutron beam is produced by the process of stripping (i.e. the proton is removed by a nuclear collision, allowing the neutron to continue its original direction of motion). Calculate approximately the distribution in angular spread of this neutron beam due to the internal motion in the deuteron. Use an approximate form for the deuteron wave function (limit of zero range) and take the binding energy of the deuteron to be 2.18 MeV.

39. Study the vibrational excitations of a nucleus of atomic weight A and charge Z, in the liquid-drop model. Include the effects of Coulomb repulsion and surface tension. Develop a criterion for instability against nuclear fission. To find the surface tension, use the fact that it contributes a term $M_s = U_0 A^{2/3}$ to the semi-empirical mass formula with $U_0 = 14$ MeV. [*Hint:* Use the methods of Problem (2–35).]

FOR THE EXPERIMENTALIST

We include here a set of problems concerning experimental physics. Most of these problems have no unique answer, with solutions of varying degrees of sophistication possible. Furthermore, because of technical developments, the appropriate response to some questions may depend strongly on time. We have, therefore, not attempted solutions to these problems.

1. Design an efficient furnace and accurate temperature-measuring equipment for studying the electrical properties of solid materials in the temperature range 1800–2300°C. The volume of the furnace is to be 10 cm³. Include methods for end-effect corrections for electrical connections to the sample in question.

2. Design a spectrophotometer for use in the range 1500–2000 Å. The design is to include a source or sources of light, dispersive agent, and method or methods of detection.

3. Devise a set of experiments that will measure the mechanical properties of so-called "Bouncing Putty" or "Krazy Klay", a silicon compound which bounces when it strikes a hard surface suddenly but flows under its own weight.

4. Design a strong focusing synchrotron to produce 50-BeV protons, paying special attention to size of magnet, gradients, ratio of straight sectors to curved sectors, energy storage in the magnetic field, radio-frequency system, and orbital stability.

5. Design a radio-frequency magnetic resonance apparatus for atomic beams to measure the nuclear spin and magnetic moment of any one of the following isotopes: Rb^{86}, Na^{24}, or Cs^{134}.

6. Design a magnet to produce a field whose homogeneity and stability are $\pm.01\%$ over a range of 1000–20,000 G. The region where B is to be constant is 5 in. in diameter and 1 in. high. Include measuring equipment to determine B to $\pm.05\%$.

7. Given the problem of measuring minute normal displacements of a surface, describe three of the most sensitive methods you can devise, and give an estimate of the minimum displacement which each can detect.

8. Discuss the relative merits of a quartz prism and a vacuum reflection-grating spectrograph for the following applications:
(a) The observation of the fine structure due to the Lamb-Retherford effect.
(b) To study the UV bands of CO between 1300 and 1550 Å.
(c) To study the Doppler effect in the aurora due to particles coming from the sun.
(d) Measuring spectral intensities by the rotating-sector disk method.

9. Suppose that you suspected the existence of gamma-rays of about 0.5 MeV energy in the radiation from the sun. Assume a flux of about 5 per cm² per sec at the earth's orbit. Assume that the absorption coefficient of air for gamma-rays is 0.1 cm²/gm. Devise an experiment to prove that such radiation exists, that it is not corpuscular, and that it really comes from the sun. Make the experiment a practical one. What would be the sources of background for your experiment? Estimate the sensitivity of your apparatus.

10. Describe any set of experiments from which the fundamental constants e, m, c, h can be determined, and estimate the accuracy obtainable with these experiments.

11. What techniques would you use for measuring
(a) Magnetic fields in the ranges: (b) Temperatures in the ranges:
   (1) 10^{-3} to 10^{-1} G,   (1) 10^{-2} to 1° K,
   (2) 1 G to 10 kG,   (2) 1° to 4° K,
   (3) 100 to 250 kG (1 kG $= 10^3$ G).   (3) 1000° to 5000° K.
In case you can think of several methods suitable for a given range, state the experimental circumstances on which you would base your choice of any one method.

12. Discuss a method for determining Avogadro's number with better than 5% accuracy.

13. The "classical" experiment for the determination of the neutron half-life is to place an evacuated container in a pile and measure the rate of accumulation of hydrogen. List three factors that make this a difficult experiment to perform.

14. Discuss nuclear reactions which are observed when lithium is bombarded by protons and when it is bombarded by neutrons. How are these reactions best detected?

15. Discuss methods suitable for measuring high, medium, and low resistance. In what range of resistances is each method the best? What is the approximate accuracy of each method?

16. Discuss the theory of operation of the high-energy machines which are now being built throughout the United States. What do you consider the main difficulties in constructing each of these machines? To what kind of experiment is each best adapted? Suggest brief programs of research for some of these machines.

Elaborate in detail one such program. What in your opinion will be the probable effect of these machines on our knowledge in physics? (This problem was given in 1948.)

17. How do you measure the wavelength of electromagnetic radiation in all different frequency ranges? Describe briefly the type of equipment required in each region.

18. Design an audio oscillator and power supply to operate at a nominal frequency of 50 kc which does not contain any inductances. The oscillator is to provide 3.0 W of power to a 200-ohm load.

19. The nuclear magnetic resonance method for measuring nuclear magnetic moments makes it desirable to measure the magnetic field between the poles of a magnet to as high a degree of absolute accuracy as possible. Propose a means for carrying out such a determination for a field of the order of 10,000 G and discuss the limits of accuracy. Take the diameter of the poles as 6 in. and the gap as 1 in.

20. For an experiment on the propagation of light in a moving medium, it is desired to have water moving as fast as possible in a pipe of at least 1 in. inside diameter and of length 5 ft. The pipe is to be provided with end windows for observation. Design the pipe and the subsidiary equipment and estimate the velocity that may be achieved.

21. Design an experiment for measuring the lifetime of an artificial beta-active substance produced by slow neutron bombardment, assuming a lifetime of the order of $\frac{1}{10}$ sec, and allowing for the possibility that fairly intensive longer lives are also excited. Estimate the accuracy.

22. The bathysphere is a hull which can be lowered to great depths below the surface of the sea and can safely accommodate an observer for the purpose of exploration of the ocean depths. Design a bathysphere for a depth of 1000 m, giving particular attention to mechanical details such as the observation windows, illumination facilities, devices for renewing the air, etc. Give as complete a design as possible.

23. Design a seismograph for detecting three-dimensional seismic motion of maximum amplitude 1 mm and maximum period 30 sec.

24. Starting with a 3-phase, 440 (± 5 percent)-volt line, specify a power supply and design a servomechanism for controlling the magnetic field of a 40-in. cyclotron. Assume an ambient temperature of $30°C \pm 10°C$, a maximum magnetic-field intensity of 18,000 G and a minimum field intensity of 15,000 G. Specify soft iron for the magnet. The controller should keep the field distribution constant as a function of time to within 0.1 percent.

25. In a synchrotron electrons are accelerated to an energy of 3×10^8 eV. Devise an apparatus for measuring the shape of the γ-ray spectrum originating from an internal thin target of tungsten struck by the electron beam.

26. Construct a spectrometer for measuring the diffraction, refraction and reflection of x-rays of a wavelength of 40 Å.

27. Heavy nuclei have been observed at the top of the atmosphere by means of nuclear photo-emulsions. It is important that the properties of these particles be determined experimentally using methods other than photographic emulsions. It is further important to design the experiments in such a way as to indicate the origin of these particles. Develop an experimental method which can successfully attack this problem and devise the methods of analysis to be used to interpret the data. In particular, describe in detail all aspects of the experiment and estimate intensities expected.

28. Design a practical experiment to detect the neutral mesons which are supposed to emerge from a beryllium target which is bombarded with 1 μA of 450 MeV protons. Calculate the minimum cross section for the production of the neutral mesons which your method would be able to detect. Show how you would distinguish these mesons from neutrons and gamma-rays which would be present.

29. What is the deepest hole which can be dug into the earth? Explain the limitations and design roughly the practical equipment necessary to dig and maintain the hole.

30. A high-impedance pulse generator produces periodic pulses of the shape shown in the diagram on p. 63. The pulse rate is 10^4/sec and the pulse height is 0.1 V. Work out the basic design of an oscilloscope to observe these pulses so that the properties of the pulse described may be measured to ± 5 percent. The deflection sensitivity of the cathode ray tube is 20 V/in. Basic circuit diagrams and schematic drawings should be used to designate electronic components. The properties of each circuit shown should be calculated or carefully estimated.

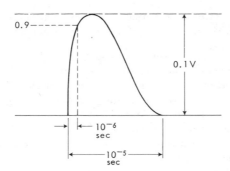

31. The neutron resonances in the range 1–100 eV for various nuclei are to be measured. A nuclear reactor is available as a source of neutrons. The neutrons may be brought through the reactor shielding by means of a cylindrical pipe measuring a cm in diameter, so that neutrons ranging from thermal to over 10^4 eV, are present at the end of the pipe. The resonances are to be determined by reflecting a collimated beam of neutrons from a crystal spectrometer. The reflected neutrons are to be detected using a proportional counter filled with BF_3 gas.

(a) Draw a plan view of the arrangement of the experimental apparatus showing all components, giving approximate dimensions for the apparatus, sample to be measured, etc.

(b) Develop the theory of the crystal spectrometer, discussing resolution, limitations on beam solid angles, beam flux requirements, effective penetration of neutrons into the crystal, and the true measurement of resonance width.

32. Design apparatus for measurement of the Lamb-Retherford shift in the 4868-Å, line of HeII (He^+) to an accuracy of 10 percent. The Lamb-Retherford shift for the $n = 2$ state of hydrogen is approximately 1000 mc.

33. Design apparatus for determining the velocity of light to an accuracy of 1 part in 10^6 by measuring the velocity of propagation of microwave radiation across a cavity. Show how microwave absorption of ammonia can be used for frequency standardization.

(a) Give the overall design and show what measurements are to be made.

(b) Give details and precautions necessary to obtain the required accuracy.

34. A few rough measurements have been made of the decay of the neutron giving a half-life of 9–18 min. What difficulties may arise in the Chalk River experiment, whose description follows? Show how a coincidence measurement would improve the experimental evidence for the decay process. Show how this experiment is related to neutrino emission.

THE RADIOACTIVE DECAY OF THE NEUTRON

J. H. Robson, Chalk River Laboratory.

The positive particle from the radioactive decay of the neutron has been identified as a proton from a measurement of charge-to-mass. A collimated beam of neutrons emerging from the Chalk River pile passes between two electrodes in an evacuated tank. One electrode is held at a positive potential, up to 20 keV, while the other electrode is grounded and forms the entrance aperture to a thin-lens magnetic spectrometer, the axis of which is perpendicular to the beam of neutrons. The positive decay particles can be focused on the first electrode of an electron multiplier. The background counting rate is 60 cpm. A peak of 80 cpm is observed above background when the magnetic field is adjusted for protons of energy expected from the electrostatic field. When a thin boron shutter is placed in the neutron beam, the proton peak disappears. Preliminary estimates of the collecting and focusing efficiency and the neutron flux indicate a minimum half-life of 9 min and a maximum of 18 min for the neutron.

35. Individual particles of single charge and unknown mass with energies of 1×10^9 eV are present in the cosmic radiation. Design apparatus for determining their mass—with a precision of 5 percent—by deflecting them in an electric and magnetic field. Discuss sources of error. The mass is assumed to be around 200 electron masses.

36. A penetrating shower of 10^{10} eV total energy consists of 5 charged mesons, 5 neutral mesons and 2 protons. Describe in detail apparatus by which (a) each individual particle can be identified, and (b) its energy can be determined.

37. A new photoconductive material, PbTe has a long-wavelength threshold for light absorption near 6 μ. Describe the expected photo-response signal and the noise as a function of wavelength and temperature. Compare the highest expected signal-to-noise ratio at room temperature with that of a thermocouple, at dry ice temperature; and at liquid air temperature. What experiments would be necessary to determine whether the photo-conductivity is intrinsic or due to an impurity; whether it is a bulk effect, surface effect, or contact effect; to determine the sign of the charge carriers, number/cm³, mean free path, and other important parameters? What would be the activation energy and other properties of this material used as a semiconductor?

38. A beam of 2-MeV neutrons of flux N cm^{-2} sec^{-1} is incident normally on a 100 μg cm^{-2} film of glycerol tristearate ($C_{57}H_{110}O_6$) (density 0.862 g/cm³) producing hydrogen recoils which enter a gas. Develop the theory for and design a detector and recording system utilizing the ionization from the recoils to measure the incident neutron flux N and the maximum energy E_{max}. Call the bias energy of the detector system E_b and the incident neutron energy E_n.

Describe the properties of the detector selected for the measurements. Calculate the expected response of the detector for recoil protons. Draw a cross-section view of the detector, indicating all components with suitable specifications such as dimensions, materials, etc.

Draw block diagrams showing all functions to be performed by electronic circuits. Tabulate the estimated specifications for each of these circuits.

What is the efficiency E of the film? What is the energy of a proton recoiling at an angle of zero degrees with respect to the incident neutron in the detector? What is the pulse size distribution in the detector?

What is the observed counting yield of the detector? Show how any spurious background proton recoils may be determined in the above measurements.

39. Consider the reaction $Au^{197} + n^0 \longrightarrow Au^{198} \xrightarrow{\beta} Hg^{198*} \xrightarrow{\gamma} Hg^{198}$. The half-life for the decay of the excited state of mercury is 10^{-8} sec. Outline experimental equipment with which it would be possible to measure this half-life, giving limitations and sources of error in the measurement.

40. Given the reaction $I^{127} + n^0 \longrightarrow I^{128} \longrightarrow$? The latter decay, for which the daughter nuclei are unknown, occurs with a half-life 25 min. Outline experimental procedures by which you could obtain the complete decay scheme of the I^{128} produced, including limitations and errors of the method you propose.

41. Discuss critically at least three pieces of evidence for a finite age of the visible universe.

42. Discuss the possibility of producing a magnetic field of 10^6 G in the laboratory. Explain the limitations, indicate the major design considerations, and describe the main features and components of apparatus to reach this field or to reach as near to this field as you think feasible.

43. Design a cryostat for making measurements of specific heat at liquid helium temperatures.

44. Design apparatus for the measurement of viscosity as a function of temperature and pressure in nitrogen gas at about 5000 atm pressure.

45. Oxygen is paramagnetic, a property which is unique among the common gases in the atmosphere. Design an apparatus using this property for analyzing the oxygen consumption in human respiration.

46. The velocity of light is known, at present, only to within a few parts in one million. Devise an experiment to measure the velocity of light to one part in 10^7. Discuss the equipment needed and the necessary corrections to be made. Try to make the experiment feasible.

47. Suppose that the neutron has a permanent electric-dipole moment. Design an experiment to detect such a moment if it is larger than that produced by two elementary charges ($+$ and $-$) separated by 10^{-14} cm. Why would the result be of interest?

48. Dyson, in an interview after returning from the U.S.S.R. (June, 1956) said, "Since World War II there have been six fundamental experiments done in physics; one in Italy, one in England, and four in the United States." If you were to select six such experiments with the boundary conditions as specified, which six would you select? Tell the significance of each experiment.

49. What is the experimental evidence for the assertion that the atomic number of a chemical element is equal to its nuclear charge?

50. State, and discuss in a few sentences, one good method for detecting beams of each of the following entities:

(a) Electrons of 100 eV energy

(b) Photons of 10^{-7} eV energy

(c) Photons of 10^{-3} eV energy

(d) Photons of 10^{+9} eV energy

(e) Metastable atoms

(f) Thermal neutrons

(g) Antineutrinos

(h) Positrons of 10^6 eV energy

(i) Free radicals

(j) Alkali atoms

(k) Phonons

(l) Neutrons of 10^7 eV energy

(m) Λ particles

(n) Spin waves

51. Consider the partial circuit diagram to the right: Estimate the total rms noise generated in the grid circuit of this high-impedance DC amplifier at 300°K. Assume usual values for the characteristics of a suitable tube. What simplifying assumptions have you made? What precautions are necessary in the use of such a circuit? List other possible sources of noise in the first stage.

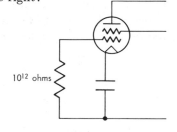

10^{12} ohms

52. Design a coincidence–anticoincidence circuit with the following properties:

(a) There is *one* output channel.

(b) There are four input channels I, II, III, IV.

(c) An output signal results when there is input to either channel I or channel II, or both; but no output pulse results if there is input to either channel III or channel IV, or both.

53. The rotor of a Beams-type ultracentrifuge is spinning at 150,000 rpm. The speed is expected to be constant to within 0.1% over any 5-min interval. Describe how you would design, construct, and use an apparatus to check the speed regulation to this precision.

54. A television commercial shows an automobile being held up by an electromagnet powered by a single flashlight cell. After making some estimates of what is needed to perform this stunt, comment on whether this is a stringent test of such a cell. Note that the circumstances allow the presence of a supplementary piece of soft iron concealed in the roof of the car.

55. What type of detector would you use for each of the following purposes?
(a) Detection of protons and β-rays with equal efficiency and pulse height.
(b) Efficient detection of 20 MeV β-rays with zero efficiency for 50-MeV protons.
(c) Efficient detection of 50-MeV protons with zero efficiency for 20-MeV β-rays.
(d) Efficient detection of 5-MeV γ-rays with zero efficiency for detecting 5-MeV α-particles.
(e) Efficient detection of fission fragments with zero efficiency for detecting α-, β-, or γ-rays.

56. Design a permanent magnet for use in cosmic-ray research. The magnet is to produce a field of 10,000 G over a 1 in. gap. The pole pieces are to be circular with a diameter of 10 in. The basic form of the magnet is to be as shown in the following diagram:

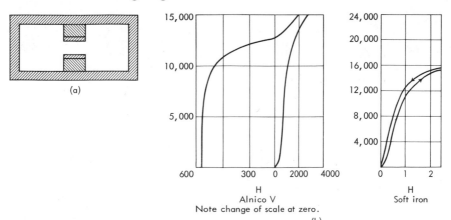

The particular shape and dimension of each piece, with the exception of the pole face itself, are completely variable. The hysteresis loops for Alnico and soft iron are given in the figure.

The design should include:

(a) Length and cross section of Alnico piece,

(b) Cross section of soft iron yoke,

(c) Apparatus and procedure for magnetization.

Discuss the feasibility of constructing each item. *Note:* Cost of Alnico = $4.00/lb; cost of soft iron = $0.05/lb; cost of copper = $0.50/lb.

SOLUTIONS

1

MATHEMATICAL PHYSICS

1. This particular method of birth control does not affect the ratio of the sexes. At each birth, the probability of having a male child is 51%, and each birth is an independent event; the ratio for the community is hence 51:49.

2. (a) Choose a specific color, say color A, and use it for the upper surface of the die. Choose any color B for the lower surface. There are five choices for B. The cube may now be rotated so that some color C is facing us. Now the remaining sides are fixed with respect to these three. We can distribute the three remaining colors over these remaining three sides in $(3!) = 6$ ways. Therefore there are a total of 30 possibilities.

We can also obtain the result by considering the rotation symmetries of the die. If there were no rotation degeneracies, we would have $(6!)$ possibilities. However, there are six sides that can be chosen facing up, and once such a choice is made, any one of four sides may be taken as facing forward. This exhausts all degeneracies and gives $6!/(6 \times 4) = 30$ ways. In general, for a figure with N faces and painted with N different colors the number of possibilities is $N!/R$, where R is the number of discrete rotations which leave the figure invariant.

(b) Here one must take account of the fact that the individual dice are indistinguishable. Thus if each die may be colored in N different ways [from part (a), $N = 30$] the number of *distinguishable* pairs of dice is

$$N(N + 1)/2 = 465.$$

3. We treat the regular octahedron as we did the die in the previous problem. The symmetry viewpoint is particularly useful. If we take one of the six corners pointing up, we get 6 possibilities; there is still a four-fold rotational degeneracy, giving 24 possibilities. Alternatively, if one of the eight sides is fixed in space, the octahedron will still entertain a three-fold rotational degeneracy. Either way, there are 24 degenerate positions. We obtain therefore, $8!/24 = 7!/3$ possibilities.

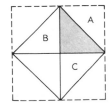

Another way of counting requires more care. We draw the octahedron flattened; the dashed lines join when the figure is folded into 3D. Color any surface with some color (shaded in the figure), and face that side up. Now color all surfaces but A, B, C. Because all other surfaces are uniquely specified with reference to these, the number of possibilities is $7 \cdot 6 \cdot 5 \cdot 4$. There are now only 2 independent choices for A, B, C. For if a given color is chosen for any of these, a rotation will bring these surfaces into one another. Thus, the first of the three to be filled gives no extra possibilities, while the remaining two leave 2 possibilities. So the answer here is

$$7 \cdot 6 \cdot 5 \cdot 4 \cdot 2 = 7!/3.$$

4. The total number of distinct ways the 52 cards may be dealt is $(52!)$.

The total number of ways a particular team may be dealt two *particular* complete suits is $a = (26!) \times (26!)$. The number of ways they may be dealt one particular suit regardless of whether they have another complete suit is $b = (39!) \times (26!)/(13!)$.

Thus the number of ways they may be dealt only one particular complete suit is $(b - 3a)$. The term $3a$ is the number of ways the other three suits may be obtained in addition and must be subtracted out. If we are not interested in which suit they may have, the probability of being dealt one and only one complete suit is $P_1 = 4(b - 3a)/52!$.

The probability of obtaining any two complete suits is $P_2 = 6a/52!$. (The factor 6 arises from the 6 different pairs of suits possible.)

The probability of obtaining at least one complete suit and possibly two is $P_1 + P_2 = (4b - 6a)/52!$. Note that it is a very good approximation to neglect the probability, a, of obtaining two suits in the latter probability.

5. The probability that in n specified intervals of length h an event has taken place, while in the remainder an event has not, is $(\lambda h)^n (1 - \lambda h)^{N-n}$, where $N = t/h$ is the total number of intervals. The number of ways we can specify n intervals out of N is $N!/n!(N - n)!$ and therefore the probability of n events is

$$P_n = \frac{N!}{n!(N - n)!}(\lambda h)^n (1 - \lambda h)^{N-n}.$$

Upon letting N go to infinity, and using Stirling's formula for large N, $N! \sim \sqrt{2\pi N}\, N^N e^{-N}$, together with the limit $(1 - \lambda t/N)^N \longrightarrow e^{-\lambda t}$, we obtain

$$P_n = \frac{(\lambda t)^n}{n!} e^{-\lambda t}.$$

Note that $\sum_{n=0}^{\infty} P_n = 1$. Also

$$\langle n \rangle = e^{-\lambda t} \sum_{n=0}^{\infty} \frac{n(\lambda t)^n}{n!} = e^{-\lambda t}(\lambda t)e^{\lambda t} = \lambda t$$

and

$$\langle n^2 \rangle = e^{-\lambda t} \sum_{n=0}^{\infty} \frac{n^2 (\lambda t)^n}{n!} = e^{-\lambda t} (\lambda t) \frac{\partial}{\partial (\lambda t)} [\lambda t e^{+\lambda t}] = (\lambda t)^2 + \lambda t.$$

This distribution function P_n is referred to as a Poisson distribution. For $\lambda t \gg 1$ this distribution becomes Gaussian in x, where

$$n = \lambda t + x \qquad \text{and} \qquad |x| \ll \lambda t,$$

for which

$$dP = \frac{dx}{\sqrt{2\pi \lambda t}} \exp \left[\frac{-x^2}{2\lambda t} \right].$$

6. (a) Two stars whose angular separation is $\theta = 1'$ lie within a solid angle $\omega \approx \pi \theta^2$. Thus the probability of the stars in a particular pair being within $1'$ of each other is $p = \omega/4\pi \approx 2.1 \times 10^{-8}$. If a double star is defined as a pair which is isolated, i.e. is not part of a triplet, quartet, etc., then the probability that some particular two stars form a doublet is

$$p(1 - p)^{N-2} \qquad \text{where } N = 6500.$$

The factor $(1 - p)^{N-2}$ represents the probability of the remaining stars not combining with the pair to form a triplet, quartet, etc. If, however, one includes as double stars those which are also part of a triplet, quartet, etc., the probability is simply p. In this problem $Np \ll 1$, and thus both definitions yield a probability p.

The number of independent pairs one can form from the N stars is $N(N - 1)/2$. Thus the number of double stars expected is

$$\frac{N(N - 1)}{2} p \approx \frac{N^2}{2} p \equiv \lambda = 0.45.$$

(b) The probability that some two pairs of stars, e.g., $(\alpha\beta)$ and $(\gamma\delta)$ where $\alpha, \beta, \gamma, \delta, \ldots$ label the stars, form two distinct double stars, while the remaining stars do not form multiple stars, is

$$P_{22} = p^2 (1 - p)(1 - 2p) \cdots [1 - (N - 3)p].$$

This is because, when adding the nth star to the celestial sphere where $n > 4$, the fraction of the sphere which it may occupy without combining with $\alpha, \beta, \gamma, \delta$, or the other stars, is $1 - (n - 3)p$. Note that, if $(N - 3)p \geq 1$, the above probability is zero. This is because in this case the stars are so dense that it is impossible to keep them separated so as to form only two double stars. In the limit $Np \ll 1$, one may obtain an interesting expression for

$$X \equiv (1 - p)(1 - 2p) \cdots [1 - (N - 3)p].$$

Form $\log X = \sum_{j=1}^{N-3} \log (1 - jp)$. Then, since $jp \ll 1$ for all $j < N$, one may expand $\log (1 - jp) \approx -jp$. Hence

$$\log X \approx -p \sum_{j=1}^{N-3} j = -p\frac{(N-3)(N-2)}{2} \approx -\frac{pN^2}{2} = -\lambda,$$

so that one obtains $X = e^{-\lambda}$ when terms of the order $N^3 p^2$ are negligible (for this problem $N^3 p^2 \approx 1.2 \times 10^{-4}$).

The number of independent ways two pairs may be formed from the N stars is

$$D = \left(\frac{1}{2}\right)\frac{N!}{2!\,2!\,(N-4)!} \approx \frac{1}{2}\left(\frac{N^2}{2}\right)^2.$$

Thus the probability of seeing any two double stars and no more is the Poisson distribution

$$DP_{22} = \frac{1}{2}\left(\frac{N^2 p}{2}\right)^2 X = \frac{1}{2}\lambda^2 e^{-\lambda} = 0.063.$$

(c) Consider three definite stars labeled α, β, γ. The probability that stars β, γ lie within $1'$ of arc of α is p^2. If a triplet is defined only when it is isolated from the other stars, i.e. is not part of some larger multiplet, then the probability of these three forming a triplet is $p^2(1-p)^{N-3}$.

If, however, one does not care whether the triplet happens to be part of a larger multiplet, then the probability is simply p^2. Since in this problem $Np \ll 1$, both these probabilities are p^2 for all practical purposes.

The number of independent triplets one may form from the N stars is

$$\frac{N!}{3!\,(N-3)!} \approx \frac{N^3}{3!}.$$

Thus the probability of observing some triplet is $N^3 p^2/3! = 2 \times 10^{-5}$.

7. (a) $M = M^+$ and $M^+ = M^{-1}$, i.e. M is both Hermitian and unitary. Hence eigenvalues μ_k are ± 1 only. Further, $\mathrm{Tr}\, M = \sum M_{ii} = 0$, so that $\sum \mu_k = 0$. Hence the four possible μ's are $+1, +1, -1, -1$. Write, for any eigenvector $\mathbf{x}^{(k)}$,

$$\mathbf{x}^{(\mu)} = \begin{pmatrix} x_1 \\ x_2 \\ x_3 \\ x_4 \end{pmatrix}, \qquad \text{for } \mu = +1,\ x_1 = x_4, \text{ and } x_3 = x_2;$$

hence

$$\mathbf{x}^{(1)} = \frac{1}{\sqrt{2}}\begin{pmatrix} 1 \\ 0 \\ 0 \\ 1 \end{pmatrix} \quad \text{and} \quad \mathbf{x}^{(2)} = \frac{1}{\sqrt{2}}\begin{pmatrix} 0 \\ 1 \\ 1 \\ 0 \end{pmatrix}$$

are two convenient choices. Similarly, for $\mu = -1$, the corresponding choices are

$$\mathbf{x}^{(3)} = \frac{1}{\sqrt{2}} \begin{pmatrix} 0 \\ 1 \\ -1 \\ 0 \end{pmatrix} \quad \text{and} \quad \frac{1}{\sqrt{2}} \begin{pmatrix} 1 \\ 0 \\ 0 \\ -1 \end{pmatrix}.$$

(b) The pedestrian approach is to compute

$$0 = \det|M - \mu I| = \mu^4 - 2\mu^2 + 1 = (\mu^2 - 1)^2 = 0.$$

8. As H and H^2 are diagonalized simultaneously, (b) is the sum of the eigenvalues of H^2. The traces of H and H^2 can be computed without diagonalization, as traces are invariants. Thus

(a) $\sum_{i=1}^{3} \lambda_i = \operatorname{Tr} H(\text{diag}) = \sum_{i=1}^{3} H_{ii} = 6,$

and

(b) $\sum_{i=1}^{3} \lambda_i^2 = \operatorname{Tr} H^2(\text{diag}) = \sum_{i=1}^{3} H_{ij}H_{ji} = \sum_{i,j}^{3} H_{ij}^2 = 42.$

The last step is true because the given H is symmetric.

9. We use the properties (a) $\sigma_i^2 = 1$, and (b) $\operatorname{Tr}(\sigma_i\sigma_k) = 2\delta_{ik}$. From the first of these, expanding the exponential

$$e^{i\boldsymbol{\sigma}\cdot\mathbf{a}} = \cos a + i\left(\frac{\boldsymbol{\sigma}\cdot\mathbf{a}}{a}\right)\sin a, \qquad \text{where } a = |\mathbf{a}|,$$

we have

$$e^{i\boldsymbol{\sigma}\cdot\mathbf{a}}e^{i\boldsymbol{\sigma}\cdot\mathbf{b}} = \cos a \cos b - \frac{(\boldsymbol{\sigma}\cdot\mathbf{a})(\boldsymbol{\sigma}\cdot\mathbf{b})}{ab}\sin a \sin b$$

$$+ i\left(\frac{\boldsymbol{\sigma}\cdot\mathbf{a}}{a}\right)\cos b \sin a + i\left(\frac{\boldsymbol{\sigma}\cdot\mathbf{b}}{b}\right)\sin b \cos a.$$

Because of (b), we drop terms linear in $\boldsymbol{\sigma}$. Further,

$$\operatorname{Tr}[(\boldsymbol{\sigma}\cdot\mathbf{a})(\boldsymbol{\sigma}\cdot\mathbf{b})] = \operatorname{Tr}[\sigma_i\sigma_j a_i b_j] = 2\delta_{ij}a_i b_j = 2\mathbf{a}\cdot\mathbf{b}.$$

Therefore, $T = 2\cos a \cos b - (2\mathbf{a}\cdot\mathbf{b}/ab)\sin a \sin b$.

10. (a) From the property of determinants, $\det(AB) = \det(A)\cdot\det(B)$ where A and B are arbitrary square matrices, it follows that the determinant of a second-rank tensor is invariant under coordinate transformations. If $A' = UAU^T$ where $\det(U) = \det(U^T) = \pm 1$, then $\det(A') = \det(A)$. U^T is the transposed matrix of U and the coordinate transformation is $X_i' = \sum_j U_{ij}X_j$. The cubic form in λ, $\det(T - \lambda I)$, is invariant in view of the above discussion. What is more, each coefficient of the various powers of λ is invariant because these powers are linearly independent. The coefficients are

$$I_0 = \det (T)$$

$$I_1 = (T_{22}T_{33} + T_{33}T_{11} + T_{22}T_{11} - T_{12}T_{21} - T_{13}T_{31} - T_{23}T_{32})$$

$$I_2 = \mathrm{Tr}\ (T) = T_{11} + T_{22} + T_{33}.$$

(b) The geometrical significance of I_0 is best determined in the coordinate system where T is diagonal. Denoting these diagonal elements by

$$T_{11} = \frac{1}{a^2}, \qquad T_{22} = \frac{1}{b^2}, \qquad T_{33} = \frac{1}{c^2},$$

we see that the equation for the surface

$$1 = \sum T_{ik}X_iX_k$$

defines an ellipsoid with principal axes a, b, and c. The volume of this ellipsoid is

$$\Omega = \frac{4\pi}{3}abc = \frac{4\pi}{3}\left[\frac{1}{T_{11} \cdot T_{22} \cdot T_{33}}\right]^{1/2}.$$

Hence $I_0 = (4\pi/3\Omega)^2$.

A plane which lies in the X_1, X_2 plane defines an ellipse through the equation

$$1 = T_{11}X_1^2 + T_{22}X_2^2 + 2T_{12}X_1X_2 \qquad (X_3 = 0).$$

This ellipse has an area A_3 where $\pi^2/A_3^2 = T_{11}T_{22} - T_{12}^2$. Thus

$$I_1 = \pi^2\left(\frac{1}{A_1^2} + \frac{1}{A_2^2} + \frac{1}{A_3^2}\right),$$

where A_1 and A_2 have been defined correspondingly. The geometrical significance is that if an ellipsoid is cut by an orthogonal trihedron (an open surface formed by the intersection of three planes) whose apex coincides with the center of the ellipsoid, the sum of the reciprocal squares of the areas cut out is independent of the orientation of the trihedron.

Similarly the invariance of I_2 is equivalent to the statement that for an orthogonal triad whose origin coincides with the center of the ellipse, the sum of the reciprocal squares of the lengths of the intercepts with the surface is independent of the orientation of the triad.

11. A function, $f(z)$, whose singularity at a point z_0 is an nth-order pole, may be expanded in a Laurent series:

$$\sum_{k=0}^{\infty} a_k(z - z_0)^{k-n},$$

where

$$a_k = \frac{1}{k!}\frac{d^{(k)}}{dz^k}[(z - z_0)^n f(z)]_{z=z_0}.$$

When integrating $f(z)$ about a closed contour which encloses this point and no other singularity, only the term in $1/(z - z_0)$ survives and the integral has the value

$$\frac{2\pi i}{(n-1)!}\frac{d^{(n-1)}}{dz^{(n-1)}}[(z - z_0)^n f(z)]_{z=z_0}.$$

The residue is defined to be

$$\frac{1}{(n-1)!}\frac{d^{(n-1)}}{dz^{(n-1)}}[(z - z_0)^n f(z)]_{z=z_0}.$$

(a) The residue is

$$\frac{1}{4!}\left[\frac{d^4}{dz^4}e^{\alpha z}\right]_{z=0} = \frac{\alpha^4}{4!}.$$

(b) The residue is

$$\frac{1}{2!}\left[\frac{d^2}{dz^2}\left(\frac{z^3}{\sin^3 z}\right)\right]_{z=0} = \frac{1}{2!}.$$

12. The integrand has poles as indicated in the diagram, and the path of integration may be deformed as shown, because no singularities are crossed in making such a deformation. As $\epsilon \longrightarrow 0$ the poles appear at $k = \pm a$. In closing the contour at infinity in the upper half-plane, only the pole at $k = +a$ is enclosed and the integral becomes

$$(2\pi i)\cdot\frac{1}{2!}\frac{d^2}{dk^2}\left[\frac{1}{(k+a)^3}\right]_{k=a} = \frac{3\pi i}{8a^5}.$$

13. Since the integrand is analytic at $z = 0$ we may deform the path of integration to that shown:

$$I = \int_c \frac{\sin^3 z}{z^3}dz \qquad \text{where } C =$$

On this contour expand

$$\sin z = \left(\frac{1}{2i}\right)(e^{iz} - e^{-iz}) \quad \text{and} \quad \sin^3 z = \left(\frac{1}{(2i)^3}\right)(e^{3iz} - 3e^{iz} + 3e^{-iz} - e^{-3iz}).$$

Those terms which have a positive imaginary exponent are evaluated by closing the contour above with a semicircle of infinite radius, while those with negative imaginary exponent are evaluated along the contour closed below. Those terms evaluated along the contour closed below give zero contribution because they enclose no singularities. Those evaluated above have a residue at $z = 0$ and I becomes

$$I = \frac{1}{(2i)^3}\int_c\left(\frac{e^{3iz} - 3e^{iz}}{z^3}\right)dz = \frac{2\pi i}{(2i)^3}\cdot\frac{6(i)^2}{2} = \frac{3\pi}{4}.$$

14. (a) Consider the integral of $z/(e^z - 1)$ over the contour shown below. The contour is closed at ∞, and since it encloses no singularities of $z/(e^z - 1)$

we obtain

$$\int_0^\infty \frac{x\,dx}{e^x - 1} + \int_0^\infty \frac{(x + \pi i)}{e^x + 1}\,dx + \int_0^\pi \frac{y\,dy}{e^{iy} - 1} = 0.$$

Upon taking the real part of this equation we obtain

$$\int_0^\infty \frac{x\,dx}{e^x - 1} + \int_0^\infty \frac{x\,dx}{e^x + 1} = \int_0^\pi \tfrac{1}{2} y\,dy.$$

(*Note:* The fact that the real part of $\int_0^\pi y\,dy/(e^{iy} - 1)$ is a simple integral serves as motivation for choosing this contour.) One should now prove the simple relationship

$$\int_0^\infty \frac{x^n\,dx}{e^x + 1} = (1 - 2^{-n}) \int_0^\infty \frac{x^n\,dx}{e^x - 1}, \qquad n > 0,$$

from which we finally conclude that $\int_0^\infty x\,dx/(e^x - 1) = \pi^2/6$.

(b) Using the same contour and identical arguments together with the results of I_1, one obtains $I_3 = \pi^4/15$. It is worthwhile to remark that the integral I_3 is needed to calculate the energy density of blackbody radiation.

15. The Fourier inversion of the equation

$$f(x) = \int_{-\infty}^{+\infty} F(k)e^{ikx}dk$$

yields

$$F(k) = \frac{1}{2\pi} \int_{-\infty}^{+\infty} e^{-ikx} f(x)dx.$$

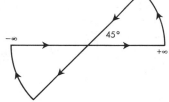

Writing $f(x) = (e^{ix^2} + e^{-ix^2})/2$, we have for $F(k)$:

$$F(k) = \frac{Je^{-ik^2/4} + J^*e^{ik^2/4}}{4\pi}, \qquad \text{where} \quad J = \int_{-\infty}^{+\infty} e^{iy^2}\,dy.$$

The integral may be written in the form

$$J = \frac{(1 + i)}{\sqrt{2}} \int_{-\infty}^{+\infty} e^{-s^2}\,ds = (1 + i)\sqrt{\frac{\pi}{2}}$$

through the substitution $y = (1 + i)s/\sqrt{2}$. The original path of integration is deformed to the one shown in the accompanying diagram. The circular arcs give zero contribution in the limit of large s. Finally we have

$$F(k) = \sqrt{\frac{1}{8\pi}}\left(\cos \frac{k^2}{4} + \sin \frac{k^2}{4}\right).$$

16. The function $F(p) = a^2/(p^2 + a^2)$ is analytic except for poles at $p = \pm ia$. This allows us to write

$$F(\sigma + ik) = \int_0^\infty e^{-(\sigma+ik)t}f(t)dt, \qquad \text{with } \sigma > 0.$$

Then by Fourier inversion with respect to k, we have

$$e^{-\sigma t}f(t) = \frac{1}{2\pi}\int_{-\infty}^\infty e^{ikt}F(\sigma + ik)dk, \qquad \text{for } t > 0.$$

The integral may be done by contour integral methods after closing the contour with a large semicircle in the upper half-plane. The function $F(\sigma + ik)$ has poles at $k = \pm a + i\sigma$. We then have

$$e^{-\sigma t}f(t) = \frac{-a^2}{2\pi}\cdot\frac{2\pi i}{2a}[e^{i(a+i\sigma)t} - e^{+i(-a+i\sigma)t}],$$

or $f(t) = a \sin (at)$.

17. Let $z = e^{i\phi}$; then

$$I = \int_0^{2\pi} \frac{d\phi}{\alpha + \cos\phi} = \frac{2}{i}\oint\frac{dz}{z^2 + 2\alpha z + 1}$$

where the integral is to be taken over the unit circle, $|z| = 1$, in the complex z-plane. The roots of the denominator are at $z = -\alpha \pm \sqrt{\alpha^2 - 1}$.

(a) For $\alpha > 1$ the root $(-\alpha + \sqrt{\alpha^2 - 1})$ is inside the unit circle while the other lies exterior to the unit circle. Thus the only contribution to the integral comes from the pole at $(-\alpha + \sqrt{\alpha^2 - 1})$. The residue of the integrand at this pole is $(-i(\alpha^2 - 1)^{-1/2})$. Finally then

$$I = 2\pi(\alpha^2 - 1)^{-1/2}.$$

(b) For $|\alpha| < 1$ we write the roots as $z_\pm = -\alpha \pm i\sqrt{1 - \alpha^2}$. Now for $\alpha = \alpha_0 + i\epsilon$, where $0 < \alpha_0 < 1$ and $\epsilon > 0$, one has, as $\epsilon \to 0$,

$$z_\pm \approx -(\alpha_0 + i\epsilon) \pm i\sqrt{1 - \alpha_0^2} \pm \frac{\alpha_0\epsilon}{\sqrt{1 - \alpha_0^2}},$$

and to first order in ϵ,

$$|z_\pm|^2 = 1 \mp \frac{2\epsilon}{\sqrt{1 - \alpha_0^2}}.$$

Thus the root z_+ lies in the interior while z_- lies exterior to the unit circle. Thus only the pole at z_+ contributes to the integral. The residue is $-(1 - \alpha_0^2)^{-1/2}$ and $I = -2\pi i\,(1 - \alpha_0^2)^{-1/2}$.

(c) For $\alpha = -1$ we may write $-1 + \cos\phi = -2\sin^2(\phi/2)$. Thus

$$I = -2\int_0^{\pi/2}\frac{d\theta}{\sin^2\theta},$$

which diverges.

18. Making the substitution $u = \sinh x$, the integrals become

$$I_1 = \int_{-\infty}^{+\infty} \frac{du}{1 + u^2} \quad \text{and} \quad I_3 = \int_{-\infty}^{+\infty} \frac{du}{(1 + u^2)^2}.$$

This form is much easier to handle by the methods of complex integration. Both integrals may be closed in the upper half-plane with the integrand of I_1 having a pole at $u = +i$ of residue $1/2i$, while the integrand of I_3 has a double pole at $u = +i$ and thus a residue

$$\frac{d}{du}\left[\frac{1}{(u + i)^2}\right]_{u=i} = +\frac{1}{4i}.$$

Thus we finally obtain

$$I_1 = \frac{2\pi i}{2i} = \pi \quad \text{and} \quad I_3 = \frac{2\pi i}{4i} = \frac{\pi}{2}.$$

19. Let $z = e^{i\phi}$; then the integral becomes a contour integral on the unit circle:

$$I = \oint \frac{dz}{iz} \frac{a(z^2 + 1) + 2bz}{(z + b/a)(z + a/b)(2ab)}.$$

There are poles at $z = 0$, $z_1 = -a/b$ and $z_2 = -b/a$.

(a) Take $|a| > |b|$. Inside the contour there are poles at $z = 0$, $z_2 = -b/a$:

$$I = \frac{2\pi}{2ab}\left[-\frac{1}{(b/a)} \cdot \frac{a(b^2/a^2 + 1) + 2b(-b/a)}{(-b/a + a/b)} + a\right] = 0.$$

(b) Take $|b| > |a|$. Inside the contour there are poles at $z = 0$, $z_1 = -a/b$:

$$I = \frac{2\pi}{2ab}\left[-\frac{1}{(a/b)} \cdot \frac{a(a^2/b^2 + 1) + 2b(-a/b)}{(b/a - a/b)} + a\right]$$

$$= \frac{2\pi}{2ab}(a + a) = \frac{2\pi}{b}.$$

20. Integrate $z^{x-1}e^{-z}$ around the contour shown.

$$\oint z^{x-1}e^{-z}\, dz = 0$$

because no singularities are enclosed. But $\int_{\Gamma_4} = 0$ as the radius of indentation goes to zero. In addition, $\int_{\Gamma_2} \to 0$ as $R \to \infty$. On Γ_1, we have $\int_0^\infty t^{x-1}e^{-t}\, dt$; on Γ_3 the integral is

$$\int_\infty^0 (iy)^{x-1}e^{-iy}i\, dy = -e^{\pi i x/2}\int_0^\infty y^{x-1}e^{-iy}\, dy.$$

Hence

$$\int_0^\infty t^{x-1}e^{-t}\, dt = e^{\pi i x/2}\int_0^\infty y^{x-1}e^{-iy}\, dy = \Gamma(x).$$

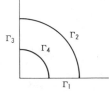

Multiply by exp $(-\pi i x/2)$ and take real and imaginary parts to get the desired results.

21. We choose the rectangle with sides $y = 0$, $y = 1$, $x = \pm R$, with indentations as shown below:

$$\oint_\Gamma \frac{dz\, e^{az}}{\sinh \pi z} = 0 = \int_{\Gamma_1 + \cdots + \Gamma_8} = I_1 + \cdots + I_8.$$

$$I_1 = \int_{-R}^{-\epsilon} \frac{e^{ax}\, dx}{\sinh \pi x} = -\int_\epsilon^R \frac{e^{-ax}\, dx}{\sinh \pi x}, \qquad I_3 = \int_\epsilon^R \frac{e^{ax}\, dx}{\sinh \pi x};$$

$$I_4 + I_8 = \int_0^1 \frac{e^{a(R+iy)}}{\sinh \pi(R + iy)}\, dy + \int_1^0 \frac{e^{-a(R-iy)}}{\sinh \pi(-R + iy)} \to 0 \text{ as } R \to \infty.$$

$$I_5 + I_7 = \int_R^\epsilon \frac{e^{a(x+i)}}{\sinh \pi(x + i)}\, dx + \int_{-\epsilon}^{-R} \frac{e^{a(x+i)}}{\sinh \pi(x + i)}\, dx$$

$$= -\int_\epsilon^R \frac{e^{ia}e^{ax}}{\sinh \pi(x + i)}\, dx - \int_\epsilon^R \frac{e^{ai}e^{-ax}dx}{\sinh \pi(-x + i)}.$$

Near $z = 0$, $e^{az}/\sinh \pi z = 1/\pi z$, and near $z = i$, $e^{az}/\sinh \pi z = -e^{ai}/\pi(z - i)$; therefore

$$I_2 = \int_\pi^0 \frac{dz}{\pi z} = -i, \qquad I_6 = -\int_{2\pi}^\pi \frac{dz\, e^{ai}}{\pi(z - i)} = ie^{ai}.$$

Now

$$\sinh \pi(i + x) = -\sinh(\pi x) = -\sinh \pi(i - x);$$

so

$$I_5 + I_7 = \int_\epsilon^R \frac{e^{ia}e^{ax}\, dx}{\sinh \pi x} - \int_\epsilon^R \frac{e^{ai}e^{-ax}}{\sinh \pi x}\, dx$$

and

$$I_1 + \cdots + I_8 = \int_0^\infty \frac{(e^{ax} - e^{-ax})}{\sinh \pi x}\, dx + \int_0^\infty \frac{e^{ia}(e^{ax} - e^{-ax})}{\sinh \pi x} + i(e^{ai} - 1) = 0$$

as $R \to \infty$ and $\epsilon \to 0$. Therefore

$$\int_0^\infty \frac{(e^{ax} - e^{-ax})}{2 \sinh \pi x}\, dx = \int_0^\infty \frac{\sinh ax\, dx}{\sinh \pi x}$$

$$= \frac{1}{2i} \frac{(e^{ai} - 1)}{(e^{ai} + 1)} = \frac{1}{2} \tan \frac{a}{2}.$$

22. There are poles of first order at $z = \pm i$; the square root branch cut is taken from 0 to ∞ on the real axis. The contour shown encloses the pole at $z = 1$, and never crosses the cut. One has

$$\oint \frac{z^{1/2}dz}{z^2 + 1} = \frac{\pi(1 + i)}{\sqrt{2}}$$

$$= (1 + i) \int_0^\infty \frac{x^{1/2}\,dx}{(x^2 + 1)}$$

from which

$$\int_0^\infty \frac{x^{1/2}\,dx}{(x^2 + 1)} = \frac{\pi}{\sqrt{2}}.$$

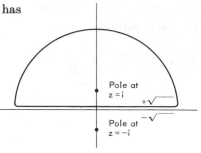

Pole at
z = i

$+\sqrt{}$

$-\sqrt{}$

Pole at
z = −i

We have used the fact that $\sqrt{i} = (1 + i)/\sqrt{2}$.

---

23. The function $1/\sin(z\pi)$ has poles at $z = n$ (n is an integer), the residues of which are $(-1)^n/\pi$. Integrating the function $1/2iz^4 \sin(z\pi)$ around a counterclockwise contour enclosing a singularity only at the point $z = n$ ($n \neq 0$) yields

$$\frac{1}{2i} \oint \frac{1}{z^4 \sin(z\pi)}\,dz = \frac{(-1)^n}{n^4}.$$

The sum over n yields

$$2 \sum_{n=1}^\infty \frac{(-1)^n}{n^4} = \frac{1}{2i} \oint_C \frac{dz}{z^4 \sin(z\pi)},$$

where C is the contour shown in the problem. This contour may be closed by adding semicircular contours above and below the real axis. These arcs add nothing to the integral. Hence

$$\sum_{n=1}^\infty \frac{(-1)^n}{n^4} = \frac{1}{4i} \oint_C \frac{dz}{z^4 \sin z\pi},$$

where C' is the clockwise contour enclosing only the singularity at $z = 0$. The function $1/(z^4 \sin \pi z)$ has a residue $14\pi^3/720$ at the point $z = 0$. Therefore

$$\sum_{n=1}^\infty \frac{(-n)^n}{n^4} = \frac{-7\pi^4}{720}.$$

---

24. (a) Possible multivaluedness of $F(z)$ arises from two sources:

(1) The multivaluedness in the definition of the logarithm, i.e. we define

$$\log \omega = \log |\omega| + i\theta + 2\pi i p,$$

where $\omega = |\omega| e^{i\theta}$ with $-\pi < \theta \leq \pi$ and p is an integer.

(2) Having made a definite choice for the branch of $\log \omega$, there is still a two-valuedness in $F(z)$ due to the two-valuedness of $\rho(z)$.

We define $\rho_1(z) = \sqrt{(z-a)/z}$, where

$$\lim_{\epsilon \to 0^+} \rho_I(x + i\epsilon) = +\sqrt{\frac{x-a}{x}} \qquad \text{for } x < 0 \text{ and } x > a;$$

then on the second sheet of $\rho(z)$ we take $\rho_{II}(z) = -\rho_I(z)$. For a fixed value of p, $F(z)$ is a two-sheeted function with

$$F_I(z) = \sqrt{\frac{z-a}{z}} \log \omega_-(z)$$

on the first sheet, and

$$F_{II}(z) = -\sqrt{\frac{z-a}{z}} \log \omega_+(z)$$

on the other. Here we have defined

$$\omega_\pm = 1 - \frac{2z}{a}\left(1 \pm \sqrt{\frac{z-a}{a}}\right).$$

One should verify that

$$\left.\begin{cases} \omega_+(z)\omega_-(z) = 1, \\ \omega_+(x) \text{ and } \omega_-(x) < 0 \qquad \text{for } x > a, \\ \omega_+(x) \text{ and } \omega_-(x) > 0 \qquad \text{for } x < 0. \end{cases}\right\} \tag{a}$$

Since the logarithmic branch cut is superimposed on that of $\rho(z)$, the only branch cuts of $F(z)$ are for $z \le 0$ and $a \le z < \infty$. The discontinuity of F_I across the left-hand cut is

$$D_l(x) = \lim_{\epsilon \to 0^+} [F_I(x + i\epsilon) - F_I(x - i\epsilon)] \qquad \text{(for } x < 0)$$

$$= \sqrt{\frac{x-a}{x}}[\log \omega_-(x) + \log \omega_+(x)]$$

$$= 4\pi i p \sqrt{\frac{x-a}{x}}.$$

The last identity follows from the first and third properties in (a) and the definition of the logarithm for fixed p. The discontinuity across the right-hand cut is

$$D_r(x) = \lim_{\epsilon \to 0} [F_I(x + i\epsilon) - F_I(x - i\epsilon)] \qquad \text{for } x > a$$

$$= \sqrt{\frac{x-a}{x}}[\log \omega_-(x + io) + \log \omega_-(x - io)].$$

Now for $x \ge a$, $\omega_-(x)$ is negative and one must calculate the sign of the imaginary part of $\omega_-(x \pm io)$ to decide whether the imaginary part of the logarithm is $(i\pi + 2\pi i p)$ or $(-i\pi + 2\pi i p)$. A simple calculation proves that the imaginary parts of both $\omega_-(x - io)$ and $\omega_-(x + io)$ are positive.

Thus one finds

$$D_r(x) = (2\pi i + 4\pi i p)\sqrt{\frac{x - a}{x}} \qquad \text{for } x \geq a.$$

The discontinuities of F_{II} are easily shown to be the negative of those for F_{I}.

(b) The function

$$G(z) \equiv \frac{F(z) - F(z_0)}{z - z_0}$$

is analytic everywhere in the finite plane except at the branch cuts of $F(z)$. Hence for a point z not on the branch cut we have

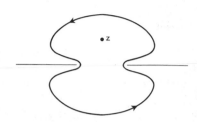

$$G(z) = \frac{1}{2\pi i} \oint \frac{G(z') \, dz'}{(z' - z)},$$

where the contour is shown at the right above. Since the function $G(z)$ goes to zero as $|z| \to \infty$, the circular portion of the contour may be expanded to $z \to \infty$ and gives zero contribution in this limit. Because of this, the contour reduces to

and

$$\frac{F(z) - F(z_0)}{z - z_0} = \frac{1}{2\pi i}\left[\int_{-\infty}^{0} \frac{D_l(s)ds}{(s - z_0)(s - z)} + \int_{a}^{\infty} \frac{D_r(s)ds}{(s - z_0)(s - z)}\right].$$

On the one sheet for which $D_l(s) = 0$ (namely the principal branch of log ω) we have

$$F(z) = F(z_0) + (z - z_0) \int_{a}^{\infty} \frac{\rho(s)}{(s - z_0)(s - z)} \, ds.$$

25. We make the substitution $y = \sqrt{n}\, x$, and the integral then becomes

$$\lim_{n \to \infty} \int_{-\infty}^{+\infty} \frac{dy}{(1 + y^2/n)^n},$$

which, since

$$\lim_{n \to \infty} \frac{1}{(1 + y^2/n)^n} = e^{-y^2},$$

becomes

$$\int_{-\infty}^{+\infty} e^{-y^2} \, dy = (\pi)^{1/2}.$$

A more difficult procedure is to evaluate $\int_{-\infty}^{+\infty} dx/(1 + x^2)^n$ first by contour integral methods, and then take the limit.

26. We note $f(a, b) = f(b/a) = -f(a/b)$. Let $y = b/a$, and take the derivative:

$$\int_0^\infty \frac{dx}{x} (e^{-x} - e^{-yx}) = f(y),$$

so

$$f'(y) = \int_0^\infty e^{-yx}\, dx = \frac{1}{y}.$$

Integrate, to get $f(y) = \ln (y) + C$. Evaluating when $y = 1$, or $a = b$, we have

$$f(1) = 0 = \ln (1) = 0 + C, \text{ or } C = 0,$$

so $f(a, b) = \ln (b/a)$.

27. $S = \left(\dfrac{d}{dx}\right)(x + x^2 + x^3 \ldots) = \dfrac{d}{dx} \sum_{n=0}^{\infty} x^n = \dfrac{d}{dx}\left(\dfrac{1}{1-x}\right) = (1 - x)^{-2}.$

28. (a) Dividing the generating equation $F(x, t) = \sum_{k=0}^{\infty} H_k(x)t^k/k!$ by t^{n+1} and integrating over a closed contour in the complex t-plane enclosing the origin, we obtain

$$H_n(x) = \frac{n!}{2\pi i} \oint \frac{\exp [x^2 - (t - x)^2]}{t^{n+1}}\, dt.$$

(b) One forms the quantity

$$\frac{\partial^2 F}{\partial x^2} + 2t \frac{\partial F}{\partial t} - 2x \frac{\partial F}{\partial x}$$

and easily verifies that it is identically zero. Using the expansion for F in terms of H_n, this identity takes the form

$$\sum_n \frac{[H_n''(x) - 2xH_n'(x) + 2nH_n(x)]t^n}{n!} = 0 \qquad \text{for all } t.$$

Hence $H_n''(x) - 2xH_n'(x) + 2nH_n(x) = 0$.

(c) Differentiating the integral representation for $H_n(x)$ we obtain

$$\frac{dH_n}{dx} = \frac{2n[(n - 1)!]}{2\pi i} \oint \frac{\exp [x^2 - (t - x)^2]}{t^n}\, dt$$

which is immediately recognized as

$$\frac{dH_n}{dx} = 2nH_{n-1}(x).$$

29. Differentiating G with respect to r, one obtains

$$\frac{(x - r)}{(1 - 2xr + r^2)^{3/2}} = \sum_l lr^{l-1}P_l(x); \tag{1}$$

differentiating G with respect to x gives

$$\frac{r}{(1 - 2xr + r^2)^{3/2}} = \sum_l r^l P_l'(x). \tag{2}$$

Eliminating $(1 - 2xr + r^2)^{-3/2}$ between (1) and (2), one obtains

$$\sum_l (x - r)r^l P_l'(x) = \sum_l lr^l P_l(x)$$

which, upon equating coefficients of equal powers of r, gives

$$xP_l'(x) = P_{l-1}'(x) + lP_l(x).$$

30. Dividing the equation

$$e^{(\mu/2)[z-(1/z)]} = \sum_{-\infty}^{\infty} J_k(\mu)z^k$$

by z^{n+1}, and integrating along the unit circle in the z-plane, we obtain

$$J_n(\mu) = \frac{1}{2\pi i} \oint z^{-n-1} e^{(\mu/2)[z-(1/z)]} \, dz.$$

On this circle however, $z = e^{i\theta}$ which, when substituted in the above expression, yields

$$J_n(\mu) = \frac{1}{2\pi} \int_{-\pi}^{\pi} \cos[\mu \sin\theta - n\theta] \, d\theta.$$

31. The Green's function for the problem is found from

$$-\nabla^2 G(\mathbf{r}, \mathbf{r}') = 4\pi\delta(\mathbf{r} - \mathbf{r}')$$

with $G(\mathbf{r}, \mathbf{r}') = 0$ for $z = 0$. The solution to $\nabla^2\psi = 0$, with ψ given on the plane $z = 0$, is

$$\psi(x, y, z) = \frac{1}{4\pi} \int dx' \, dy' \, \phi(x', y') \frac{\partial G}{\partial z'}.$$

It can be checked that

$$G = [(x - x')^2 + (y - y')^2 + (z - z')^2]^{-1/2}$$
$$- [(x - x')^2 + (y - y')^2 + (z + z')^2]^{-1/2},$$

which is found using the method of images, common in electrostatics. Thus

$$\psi(x, y, z) = \frac{z}{(2\pi)} \int dx' \, dy' \, \phi(x', y')[(x - x')^2 + (y - y')^2 + z^2]^{-3/2}.$$

32. A Bessel function of nth order, $J_n(x)$, satisfies

$$x^2 J_n''(x) + J_n'(x) + (x^2 - n^2)J_n(x) = 0.$$

Therefore $K_0(x) \equiv J_0(ix)$ satisfies $x^2 K_0'' + xK_0' - x^2 K_0 = 0$. By hypothesis

$K_0(x) = \int_0^\infty \exp(-x \cosh \phi) d\phi$, whence

$$K_0'' = \int_0^\infty \cosh^2 \phi \exp(-x \cosh \phi) d\phi$$

and

$$K_0' = -\int_0^\infty \cosh \phi \exp(-x \cosh \phi) \, d\phi.$$

When K_0' is integrated by parts, the relation $x^2 K_0'' + x K_0' - x^2 K_0 = 0$ is easily verified. To find the asymptotic form $K_0 = De^{-x}/x^{1/2}$, make the substitution $\phi = y/\sqrt{x}$ in the integral for K and obtain

$$K_0 = \frac{1}{\sqrt{x}} \int_0^\infty \exp\left[-x \cosh \frac{y}{\sqrt{x}}\right] dy,$$

and for large x make the expansion

$$x \cosh \frac{y}{\sqrt{x}} = x + \frac{y^2}{2} + O(y^4/x).$$

Therefore for $x \gg 1$ we have

$$K_0 \approx \frac{e^{-x}}{\sqrt{x}} \int_0^\infty e^{-y^2/2} \, dy \qquad \text{and} \qquad D = \int_0^\infty e^{-y^2/2} \, dy = \left(\frac{\pi}{2}\right)^{1/2}$$

33. By the divergence theorem, $\int \mathbf{r} \cdot d\mathbf{A} = \int \operatorname{div} \mathbf{r} \cdot dV = 3 \int dV = 3V$ where V is the volume of the torus. Let the inner and outer radii of the torus be R_1, R_2, respectively. We find V by using a theorem of Pappus about volumes of revolution. Consider a circle on the xy-plane, with center at $y = 0$, $x = (R_1 + R_2)/2$, and radius $(R_2 - R_1)/2$. The volume enclosed, when this is rotated about the y-axis, is, by Pappus' theorem, the area of the surface times the distance traveled by the center-of-mass of the surface, or

$$\pi \left(\frac{R_2 - R_1}{2}\right)^2 \times 2\pi \left(\frac{R_1 + R_2}{2}\right) = \frac{\pi^2}{4}(R_1 + R_2)(R_2 - R_1)^2.$$

Hence the integral is $(3\pi^2/4)(R_1 + R_2)(R_2 - R_1)^2$. *Note:* The reader should verify Pappus' theorem.

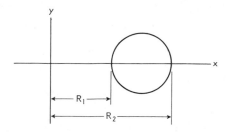

34. The volume is given by $V = \int dx_1 \, dx_2 \, dx_3 \, dx_4$ over the interior of the unit sphere. We know from three-dimensional spherical coordinates that

$dx_1\, dx_2\, dx_3 = 4\pi\rho^2\, d\rho$ where ρ is the three-dimensional radius, $\rho = r \sin\phi_2$. The volume integral then becomes $V = 4\pi\rho^2\, d\rho\, dx_4$. But by analogy with two-dimensional geometry we also know $d\rho\, dx_4 = r\, dr\, d\phi_2$. Therefore $V = \int dr\, 4\pi r^3 \int d\phi_2 \sin^2\phi_2$.

The only problem now is to find the limits of the integration over ϕ_2. These may be determined from the constraint that ρ be of only one sign (by convention $\rho \geq 0$) so that the transformation between the x's and the ϕ's is single-valued. The limits are $0 \leq \phi_2 \leq \pi$. Therefore

$$V = 4\pi \int_0^\pi d\phi_2 \sin^2\phi_2 \int_0^1 r^3\, dr = \frac{\pi^2}{2}.$$

A more general approach calculates the volume of an n-dimensional sphere, with coordinate axes $x_i \ldots x_n$ using the integral,

$$\int_{-\infty}^\infty e^{-(x_1^2 + \cdots + x_n^2)}\, dV^{(n)} = \int_{-\infty}^\infty e^{-x_1^2} \cdots e^{-x_n^2}\, dx_i \cdots dx_n = \pi^{n/2}$$

$$= \int_0^\infty e^{-R^2} A R^{n-1}\, dR,$$

with A to be determined.
Let $y = R^2$, and the integral is

$$\frac{A}{2} \int_0^\infty e^{-y} y^{(n-1)/2} y^{-1/2}\, dy = \frac{A}{2} \int_0^\infty e^{-y} y^{(n/2-1)}\, dy = \frac{A}{2}\Gamma\left(\frac{n}{2}\right).$$

Hence $A = 2\pi^{n/2}/\Gamma(n/2)$ and the volume of an n-dimensional sphere with unit radius is

$$\int_0^1 A R^{n-1}\, dR = \frac{A}{n} = \frac{2\pi^{n/2}}{n\Gamma(n/2)}.$$

Special case: $n = 2N$ (i.e. n even). Then $\Gamma(N) = (N-1)!$ and $V = \pi^N/N!$. This has the amusing property of going to zero as $N \to \infty$.

35. We take the density of air and helium at a point x in the pipe to be $n_a(x)$ and $n_h(x)$ respectively. From the assumed equilibrium conditions and lack of temperature gradients, we deduce that

$$n_a(x) + n_h(x) = N = \text{const};$$

otherwise a pressure gradient would exist in the pipe. The flux of air in the pipe is given by

$$f_x = -D\frac{dn_a}{dx} + n_a v.$$

The term $-D(dn_a/dx)$ is due to diffusion of the air in He while the latter term represents the convection of air by He. In the steady state, the continuity equation becomes $\nabla \cdot \mathbf{f} = 0$ (i.e. $(\partial f_x/\partial x) = 0$ in our case) and the solution for $n_a(x)$, which goes to zero as $x \longrightarrow -\infty$, is

$$n_a(x) = N e^{vx/d}.$$

We have taken $n_a(0) = N$ since we are told $n_h(0) = 0$.
Thus the concentration of air for $x < 0$, is

$$\frac{n_a(x)}{n_a(x) + n_h(x)} = e^{vx/D}.$$

36. (a) $\nabla^2 n + K^2 n = 0$. The lowest solution is obtained by setting $n = (\sin kr)/r$. But $n(R) = 0$, so $kR = \pi p$ (p an integer). Therefore $R = \pi/k$. Solutions with $p > 1$ lead to negative densities, and must be ruled out.
(b) The neutron density in the surface layer, $n_2(r)$, satisfying $n_2(R + t) = 0$, is

$$n_2(r) = \frac{A}{r} \sinh \left[\mu(r - R - t) \right].$$

In the interior of the pile we have a neutron density, $n_1(r)$, where $n_1(r) = (B/r) \sin (kr)$. The boundary conditions

$$n_1(R) = n_2(R) \qquad \text{and} \qquad dn_1(R)/dr = dn_2(R)/dr$$

require:
  (1) $B \sin (kR) = -A \sinh \mu t$,
  (2) $Bk \cos (kR) = A\mu \cosh \mu t$.
Therefore the radius, R, of the interior region is found from

$$\mu \tan (kR) = -k \tanh (\mu t),$$

which for $\mu \gg k$ allows the approximation $kR \approx \pi - (k/\mu) \tanh (\mu t)$. Thus the new radius is shorter by the amount $\tanh (\mu t)/\mu$.

37. Choosing a Cartesian coordinate system with z coinciding with the axis of symmetry, the solution to the steady-state diffusion equation $-D\nabla^2 n = Q\delta(\mathbf{r})$ is

$$n(x, y, z) = \frac{Q}{2Da^2} \sum_{j,k} \frac{e^{-b_{jk}|z|}}{b_{jk}} \cos \left[\frac{(2j + 1)\pi x}{2a} \right] \cos \left[\frac{(2k + 1)\pi y}{2a} \right];$$

here

$$Q = 10^6/\text{sec}; \qquad a = 75 \text{ cm}; \qquad b_{jk} = \frac{\pi}{2a}\sqrt{(2j + 1)^2 + (2k + 1)^2},$$

and the boundary condition $n \equiv 0$, for x or $y = \pm a$, is used.
  On the z-axis the flux $j_z = -D(\partial n/\partial z)$ is given by

$$j_z = \frac{Q}{2a^2} \sum_{j,k} e^{-b_{jk}|z|}.$$

It is interesting to note that if $z \ll a$, then the sum may be replaced by an integral and one finds

$$\sum_{j,k} e^{-b_{jk}|z|} \longrightarrow \frac{a^2}{2\pi z^2} \qquad \text{for } z \ll a.$$

This must be so, because for $z \ll a$ one expects the spherical symmetry of source to dominate and hence

$$j_z \longrightarrow \frac{Q}{4\pi z^2} \qquad \text{for } z \ll a.$$

However, for $z = 100$ cm, it is a good approximation to keep only the first term in the sum. Notice that the answer does not depend on the coefficient of diffusion.

2

MECHANICS

1. We want to calculate the force on a sphere of radius r, with velocity v, in a fluid of viscosity η. The dimensions of the various quantities concerned are

$$[F] = \frac{ml}{t^2}; \qquad [\eta] = \frac{m}{lt}; \qquad [v] = \frac{l}{t}; \qquad [r] = l.$$

A formula for the force which involves only these quantities must be of the form $F = C\eta^\alpha r^\beta v^\gamma$ where C is dimensionless. The dimensional equation

$$[F] = [\eta]^\alpha [r]^\beta [v]^\gamma$$

yields the solution $\alpha = \beta = \gamma = 1$ upon equating exponents of m, l, and t. Therefore $F = C\eta r v$ (actually $C = 6\pi$). If the density of the fluid is included as a parameter, then one can form the dimensionless number $R = rv\rho/\eta$, and in this case one can only argue that $F = C(R)\eta r v$ where C is a function of R. However, for streamline flow one expects the density of the fluid to be unimportant in calculating the force on the sphere. When R reaches a certain critical value, turbulent flow begins and Stokes' law is no longer valid. R is referred to as the Reynolds number. These considerations illustrate a limitation on the dimensional analysis procedure; i.e. when one has enough physical quantities to form dimensionless numbers, one must decide if all, or to what extent, these quantities are important.

2. The dimensions of the various quantities involved are

$$[T] = t; \qquad [p] = \frac{m}{lt^2}; \qquad [d] = \frac{m}{l^3}; \qquad [e] = \frac{ml^2}{t^2}.$$

Equating exponents of m, l and t in the dimensional equation

$$[T] = [p]^a [d]^b [e]^c,$$

we obtain

$$a + b + c = 0, \qquad a + 3b - 2c = 0, \qquad 2a + 2c = -1,$$

for which the solution is $a = -5/6$; $b = 3/6$; $c = 2/6$.

91

3. Consider

$$\frac{dE}{dt} = \mathbf{F} \cdot \mathbf{V} = A V^{\alpha+1} = \frac{dE}{dr}\frac{dr}{dt} = -C\frac{dE}{dr} = -C\left[\frac{GMm}{2r^2}\right] = A\left(\frac{GM}{r}\right)^{(\alpha+1)/2}.$$

This identity in r can hold only if $\alpha = 3$. Also, $-C(GMm/2) = A(GM)^2$.

Alternately we could have proceeded from the torque equation

$$\mathbf{r} \times \mathbf{F} = \frac{d\mathbf{L}}{dt}$$

which leads to $F_\theta = m(2\dot{r}\dot{\theta} + r\ddot{\theta})$. But $\dot{\theta} = (GM/r^3)^{1/2}$ and $R\ddot{\theta} = 3C\dot{\theta}/2$. Thus substituting $F_\theta = Av^\alpha$ leads to the same conditions on α and A.

4. To find the equations of motion, one may proceed via a Lagrangian, so as to eliminate consideration of the tension in the rope. A convenient coordinate is the length of cord unwound; then

$$\mathscr{L} = \frac{1}{2}mv^2 = \frac{1}{2}mL^2\dot{\phi}^2 = \frac{mL^2\dot{L}^2}{2R^2}.$$

(a) The equation of motion

$$\frac{d}{dt}\left(\frac{\partial\mathscr{L}}{\partial\frac{dL}{dt}}\right) - \frac{\partial\mathscr{L}}{\partial L} = 0$$

reduces to $(d/dt)(L\dot{L}) = 0$.

(b) The general solution satisfying $L = 0$, and $v = v_0$, at $t = 0$, is

$$L^2 = 2Rv_0 t.$$

(c) At any time the angular momentum about the axis of the cylinder is

$$J = mvL = \frac{m\dot{L}L^2}{R} = m(2Rv_0^3 t)^{1/2}.$$

Alternate solution: Tension in the cord is given by

$$T = \frac{mv^2}{L}. \tag{1}$$

Take torques about the center of the disk.

$$TR = m\frac{d}{dt}(vL). \tag{2}$$

But T is always normal to the particle trajectory, so $dv/dt = 0$. Hence, eliminating T between Eq. (1) and (2), one finds

$$\frac{dL}{dt} = \frac{v_0 R}{L}, \qquad L^2 = 2v_0 Rt.$$

Alternate solution: Because the tension T is normal to the velocity, the speed is constant. The rest is kinematics. Imagine that, after a length L of rope has unwound, an additional length dL unwinds. Then

$$\frac{dL}{R} = d\phi = \dot\phi dt = \frac{v_0 \, dt}{L}.$$

Hence

$$L \, dL = v_0 R dt \implies L^2 = 2v_0 Rt.$$

5. The range of a droplet sprayed from an angle α is $R = v_0^2 \sin(2\alpha)/g$, and its variation with α is given by $dR = 2v_0^2 \cos(2\alpha) \, d\alpha/g$. The amount of water falling on an annulus of radius R and thickness dR is proportional to $\rho(\alpha) \sin\alpha \, d\alpha = cR \, dR$, where c is a constant. If this is to be independent of α,

$$\rho(\alpha) \propto \frac{\sin(4\alpha)}{\sin(\alpha)}, \qquad 0 < \alpha < 45°.$$

6. The equation of motion for a particle on the surface is

$$m\ddot{s} = -mg \sin\theta = -mg \frac{dy}{ds}, \tag{1}$$

where s is the path length traveled and y is the height. If this motion is to be periodic, then one has

$$\ddot{s} = -\omega^2 s, \tag{2}$$

which implies

$$\omega^2 s = g \frac{dy}{ds}.$$

The solution is $y = (\omega^2/2g)s^2$ and a useful alternate equation obtained directly from Eq. (1) and (2) is

$$s = \frac{g}{\omega^2} \sin\theta.$$

Note that these equations imply that the curve has a maximum height $Y = g/2\omega^2$. That a maximum height exists is obvious without solving any equations, since a ball on the surface released at a given height cannot reach the bottom before a ball dropped vertically.

7. Let the positions of the particles be $\mathbf{r}_1$, $\mathbf{r}_2$ and $\mathbf{r}_3$ in the center-of-mass frame. The force on particle 1 is

$$F_1 = \frac{Gm_1 m_2 (\mathbf{r}_2 - \mathbf{r}_1)}{d^3} + \frac{Gm_1 m_3 (\mathbf{r}_3 - \mathbf{r}_1)}{d^3},$$

where d is the constant separation of the particles and is equal to a side of the equilateral triangle. The force $\mathbf{F}_1$ may be grouped to read

$$F_1 = \frac{Gm_1 M\mathbf{R}}{d^3} - \frac{Gm_1 M\mathbf{r}_1}{d^3},$$

where $\mathbf{R} = \mathbf{0}$ is the center-of-mass position, which is chosen as the origin, and $M = m_1 + m_2 + m_3$. Therefore

$$\frac{d^2\mathbf{r}_1}{dt^2} = -\left(\frac{GM}{d^3}\right)\mathbf{r}_1, \tag{1}$$

with identical expressions for $\mathbf{r}_2$ and $\mathbf{r}_3$. In addition, for rigid rotation about the center of mass, one has

$$\frac{d\mathbf{r}_i}{dt} = \mathbf{\Omega} \times \mathbf{r}_i$$

and

$$\frac{d^2\mathbf{r}_i}{dt^2} = \mathbf{\Omega} \times (\mathbf{\Omega} \times \mathbf{r}_i) = \mathbf{\Omega}(\mathbf{\Omega}\cdot\mathbf{r}_i) - \Omega^2\mathbf{r}_i. \tag{2}$$

For Eq. (2) to be consistent with the equation of motion, Eq. (1), one must choose $\mathbf{\Omega}\cdot\mathbf{r}_i = 0$ and $\Omega^2 = GM/d^3$. In conclusion, the axis of rotation must pass through the center of mass and be perpendicular to the plane of the triangle. The magnitude of the angular velocity must be $\Omega = (GM/d^3)^{1/2}$.

8. For a particle moving under the influence of a central force, the effective potential is:

$$V_{\text{eff}} = V(r) + \frac{L^2}{2mr^2}.$$

A circular orbit is possible at that value of r for which $\partial V_{\text{eff}}/\partial r = 0$; and the orbit is stable if $\partial^2 V_{\text{eff}}/\partial r^2$ is positive. For this problem,

$$V_{\text{eff}} = -\frac{km}{r^n} + \frac{L^2}{2mr^2}, \qquad \frac{\partial V_{\text{eff}}}{\partial r} = \frac{kmn}{r^{n+1}} - \frac{L^2}{mr^3},$$

and

$$\frac{\partial^2 V_{\text{eff}}}{\partial r^2} = -\frac{kmn(n+1)}{r^{(n+2)}} + \frac{3L^2}{mr^4}.$$

Setting $\partial V_{\text{eff}}/\partial r = 0$, one finds that $\partial^2 V_{\text{eff}}/\partial r^2 > 0$ if $3 - (n+1) > 0$, i.e. $n < 2$.

9. From energy conservation, one has, in terms of the relative coordinate r and the reduced mass $\mu = m_1 m_2/(m_1 + m_2)$,

$$\left(\frac{1}{2\mu}\right)\left(\frac{dr}{dt}\right)^2 - \frac{Gm_1m_2}{r} = -\frac{Gm_1m_2}{r_0}. \tag{1}$$

The constant r_0 is obtained from the centrifugal equation

$$F = Gm_1m_2/r_0^2 = \mu v^2/r_0;$$

that is,

$$r_0^3 = \frac{G(m_1 + m_2)\tau^2}{4\pi^2}. \qquad \text{(Kepler's Third Law)} \tag{2}$$

From (1) the time t at which the particles collide is

$$t = C \int_0^{r_0} \frac{dr}{(1/r - 1/r_0)^{1/2}} \qquad \text{where } C = \sqrt{\frac{\mu}{2Gm_1m_2}}.$$

Let $r = r_0 \sin^2 \theta$. Then

$$t = 2r_0^{3/2}C \int_0^{\pi/2} \sin^2 \theta \, d\theta = 2C \frac{\pi}{4} r_0^{3/2} = \frac{\pi}{2} \sqrt{\frac{\mu r_0^3}{2Gm_1m_2}}.$$

Using Eq. (2), this is $\tau/4\sqrt{2}$.

Alternate solution: The "fall to the center" is regarded as the limiting case of elliptical motion. As the eccentricity of the orbit approaches unity, the orbit appears as shown. The collision point is marked by X, so the collision time t is one-half the orbit time. One now uses the property of Keplerian orbits that the period is proportional to $(-E)^{-3/2}$ where $E = T + V$ is the total energy. This may be verified from Eq. (2) of the preceding solution. In addition, when the particle is stopped in circular motion and allowed to fall to the center, the energy is doubled. This occurs because, when the particle is stopped, kinetic energy is removed and the particle is more tightly bound, hence $(-E)$ increases. The fact that the energy doubles follows from the $1/r$ dependence of the potential. Thus $E_i = V_i/2$ for a circle, and when the particle is stopped the energy is all potential, and $E_f = V_i = 2E_i$. Finally, then,

$$\frac{t}{\tau} = \frac{1}{2} \left(\frac{E_i}{E_f}\right)^{3/2} = \frac{1}{2} \left(\frac{1}{2}\right)^{3/2},$$

$$t = \frac{\tau}{4\sqrt{2}}.$$

10. Assume that the man exerts constant force on the ground, until he leaves the ground. Assume also that when jumping from the surface of the moon he exerts the same constant force. Considering that the gravitational acceleration on the moon is

$$g_m = g_e \left(\frac{\rho_m}{\rho_e}\right)\left(\frac{r_m}{r_e}\right) \approx \frac{g_e}{6},$$

then, since the force is the same on earth as on the moon, the work done is the same, and $Mg_m(h_m + 50) = Mg_e(60 + 50)$. Hence $h_m = 6.1$ m.

11. Let the center of mass of the rod have downward acceleration $\ddot{x}$. Then

$$W - F = m\ddot{x}. \tag{1}$$

The angular momentum equation implies

$$\frac{WL}{2} = I\ddot{\theta}, \tag{2}$$

where I = moment of inertia about an end = $\frac{1}{3}mL^2$. One has $\ddot{\theta} = \ddot{x}/(L/2)$, a relation that is true for small θ, i.e., short times. From (1) and (2), $F = W/4$.

12. The vertical force F_v exerted by the plane on one of the cylinders is fixed at the value $3W/2$ where W is the weight of one of the cylinders. Since we wish to minimize the angle which the total force makes with the vertical, we should minimize the horizontal force, F_h, which the plane exerts on the cylinder.

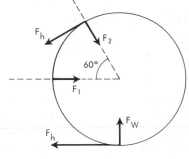

From the force diagram we see that we must have:

Vertical equilibrium: $W + F_2 \cos 30° + F_h \sin 30° = 3W/2$

Horizontal equilibrium: $F_1 + F_2 \cos 60° = F_h + F_h \cos 30°$.

F_h has a minimum when $F_1 = 0$, for which we find $F_h = W/(4 + 2\sqrt{3})$, and the tangent of the minimum angle is given by $\tan \theta = \frac{1}{3}(2 + \sqrt{3})$.

13. It rolls in the direction of the pull. When rolling without slipping, the bottom point P of the yo-yo has zero instantaneous velocity. Rotation around a fixed axis passing through P is determined by torque about that axis, but only F has a nonvanishing torque. Therefore rotation is clockwise, and the center of mass moves to the right.

14. Denote by B the point fixed on the disk; β is the angle through which the dog travels on the disk, and α the angle the disk travels. At $t = 0$ the angular momentum is zero; this quantity is conserved. The angular momentum of the disk is $(3MR^2/2)\dot{\alpha}$ (use the parallel axis theorem.), with respect to an axis through B, but fixed in space. The angular momentum of the dog is $[\dot{\alpha} - (\dot{\beta}/2)][4mR^2 \sin^2 (\beta/2)]$. Setting the sum equal to zero we get

$$\frac{d\alpha}{dt} = \frac{d\beta}{2dt} \frac{4m \sin^2 (\beta/2)}{3M/2 + 4m \sin^2 (\beta/2)},$$

or

$$\frac{1}{2} \int_0^{2\pi} d\beta \frac{4m \sin^2 (\beta/2)}{3M/2 + 4m \sin^2 (\beta/2)}.$$

We can write this in terms of $\gamma = \beta/2$. Then the angle is given by the expression

$$\int_0^\pi d\gamma \frac{4m \cos^2 \gamma}{3M/2 + 4m \cos^2 \gamma}.$$

Note that the angle depends only on the mass ratio of the disk and dog.

15. Conservation of angular momentum implies $I_0\omega_0 = I\omega$ with $I_0 = 2MR^2/5$ and $I = (2M/5 + 2m/3)R^2$. Now

$$\frac{2M}{5} = \frac{8\pi}{15} R^3 D \quad \text{and} \quad \frac{2m}{3} = \frac{8\pi}{3} dR^2 k.$$

Therefore

$$\frac{T}{T_0} = \frac{\omega_0}{\omega} = \frac{I}{I_0} = 1 + \frac{5dh}{DR},$$

and hence $(T - T_0)/T_0 = 5\, dh/DR$.

16. Since the earth is rotating with angular velocity $\boldsymbol{\Omega}$, the angular momentum expressed in a coordinate system on earth satisfies

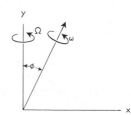

$$\frac{d\mathbf{L}}{dt} = \mathbf{N} - \boldsymbol{\Omega} \times \mathbf{L},$$

where N is the total applied torque. Looking down on the gyroscope at the equator, where $\boldsymbol{\Omega}$ is along the y-axis and the z-axis is vertical, one may write the following expression for $\mathbf{L}$

$$\left.\begin{array}{l} L_x = C\omega \sin\phi \approx C\omega\phi \\ L_y = C\omega \cos\phi \approx C\omega \end{array}\right\} \quad \text{for } |\phi| \ll 1,$$

$$L_z = -A\dot\phi.$$

In this expression terms of order $\omega/\Omega \ll 1$ have been dropped. Since there are no forces in the xy-plane, $N_z \equiv 0$, and the equation for L_z becomes $-A\ddot\phi = C\omega\Omega\phi$. Thus ϕ oscillates with angular frequency ν, where $\nu^2 = C\omega\Omega/A$.

17. When the total angular momentum is expressed in an inertial frame of reference it satisfies $(d\mathbf{L}/dt)_{\text{inertial}} = 0$. However, the principal moments of inertia are given in the body-fixed frame of reference, rotating with an angular velocity $\boldsymbol{\Omega}$ with respect to an inertial frame. The connection between $(d\mathbf{L}/dt)_{\text{inertial}}$ and $(d\mathbf{L}/dt)_{\text{body}}$ is

$$(d\mathbf{L}/dt)_{\text{inertial}} = (d\mathbf{L}/dt)_{\text{body}} + \boldsymbol{\Omega} \times \mathbf{L} = 0.$$

In the body frame we then have the equations:

$$\frac{d}{dt}(I_{zz}\Omega_z) = 0, \tag{1}$$

$$\frac{d}{dt}(I_{xx}\Omega_x) + \frac{3}{2} I_0\Omega_y\Omega_z\epsilon \cos\omega t = 0, \tag{2}$$

$$\frac{d}{dt}(I_{yy}\Omega_y) - \frac{3}{2} I_0\Omega_x\Omega_z\epsilon \cos\omega t = 0, \tag{3}$$

where $I_0 = 2mr^2/5$. Equation (1) has the solution $\Omega_z = \Omega_{0z}/(1 + \epsilon \cos\omega t)$

and Ω_z varies little with time since $\epsilon \ll 1$. From (2) and (3) we see that $\Omega_x \hat{x} + \Omega_y \hat{y}$ rotates with an angular frequency $\omega_p = \frac{3}{2} \Omega_z \cos \omega t$ when dI_{xx}/dt and dI_{yy}/dt are neglected, i.e., when $\omega \ll \Omega_z$.

18. For sufficiently small displacements, motion along the direction of the rods is decoupled from motion normal to the rods. Denoting displacements along the rod by (x_1, x_2, x_3) and those normal to the rod by (y_1, y_2, y_3), we find that the Lagrangian is

$$\mathscr{L} = \frac{m}{2}(\dot{x}_1^2 + 2\dot{x}_2^2 + \dot{x}_3^2 + \dot{y}_1^2 + 2\dot{y}_2^2 + \dot{y}_3^2) - \frac{k}{2}[(x_1 - x_2)^2 + (x_3 - x_2)^2]$$

$$- \frac{k'}{2}[(y_1 - y_2) - (y_2 - y_3)]^2.$$

The last term is constructed so as to vanish when all the particles lie on a straight line.

Motion in the x-direction is governed by the equations

$$m\ddot{x}_1 + k(x_1 - x_2) = 0, \qquad m\ddot{x}_3 + k(x_3 - x_2) = 0,$$
$$2m\ddot{x}_2 - k(2x_2 - x_1 - x_3) = 0.$$

Conservation of momentum requires that $\ddot{x}_1 + 2\ddot{x}_2 + \ddot{x}_3 = 0$. This condition implies that there are only two normal modes for vibration in the x-direction. The normal modes are

and
$$x_1 - x_3 \quad \text{with frequency} \quad (k/m)^{1/2}$$
$$x_1 - 2x_2 + x_3 \quad \text{with frequency} \quad (2\,k/m)^{1/2}.$$

Motion in the y-direction is constrained by conservation of angular momentum, as well as linear momentum. Consequently there is only one mode of vibration, with frequency $(4k'/m)^{1/2}$. The modes may be sketched as in the accompanying figure.

19. Let the vertical displacements from the equilibrium positions be X_1, X_2. For the motion of the center of mass C, $F = ma$ yields

$$\frac{M}{2}(\ddot{X}_1 + \ddot{X}_2) = -k(X_1 + X_2) - mg, \tag{1}$$

and the torque condition gives (for small X_1, X_2)

$$\frac{I_0}{L}(\ddot{X}_2 - \ddot{X}_1) = \frac{-L}{2}k(X_2 - X_1) \tag{2}$$

and I_0 = moment of inertia about $C = ML^2/12$.

The gravity term, which merely determines the unextended spring lengths, can be transformed away. It does not affect the modes. From (1) and (2) there are obviously two modes:

(a) a symmetric one, $X_1 = x_1 + x_2$, with frequency $\omega_s^2 = 2k/M$, and

(b) an antisymmetric one, $X_2 = x_1 - x_2$, $\omega_a^2 = 6k/M$.

20. This problem was first studied by Daniel Bernoulli. Denote by $\mu = m/L$, the mass per unit length of the string. Application of Newton's Second Law to a small element dx of the string gives the equation

$$\frac{\partial^2 y}{\partial t^2} = \frac{1}{\mu} \frac{\partial}{\partial x}\left[T(x) \frac{\partial y}{\partial x}\right], \tag{1}$$

where T is the tension in the string. Because the rope is in equilibrium with respect to motion along its length, $T(x) = g[m(1 - x/L) + M]$. The appropriate boundary conditions for $y(x, t)$ are

$$y(L, 0) = \delta, \qquad \frac{\partial y}{\partial t}(x, 0) = 0,$$

$$y(0, t) = 0, \qquad \frac{\partial^2 y}{\partial x^2}(L, t) = 0.$$

The second of these is satisfied by putting $y(x, t) = y(x) \cos \omega t$. Then $y(x)$ satisfies an ordinary differential equation

$$\frac{1}{\mu} \frac{d}{dx}\left[T(x) \frac{dy}{dx}\right] + \omega^2 y = 0. \tag{2}$$

With the ansatz $y = f(\sqrt{T(x)})$, this reduces to the Bessel equation of order zero:

$$f'' + \frac{1}{\sqrt{T(x)}} f' + \mu\left(\frac{2L\omega}{mg}\right)^2 f = 0,$$

with the general solution

$$y(x) = AJ_0\left(\frac{2\omega}{g}\sqrt{\frac{T(x)}{\mu}}\right) + BN_0\left(\frac{2\omega}{g}\sqrt{\frac{T(x)}{\mu}}\right),$$

where N_0 is the zeroth Neumann function (i.e., Bessel function "of the second kind", having a logarithmic singularity at the origin). The first and third boundary conditions yield the coefficients:

$$y = \delta\left[\frac{N_0\left(\frac{2\omega}{g}\sqrt{\frac{g}{\mu}(M+m)}\right)J_0\left(\frac{2\omega}{g}\sqrt{\frac{T(x)}{\mu}}\right) - J_0\left(\frac{2\omega}{g}\sqrt{\frac{g}{\mu}(M+m)}\right)N_0\left(\frac{2\omega}{g}\sqrt{\frac{T(x)}{\mu}}\right)}{J_0\left(\frac{2\omega}{g}\sqrt{\frac{Mg}{\mu}}\right)N_0\left(\frac{2\omega}{g}\sqrt{\frac{g}{\mu}(M+m)}\right) - J_0\left(\frac{2\omega}{g}\sqrt{\frac{g}{\mu}(M+m)}\right)N_0\left(\frac{2\omega}{g}\sqrt{\frac{Mg}{\mu}}\right)}\right].$$

The fourth boundary condition then gives a transcendental equation for ω.

In the case that $m \ll M$, the arguments of the Bessel function are large, and one may apply the asymptotic expansions:

$$J_0(z) \longrightarrow \frac{1}{\sqrt{\pi z}}(\cos z + \sin z),$$

$$N_0(z) \longrightarrow \frac{1}{\sqrt{\pi z}}(\cos z - \sin z),$$

as z increases without limit. In that limit, $y = \delta(x/L) \cos \omega t$.

In place of the fourth boundary condition, which is satisfied identically, we use the equivalent condition

$$\frac{\partial^2 y}{\partial t^2} = -g\frac{\partial y}{\partial x} \qquad \text{at } x = L$$

to obtain $\omega^2 = g/L$.

21. Waves on a stretched membrane obey the differential equation

$$\sigma\frac{\partial^2 y}{\partial t^2} = T\nabla^2 y.$$

Letting $y = u(x, y) \cos \omega t$ gives $-\omega^2 \sigma u = T\nabla^2 u$. Consequently,

$$\omega^2 = -\frac{T}{\sigma}\frac{\int u\nabla^2 u\, dA}{\int u^2\, dA}.$$

This may be computed for various trial functions to obtain an approximate eigenfrequency. The trial functions u must be chosen so as to satisfy the boundary conditions, in this case $u = 0$ on the boundary.

22. On the surface of the earth, the moment of inertia of the flywheel is increased by a slight amount beyond its value in vacuum because of the small viscosity of air; the flywheel tends to drag air around with it. A watch made to run on the earth's surface is adjusted to compensate for this effect. When removed to a high altitude, the moment of inertia is decreased slightly, and the watch runs slightly fast. According to Professor Allison, this effect was noticed by Professor I. Rabi, who was very proud of an extremely accurate watch, which ran slightly fast when he went to a mountain top for cosmic-ray experiments. Rabi proposed the problem to Professor Fermi while they were riding a train during the war, as a diversion for Fermi, who hated train rides. (For security reasons, they could not discuss project business on the train; they could not fly for security reasons.) Fermi explained the effect immediately, and in about an hour's time had worked out a complete quantitative theory of the phenomenon.

23. Young's Law requires $\Delta F/S = Y((\Delta L)/L)$, where ΔL is the extension of the string, and ΔF is the force felt by the string as the mass decelerates.

The maximum ΔL is determined by conservation of energy:

$$mg(L + \Delta L) = \frac{S}{2} YL\left(\frac{\Delta L}{L}\right)^2,$$

from which

$$\frac{\Delta L}{L} = \frac{mg}{SY} + \sqrt{\left(\frac{mg}{SY}\right)^2 + \frac{2mg}{SY}}.$$

In order to prevent breaking, we must have $\Delta F/S < T$; therefore

$$Y < \left(\frac{ST^2}{2mg} - T\right).$$

24. The train must be stopped before the spring is fully compressed. Otherwise, if the jump in the spring constant is large enough, the acceleration will exceed a_{max}. As the spring is compressing we have $F = k_0(l_0 - l) = Ma$, and a will be largest when l is smallest (i.e. when the train is stopped). At this point the energy equation tells us that $\frac{1}{2}k_0(l_0 - l)^2 = \frac{1}{2}Mv^2$. Eliminating k_0 we obtain

$$a_{max} = \frac{v^2}{(l_0 - l)} \approx \frac{v^2}{l_0}$$

since $l \ll l_0$. Therefore the minimum value of l_0 is given by $l_0(v^2/a_{max})$.

25. (a) We use Archimedes' principle to put $\rho\pi R^2 h\ddot{x} = -\rho_0 g\pi R^2 x$. Therefore $\omega^2 = \rho_0 g/\rho h$.
(b) Consider a slab of fluid of thickness dr and of area A parallel to the cylinder. The net force on this slab due to viscosity is

$$f = A\eta \frac{\partial^2 v}{\partial r^2} dr,$$

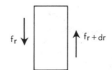

which must equal $(\rho_0 A\, dr)(\partial v/\partial t)$. For a sinusoidal time dependence $e^{+i\omega t}$ one has

$$\eta \frac{\partial^2 v}{\partial r^2} = i\rho_0\omega v,$$

with solution $v = v_0 e^{-(r/\delta)(1+i)}$ where the damping length is given by $\delta = \sqrt{2\eta/\rho_0\omega}$. The total viscous force acting on the cylinder is thus

$$F = \eta A \frac{dv}{dr} = \eta\left(2\pi R\frac{\rho}{\rho_0}h\right)\left[-\frac{(1+i)}{\delta}\right]i\omega x.$$

The magnitude of F is $|F| = 2\pi R h\rho(\eta\omega^3/\rho_0)^{1/2}|x|$.

26. Consider a section of the surface between the planes $Z = Z_1$ and $Z = Z_2$. In order to have equilibrium in the vertical direction we must have $(2\tau \cos\theta_1)(2\pi r_1) = (2\tau \cos\theta_2)(2\pi r_2)$ or simply $r \cos\theta = \text{const}$, where

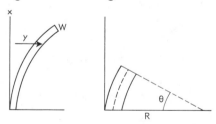

$\cos \theta = [1 + (dr/dZ)^2]^{-1/2}$. The equation for the surface is thus

$$\left(\frac{r}{r_0}\right)^2 = 1 + \left(\frac{dr}{dZ}\right)^2,$$

for which the general solution is $r = r_0 \cosh (Z/r_0)$. (The origin of Z is taken midway between the two loops.) We calculate r_0 from the condition

$$a = r_0 \cosh (d/r_0).$$

Define $x = (d/r_0)$; then x satisfies the equation $(a/d)x = \cosh x$, which has a solution for x only if (a/d) is greater than some minimum value. This can be seen by making a graph of $\cosh x$ and $(a/d)x$. In addition, at the minimum value of (a/d), a solution exists only if $a/d = \sinh x$, i.e. the graphs of $(a/d)x$ and $\cosh x$ have only a single intersection. Therefore the x corresponding to this value of (a/d) is found from $x = \coth x$, for which we find $x \cong 1.2$, and (a/d) is calculated from $(a/d)^2(x^2 - 1) = 1$. Therefore the maximum value of (d/a) allowed for stability is 0.66.

27. Let the bending be defined by y as shown in the figure. One must first calculate the bending moment through a section of the strut at a height x.

Consider a plane section of the strut at a distance Z from the mid-section. The length of this section, when the cylinder is bent through an angle θ, is $(R + z)\theta$. The strain is therefore $(R + z)/R$, and the restoring moment

$$N = \frac{Ya}{R} \int_{-a/2}^{a/2} (R + z)z \, dz = \frac{Ya^4}{12R}.$$

The radius of curvature R is

$$\frac{1}{R} = \frac{d^2y}{dx^2}\left[1 + \left(\frac{dy}{dx}\right)^2\right]^{-3/2}.$$

We shall consider only infinitesimal deformations for which we may neglect $(dy/dx)^2$ as compared to 1. Thus the restoring moment at height x becomes

$$N = \frac{Ya^4}{12}\frac{d^2y}{dx^2}.$$

In order to have equilibrium of the portion of the strut above this height, the moment must equal the torque of the weight about this point, $W(\epsilon - y)$, where $\epsilon = y(l)$. We therefore have the equation

$$\frac{Ya^4}{12}\frac{dy^2}{dx^2} = W(\epsilon - y),$$

for which the general solution is $y = A \sin(\omega x) + B \cos(\omega x) + \epsilon$, where $\omega^2 = (12W/Ya^4)$. The boundary conditions give conditions on A, B, and ω:

$$y(0) = 0 \longrightarrow B = -\epsilon,$$

$$y'(0) = 0 \longrightarrow A = 0,$$

$$y(l) = \epsilon \longrightarrow A \sin \omega l + B \cos \omega l = 0.$$

The solution is thus:

$$y = \epsilon(1 - \cos \omega x) \qquad \text{with } \epsilon \cos \omega l = 0.$$

For $\omega l < \pi/2$, the condition $\epsilon \cos \omega l = 0$ can only be satisfied by $\epsilon = 0$, i.e. the vertical position is stable. However, when W increases to the point where $\omega l = \pi/2$, an infinitesimal deformation is possible. Hence

$$W = (\pi^2 Ya^4/48l^2)$$

is the beginning of instability.

28. The notation is as defined in the figure. By Young's Law, the force required to bend a segment of area $a(du)$ and length dz through an angle $(d\theta)$ is $Yau(d\theta)\,du/dz$, and the total moment about a point P of those forces acting across the AB-plane is

$$M = \int \frac{Y\,au^2(d\theta)\,du}{dz} = \left(\frac{Ya^3}{12}\right)\frac{d\theta}{dz}.$$

For small bending, $(d\theta) = dz(\partial^2 h/\partial z^2)$, so

$$M = \frac{Ya^3}{12}\left(\frac{\partial^2 h}{\partial z^2}\right).$$

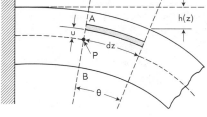

In order to have rotational equilibrium for the section of the beam to the right of a point z, one must have

$$M(z) = \int_z^L \rho ga^2\,(L - z')\,dz';$$

hence

$$\frac{\partial^2 h}{\partial z^2} = \frac{6\rho g}{Ya}(L^2 - 2Lz + z^2).$$

The boundary conditions are $h(0) = h'(0) = 0$. Therefore $h(z) = (\rho g/2Ya)$ $[6L^2z^2 - 4Lz^3 + z^4]$ and $h(L) = (3\rho g L^4/2Ya)$. Note that the moment M and force $\partial M/\partial z$ vanish at the free end.

29. Imagine the chimney separated at a distance x above the ground. The rotation of the chimney as a whole about O satisfies the equation

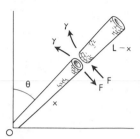

$$\frac{ML^2\ddot{\theta}}{3} = \frac{MgL}{2}\sin\theta, \quad \text{or} \quad \ddot{\theta} = \frac{3g\sin\theta}{2L}. \quad (1)$$

Likewise the lower portion satisfies the equation

$$\frac{Mx^3\ddot{\theta}}{3L} = \frac{Mx^2g\sin\theta}{2L} + xF - \gamma, \quad (2)$$

where F is the shear force and γ the internal flexion torque at the point x. Similarly, the rotation of the upper portion about its center of mass satisfies the equation

$$\frac{M(L-x)^3}{12L}\ddot{\theta} = \frac{(L-x)F}{2} + \gamma. \quad (3)$$

Equations (1) through (3) may be solved to yield

$$F = \frac{Mg\sin\theta(L-x)}{L^2(L+x)}\left[\frac{(L-x)^2}{4} - x^2\right],$$

and

$$\gamma = \frac{Mx(L-x)^2g\sin\theta}{4L^2}.$$

Since the chimney is thin, i.e. $w \ll L$, where w is its width, the forces responsible for γ are very much greater than F. That is to say, $\gamma/w \gg F$. Thus the breaking is dominated by the maximum of γ. This occurs at $x = L/3$.

30. Consider a small volume element of liquid near the surface. The potential energy/volume is

$$V = \rho g y - \tfrac{1}{2}\rho\omega^2 x^2.$$

The equation of the surface is obtained by setting the potential energy equal to a constant (in this case zero because of the choice of coordinates). Thus

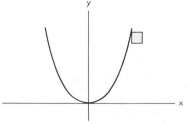

$$y = \omega^2 x^2 / 2g.$$

31. From the expression for ϕ one finds the components of the velocity

$$v_r = v_\infty(1 - R^2/r^2) \cos \theta$$
$$v_\theta = -v_\infty(1 + R^2/r^2) \sin \theta.$$

Because the door is open, the pressure P inside and outside must be equal, with value P_0. From Bernoulli's equation for streamline flow, at any other point on the exterior surface,

$$P + \frac{1}{2}\rho v^2(\theta) = P_0 \Longrightarrow (P - P_0) = -\frac{1}{2}\rho v^2(\theta)$$

gives the pressure at any angle θ. Then

$$F = \int (P_0 - P) \cdot dA = \frac{1}{2}\rho v_\infty^2 \left(\frac{1 + R^2}{R^2}\right)^2 \int_0^\pi \sin^2 \theta R \, d\theta L$$

$$= 2\rho v_\infty^2 LR \cdot \frac{\pi}{2} = \rho \pi v_\infty^2 LR = 2.1 \times 10^7 \text{ N}.$$

The force is upward.

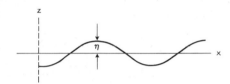

32. Let the upper medium have density ρ_1 and the lower one density ρ_2. Since the oscillations are incompressible, the pressure satisfies $\nabla^2 P = 0$ everywhere. For a boundary oscillation of the form $\eta = \alpha \exp [i(kx - \omega t)]$, the oscillations cause slight changes in the pressure from its static value which damp out as one travels away from the interface. Thus

$$P_1 = -\rho_1 gz + \beta_1 \exp [-kz + i(kx - \omega t]$$

and

$$P_2 = -\rho_2 gz + \beta_2 \exp [kz + i(kx - \omega t)],$$

where β_1 and β_2 are small constants which go to zero as η goes to zero. The boundary conditions are

(1) $P_1 = P_2$ at $z = \eta$
(2) From $\rho(\partial v/\partial t) = -\nabla P$ we have $\rho_1 \ddot{\eta} = -\partial P_1/\partial z$ and $\rho_2 \ddot{\eta} = -\partial P_2/\partial z$.

Remembering that η is small, we find that these yield the equations

(1) $-\rho_1 g\alpha + \beta_1 = -\rho_2 g\alpha + \beta_2$
(2) $-\omega^2 \rho_1 \alpha = k\beta_1$ and $\omega^2 \rho_2 \alpha = k\beta_2$.

These equations are easily solved to yield

$$\omega^2 = \frac{(\rho_2 - \rho_1)kg}{(\rho_2 + \rho_1)}.$$

Note that if $\rho_1 > \rho_2$ the frequency is imaginary, and thus the oscillations are unstable. This is expected, since in this case the heavier air should fall. The phase velocity $v = \omega/k$ is given by

$$v = \left[\frac{(\rho_2 - \rho_1)g}{(\rho_2 + \rho_1)k}\right]^{1/2} = \left[\frac{(\rho_2 - \rho_1)g\lambda}{2\pi(\rho_2 + \rho_1)}\right]^{1/2}.$$

To relate ρ to the temperature one may use the ideal gas law, i.e. $\rho = K/T$, hence

$$v = \left[\frac{(T_1 - T_2)g\lambda}{2\pi(T_2 + T_1)}\right]^{1/2}.$$

33. The velocity will be taken as irrotational, for which we have $\mathbf{v} = \nabla\phi$ and $\nabla\cdot\mathbf{v} = K\delta((x - a), (y - b))$. Due to the presence of the walls the flow must be such that $v_x(0, y) = 0$ and $v_y(x, 0) = 0$. This boundary condition will be satisfied if we consider the flow in the region of interest to be due to the original source and image sources of strength K located at $(x, y) = (-a, b)$, $(-a, -b)$ and $(a, -b)$. For a source of strength K the solution to $\nabla^2\phi = K\delta((x - a), (y - b))$ is

$$\frac{K}{2\pi} \log \sqrt{(x - a)^2 + (y - b)^2}.$$

The velocity potential for the flow with boundaries (which has a velocity which goes to zero as the distance from (a, b) increases) is

$$\phi = \frac{K}{2\pi}(\log \sqrt{(x - a)^2 + (y - b)^2} + \log \sqrt{(x - a)^2 + (y + b)^2}$$
$$+ \log \sqrt{(x + a)^2 + (y - b)^2} + \log \sqrt{(x + a)^2 + (y + b)^2}).$$

Use Bernoulli's equation to relate the velocity to pressure; that is,

$$\frac{\rho}{2}(\nabla\phi)^2 + P = P_0,$$

where P_0 is the pressure as $\mathbf{v} \longrightarrow 0$. Thus the pressure is given by

$$P = P_0 - \tfrac{1}{2}\rho\left(\frac{K}{\pi}\right)^2\left[\frac{(x - a)}{b^2 + (x - a)^2} + \frac{(x + a)}{b^2 + (x + a)^2}\right]^2 \quad \text{on } OB;$$

$$P = P_0 - \tfrac{1}{2}\rho\left(\frac{K}{\pi}\right)^2\left[\frac{(y - b)}{a^2 + (y - b)^2} + \frac{(y + b)}{a^2 + (y + b)^2}\right]^2 \quad \text{on } OA.$$

34. Consider a body such as the moon, located as shown in the accompanying figure. The shape of the water surface is defined by the spherical coordinates (ρ, θ). In the center-of-mass coordinate system, which is rotating with angular velocity ω, a unit mass of water on the surface is acted upon by gravitational and inertial forces obtainable from the potential

$$V(\rho, \theta) = V_e + V_m + V_c,$$

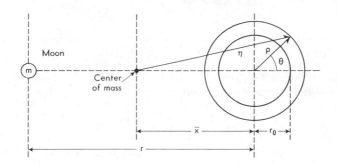

where

$$V_e = \frac{-GM_e}{\rho} \qquad \text{(gravitational potential due to earth)},$$

$$V_m = -\frac{Gm}{(r^2 + \rho^2 + 2r\rho \cos \theta)^{1/2}} \qquad \text{(gravitational potential due to moon)},$$

$$V_c = -\frac{\omega^2 \eta^2}{2} \qquad \text{(centrifugal force potential)},$$

with $\eta^2 = \rho^2 + \bar{x}^2 + 2\bar{x}\rho \cos \theta$. In order to obtain ω in terms of the given constants, note that

$$M_e \omega^2 \bar{x} = \frac{GM_e m}{r^2} \qquad \text{and} \qquad \bar{x} = \frac{m}{M_e + m} r.$$

It should be pointed out that in the above, the axis of rotation of the earth-moon system and the axis of the earth's 24-hour rotation are taken perpendicular to the plane of the drawing. The latter choice is made so that maximum tidal changes will be seen in a 24-hour period as defined by the angle θ. The tides are also calculated at the equator for simplicity.

Upon expanding V_m in powers of (ρ/r) and neglecting terms of higher order than $(\rho/r)^2$, one has

$$V_m \simeq \frac{-Gm}{r} \left[1 - \frac{\rho}{r} \cos \theta + \frac{\rho^2}{2r^2} (3 \cos^2 \theta - 1) + \cdots \right].$$

In order to calculate $\rho(\theta)$, one notes that the surface must be an equipotential. Thus,

$$V(\rho, \theta) = -\frac{Gm(2M_e + 3m)}{2r(M_e + m)} - \frac{Gm\rho^2}{2r^3}(3 \cos^2 \theta - 1) - \frac{G(M_e + m)}{2r^3} \rho^2 - \frac{GM_e}{\rho}$$

$$= \text{const on the surface}.$$

Setting $\rho = r_0 + h$ and neglecting terms quadratic and higher in h and terms of order $[(M_e + m)/M_e](r_0^3/r^3)$, one obtains the equation for h

$$-\frac{Gmr_0^2}{2r^3}(3 \cos^2 \theta - 1) - \frac{G(M_e + m)r_0^2}{2r^3} + \frac{GM_e h}{r_0^2} = \text{const}.$$

Hence

$$h = h_0 + \frac{3mr_0^4 \cos^2 \theta}{2M_e r^3}.$$

The tidal variation is given by $\Delta h = h(0) - h(\pi/2) = 3mr_0^4/2M_e r^3$. The ratio of the tides due to the sun and moon is thus

$$\frac{(\Delta h)_{\text{sun}}}{(\Delta h)_{\text{moon}}} = \frac{Mr^3}{mR^3} \approx 0.4.$$

The presence of the $\cos^2 \theta$ in $h(\theta)$ shows that there are two tides in a 24-hour period.

35. A small but otherwise arbitrary deformation of the surface from a sphere will be represented by $h(\hat{\mathbf{n}})$ as shown in the figure. The function $h(\hat{\mathbf{n}})$ may be expanded in spherical harmonics to $h(\hat{\mathbf{n}}) = R \sum_{lm} A_{l,m} Y_{lm}(\hat{\mathbf{n}})$. The gravitational energy is

$$U = -\frac{1}{2} \int dV \, dV' \frac{G\rho(\mathbf{r})\rho(\mathbf{r}')}{|\mathbf{r} - \mathbf{r}'|}.$$

We write $\rho(\mathbf{r}) = \rho_0(\mathbf{r}) + \delta\rho(\mathbf{r})$, where ρ_0 is the density for a homogeneous sphere of radius R, and the energy becomes

$$U = U_0 + U_1 + U_2,$$

where

$$U_0 = -\frac{G}{2} \int dV \, dV' \frac{\rho_0(\mathbf{r})\rho_0(\mathbf{r}')}{|\mathbf{r} - \mathbf{r}'|}$$

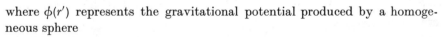

is independent of h, and

$$U_1 = -G \int \frac{dV \, dV' \rho_0 \, \delta\rho(\mathbf{r}')}{|\mathbf{r} - \mathbf{r}'|} = -\int \delta\rho(\mathbf{r}')\phi(\mathbf{r}') \, dV',$$

where $\phi(\mathbf{r}')$ represents the gravitational potential produced by a homogeneous sphere

$$\phi(\mathbf{r}') = G \int \frac{\rho_0(\mathbf{r}) \, dV}{|\mathbf{r} - \mathbf{r}'|}$$

and

$$U_2 = -\frac{G}{2} \int \frac{dV \, dV' \, \delta\rho(\mathbf{r}) \, \delta\rho(\mathbf{r}')}{|\mathbf{r} - \mathbf{r}'|}.$$

At this point note that $\delta\rho(\mathbf{r})$ is nonzero only in the shaded regions, and that in these regions $\delta\rho = \pm\rho_0$ depending, respectively, on whether h is positive or negative. Thus, for instance, the expression for U_2 is

$$U_2 = -\frac{G}{2} \int_R^{R+h(\hat{\mathbf{n}})} r^2 \, dr \int_R^{R+h(\hat{\mathbf{n}}')} (r')^2 \, dr' \, \rho_0^2 \int \frac{d\Omega \, d\Omega'}{|\mathbf{r} - \mathbf{r}'|}.$$

By a Taylor expansion of the integral, keeping terms up to second order in h, U_2 becomes

$$-\frac{G\rho_0^2 R^4}{2} \int d\Omega \, d\Omega' \frac{h(\hat{\mathbf{n}})h(\hat{\mathbf{n}}')}{|\mathbf{r}-\mathbf{r}'|}.$$

Similarly the expression for U_1 to second order in h is

$$U_1 = -\frac{4\pi G\rho_0^2 R^3}{3} \int d\Omega\left(Rh + \frac{h^2}{2}\right).$$

Finally, the total gravitational energy to second order in h is

$$U = U_0 - \rho_0 GM \int d\Omega\left(Rh + \frac{h^2}{2}\right) - \frac{G\rho_0^2 R^4}{2} \int d\Omega \, d\Omega' \frac{h(\hat{\mathbf{n}})h(\hat{\mathbf{n}}')}{|\mathbf{r}-\mathbf{r}'|}. \quad (1)$$

Since the water is incompressible, one also has the equation

$$\int_R^{(R+h)} d\Omega \, dr r^2 = 0,$$

which, to second order in h, becomes

$$R \int d\Omega \, h(\hat{\mathbf{n}}) = -\int d\Omega \, h^2(\hat{\mathbf{n}}). \quad (2)$$

When this expression is substituted in Eq. (1) and the identity

$$\frac{1}{|\mathbf{r}-\mathbf{r}'|} = \frac{1}{R} \sum \frac{4\pi}{2l+1} Y_{lm}(\hat{\mathbf{n}}) \, Y_{lm}^*(\hat{\mathbf{n}}')$$

is used, one finally obtains

$$U = U_0 + \frac{4\pi G\rho_0^2 R^5}{3} \sum \frac{(l-1)}{(2l+1)} A_{l,m}^* A_{l,m}.$$

It should be noted that because of incompressibility (Eq. (2)), the mode with $l = 0$ is not independent. If all $(l \neq 0)$-modes vanish, so does the $(l = 0)$-mode. Otherwise, the $(l = 0)$-mode does occur; however, it is of second order in smallness. In addition, for the mode with $l = 1$, corresponding to uniform displacement of the sphere, there is no change in the gravitational energy.

The kinetic energy $(\rho_0/2) \int dV |\mathbf{v}|^2$ becomes $(\rho_0 R^2/2) \int \phi(\partial\phi/\partial r) \, d\Omega$ when $\mathbf{v} = \nabla\phi$, as in irrotational flow, and $\nabla \cdot \mathbf{v} = 0$ as for an incompressible fluid. To find ϕ we note that

$$\dot{h}(\hat{\mathbf{n}}, t) = \mathbf{v} \cdot \hat{\mathbf{n}} = \frac{\partial\phi}{\partial r}(R),$$

and expanding $\phi = \sum b_{lm}(r/R)^l Y_{lm}(\hat{\mathbf{n}})$ one finds

$$b_{lm} = R^2 \dot{A}_{lm}/l.$$

The kinetic energy thus becomes

$$\frac{\rho_0 R^5}{2} \sum \frac{1}{l} \dot{A}_{l,m}^* \dot{A}_{l,m}.$$

The total Hamiltonian for small oscillations is the kinetic plus potential energies,

$$H = \sum_{l,m} \left\{ \left(\frac{\rho_0 R^5}{2l} \right) \dot{A}^*_{l,m} \dot{A}_{l,m} + \left[\frac{4\pi \rho_0^2 G R^5 (l-1)}{3(2l+1)} \right] A^*_{l,m} A_{l,m} \right\}.$$

The frequency of the mode $A_{l,m}$ is thus

$$\omega_l^2 = \frac{8\pi \rho_0 G l(l-1)}{3(2l+1)},$$

or, in terms of the acceleration of gravity,

$$g = \left(\frac{4\pi}{3} \rho_0 R^3 \right) \frac{G}{R^2}, \qquad \omega_l^2 = \frac{2l(l-1)g}{R(2l+1)}.$$

The period of the lowest mode ($l=2$) is $\tau = 2\pi/\omega_2 = \pi (5R/g)^{1/2} = 5640$ sec.

36. From the relativistic transformation of coordinates, one may write

$$t = \gamma_1 (t_1 + x_1 v_1); \qquad x = \gamma_1 (x_1 + v_1 t_1); \qquad \text{and } t_2 = \gamma_2 (t - x v_2).$$

But $x_1 \equiv 0$ since the clock is at rest in S_1. These equations are easily solved to yield $t_2 = \gamma_2 (1 - v_1 v_2)t$.

37. At time t, as measured in the rest frame of the star, let the rocket have a velocity v. Consider the Lorentz transformation from this frame to an inertial frame moving with velocity v: $x = \gamma(x' + vt')$ and $t = \gamma(t' + vx')$. Then

$$\frac{dx}{dt} = \frac{v + (dx'/dt')}{\left[1 + v\left(\frac{dx'}{dt'} \right) \right]} \quad \text{and} \quad \frac{d^2x}{dt^2} = \frac{d^2x'/dt^2}{\gamma^3 \left[1 + v\left(\frac{dx'}{dt'} \right) \right]^3}.$$

For $(dx'/dt') = 0$ we have the transformation law for accelerations along the direction of motion, $a = a'\gamma^{-3}$. We also have the transformation law for time intervals $dt' = dt/\gamma$. The total time elapsed as seen by the moving clock is

$$\int dt' = \int \frac{1}{\gamma} dt = \int \frac{1}{\gamma a} dv = \frac{1}{a'} \int \frac{dv}{\left(1 - \frac{v^2}{c^2} \right)},$$

or

$$T' = \frac{1}{2a'} \log \frac{(1 + v_f/c)}{(1 - v_f/c)},$$

where v_f is the final velocity as seen by the star and a' is the constant acceleration as seen by the rocket. To calculate v_f we notice that the total distance traveled is given by

$$D = \int v\, dt = \int \frac{v\, dv}{a} = \frac{1}{a'} \int \frac{v\, dv}{(1 - v^2)^{3/2}} \quad \text{or} \quad D = \frac{1}{a'} \left[\frac{1}{\sqrt{1 - v_f^2}} - 1 \right].$$

Elimination of v_f between these two equation gives T' in terms of D and a'

38. The period τ is given by $\tau = 4 \int_0^a dx/v$, where the velocity v is calculated from the energy equation

$$mc^2 \left(1 - \frac{v^2}{c^2}\right)^{-1/2} + \frac{m\omega^2 x^2}{2} = mc^2 + \frac{m\omega^2 a^2}{2}.$$

In this manner the expression for the period becomes

$$\tau = \frac{4}{\omega} \int_0^a \frac{[1 + \omega^2 (a^2 - x^2)/2c^2]}{(a^2 - x^2)^{1/2} \left[1 + \frac{\omega^2}{4c^2}(a^2 - x^2)\right]^{1/2}}.$$

We expand the integrand in powers of $\omega^2 a^2/c^2$, and get, as the leading terms in the expression for the period, $\tau = (2\pi/\omega)[1 + 3\omega^2 a^2/16c^2 + \cdots]$.

39. Define the four-vector $p = p_\pi + p_n = \bar{p}_p + p_D$, where, in the lab frame, $p = (m_p + m_D, \mathbf{0}) \equiv (M, \mathbf{0})$. Solving for $p_n = p - p_\pi$ and squaring, one finds

$$m_n^2 = p_n^2 = (p - p_\pi)^2 = (M - E_\pi)^2 - \mathbf{p}_\pi^2 = M^2 - 2E_\pi M + m_\pi^2.$$

Therefore

$$E_\pi = \frac{M^2 + m_\pi^2 - m_n^2}{2M} \approx 1.24 \text{ BeV}.$$

The lab frame was chosen to evaluate the scalar product $p \cdot p_\pi$; since it is a Lorentz invariant quantity, any convenient inertial frame may be chosen.

40. (a) Any collection of particles has a center-of-mass velocity

$$\mathbf{v} = \frac{\mathbf{p} \text{ (total)}}{E \text{ (total)}},$$

as can be seen from the Lorentz transformation of momentum $p'_\parallel = \gamma(p_\parallel - Ev)$; $p'_\perp = p_\perp$. Choosing $\mathbf{v}$ parallel to $\mathbf{p}$, we see $\mathbf{p}' = \mathbf{0}$ when $\mathbf{v} = \mathbf{p}/E$. In our case the result is $\mathbf{v} = (\mathbf{p}_+ + \mathbf{p}_-)/(E_+ + E_-)$.

(b) The total energy and momentum constitute a Lorentz four-vector; hence the quantity $[(E_+ + E_-)^2 - (\mathbf{p}_+ + \mathbf{p}_-)^2]$ is an invariant. The barycentric frame is the one in which $\mathbf{p}'_+ + \mathbf{p}'_- = \mathbf{0}$; as $m_+ = m_-$, one also has $E'_+ = E'_-$ in this frame. Thus

$$4(E'_+)^2 = (E_+ + E_-)^2 - (\mathbf{p}_+ + \mathbf{p}_-)^2$$

or finally

$$E'_+ = E'_- = \frac{\sqrt{(E_+ + E_-)^2 - (\mathbf{p}_+ + \mathbf{p}_-)^2}}{2}.$$

(c) Consider the invariant $I = (\mathbf{p}_+ - \mathbf{p}_-)^2 - (E_+ - E_-)^2$. In the rest frame of the electron, $\mathbf{p}_- = 0$, $E_- = m$, we find that $E_+ = m/\sqrt{1 - v_{\text{rel}}^2}$ where

v_{rel} is the relative velocity. Thus

$$I = \frac{2m^2}{\sqrt{1 - v_{rel}^2}} - 2m^2, \text{ and we have } v_{rel} = \left[1 - \frac{1}{\left(1 + \frac{I}{2m^2}\right)^2}\right]^{1/2}.$$

41. Choose a coordinate system with x-axis to the right and y-axis vertical. The force equations are

$$\frac{d}{dt}(p\cos\theta) = 0 \quad \text{and} \quad \frac{d}{dt}(p\sin\theta) = \frac{-eV}{d}.$$

Thus $p\cos\theta = p_0\cos\alpha$ is a constant of motion, while the second equation gives $(d/dt)(p_0\cos\alpha\tan\theta) = -eV/d$. We use $\tan\theta = dy/dx$ and the approximation $d/dt = c(d/ds)$ where c is the speed of light. This is a good approximation for a relativistic electron since its speed is approximately c throughout its motion. Thus we obtain

$$\frac{d}{ds}\left(\frac{dy}{dx}\right) = \frac{-eV}{dp_0c\cos\alpha} \quad \text{or} \quad \frac{d^2y}{dx^2} = \frac{-eV}{cp_0d\cos\alpha}\left[1 + \left(\frac{dy}{dx}\right)^2\right]^{1/2}.$$

The general solution to this differential equation is

$$y = A - \frac{1}{\beta}\cosh\beta(x - \alpha),$$

where $\beta = eV/cp_0d\cos\alpha$. The constants A and a are calculated from $y(0) = 0$ and $dy(0)/dx = \tan\alpha$.

42. Let one photon have momentum $\mathbf{p}$ in the lab, and $\mathbf{p}'$ in the pion rest frame. For a photon $|\mathbf{p}| = E$. The Lorentz transformation connects $\mathbf{p}$ and $\mathbf{p}'$:

$$p'\sin\theta' = p\sin\theta,$$
$$p'\cos\theta = p(\cos\theta - v),$$
$$p' = p(1 - v\cos\theta).$$

By division,

$$\tan\theta' = \frac{1}{\gamma}\frac{\sin\theta}{\cos\theta - v} \quad \text{and} \quad \cos\theta' = \frac{\cos\theta - v}{1 - v\cos\theta}.$$

Probability is conserved, so $P(\theta)\,d\Omega = P'(\theta)\,d\Omega'$. Then

$$P(\theta) = P'(\theta)\frac{d(\cos\theta')}{d(\cos\theta)} = P'(\theta)\frac{1}{\gamma^2(1 - v\cos\theta)^2} = \frac{1}{4\pi\gamma^2(1 - v\cos\theta)^2}$$

is the normalized probability density.

43. (a) One year contains 3.15×10^7 sec. Consequently $t_{earth} = \gamma t_{neutron} = 10^3\,\gamma$ sec $= 3.15 \times 10^7$ sec $\implies \gamma = 3.15 \times 10^4$. As seen from the earth, the neutron energy is 3.15×10^4 times the neutron rest mass, or about 30,000 BeV.

(b) Let variables in the neutron rest frame be primed, and those in the earth frame be unprimed. The angles made by the decay product with respect to the neutron velocity (designated as along the z-axis) are θ and θ' in the two systems. The relation

$$\tan \theta = \frac{1}{\gamma} \frac{u' \sin \theta'}{u' \cos \theta' + v},$$

where u' is the velocity of the decay product in question, is derived in the preceding problem. This relation can also be found from the Einstein addition law for velocities.

Both the neutrino and electron are ultrarelativistic in the primed frame. For the electron, $d(\tan \theta)/d\theta' = 0$ implies $\cos \theta' = -u'/v$. Then

$$\tan \theta_{\max} = \frac{u'/v}{\gamma_{\text{C.M.}}} \frac{1}{\sqrt{1 - (u'/v)^2}} \approx \frac{\gamma_{\text{electron}}}{\gamma_{\text{C.M.}}}.$$

Now $\gamma_{\text{C.M.}} = 3.15 \times 10^4$, while $\gamma_{\max}$ (electron) ≈ 2.6. We see $\theta_{\max} \approx 10^{-4}$ rad.

(c) The method of (b) breaks down when $(u' = 1) > v$, since this would lead to $\cos \theta' < -1$. The largest angle θ is π; we obtain this for backward motion in the neutron rest frame.

(d) In the neutron frame, the neutrino has maximum energy

$$E_\nu \approx M_{\text{N}} - M_{\text{p}} - M_{\text{e}} = 0.8 \text{ MeV}.$$

And for a neutrino emitted backwards in the lab frame,

$$E^{\max} = \gamma[E' + \mathbf{v} \cdot \mathbf{p}']$$

$$= E'\gamma(1 - v) \approx E' \sqrt{\frac{1-v}{2}}$$

$$\approx \frac{E'}{2\gamma_{\text{C.M.}}} \approx 12.7 \text{ eV}.$$

44. We begin with a crude order-of-magnitude estimate. From Kepler's Laws, $\nu \sim E^{3/2}$, where E is the energy excluding the rest mass. The first-order change in ν due to relativistic effects is

$$\frac{d\nu}{\nu} = \frac{3}{2} \frac{dE}{E} = \frac{3}{2} \frac{\frac{3}{8}(v^4/c^2)}{GM/2R} = \left(\frac{9}{16c^2}\right) \frac{(GM/R)^2}{(GM/2R)}$$

$$= \frac{9}{8} \frac{GM}{Rc^2} = \frac{\delta\theta_0}{2\pi} \Longrightarrow \delta\theta_0 \approx \frac{2\pi GM}{Rc^2}.$$

This result is not rigorous. The relativistic effects may be manifested through a speeding up of the particle in the nonrelativistic orbit, as well as through a rotation of the perihelion. In fact, this is just what happens for a circular orbit. For an orbit with eccentricity, the perihelion does precess, the rate of precession being roughly that given above. A more exact treatment follows.

The radial part of the force law is

$$F_r = \frac{d}{dt}(m\gamma\dot{r}) - m\gamma r\dot{\theta}^2 = -\frac{\partial V}{\partial r}.$$

The time is eliminated through the angular momentum equation $L = m\gamma r^2\dot{\theta}$, with L a constant of motion, from which

$$\frac{d}{dt} = \frac{L}{m\gamma r^2}\frac{d}{d\theta}.$$

In terms of the variable $u = 1/r$, the force law becomes

$$-\frac{L^2 u^2}{m\gamma}\frac{d^2 u}{d\theta^2} - \frac{L^2 u^3}{m\gamma} = u^2\frac{\partial V}{\partial u},$$

which is

$$\frac{d^2 u}{d\theta^2} + u = -\frac{(E-V)}{c^2 L^2}\frac{\partial V}{\partial u},$$

when the energy equation $E = mc^2\gamma + V$ is used. In a gravitational field (neglecting general relativity), $V = -GMm/r \equiv -ku$, and the force equation becomes

$$u'' + u\left[1 - \frac{k^2}{c^2 L^2}\right] = \frac{Ek}{c^2 L^2},$$

with solution

$$u = \frac{1}{r} = \frac{Ek}{c^2 L^2 d^2}[1 + \epsilon\cos\alpha(\theta - \theta_0)],$$

where $\alpha = (1 - k^2/c^2 L^2)^{1/2} \approx 1 - k^2/2c^2 L^2$.

In one revolution θ increases by 2π; thus the perihelion advances through an angle $\delta = 2\pi(1 - \alpha) = \pi k^2/c^2 L^2$ per revolution. For a nearly circular orbit of radius R this becomes

$$\delta = \frac{\pi GM}{Rc^2}.$$

This is one-sixth the amount predicted in the general theory of relativity.

Alternate solution: In the nonrelativistic Kepler problem there are two well-known constants of the motion, namely the energy and the angular momentum. There is, however, an additional constant of the motion, namely

$$\mathbf{A} = \hat{\mathbf{r}} + \frac{\mathbf{L}\times\mathbf{p}}{km}.$$

This may be verified from the identity $d(\hat{r})/dt = \mathbf{L}\times r/mr^2$ and the force law, from which $\mathbf{p} = -k\hat{r}/r^2$. The necessity for an additional constant of the motion follows from the "accidental" degeneracy of the Coulomb potential; the Hamilton-Jacobi equation (in quantum mechanics, the Schrödinger equation) is separable in both spherical and parabolic coordinates. $\mathbf{A}$ has the property of pointing always to the perihelion of the orbit; its length is the eccentricity ϵ of the orbit; $\mathbf{A}$ is called the Runge-Lenz vector.*

In special relativity, $\mathbf{A}$ is no longer conserved. If K is a frame moving with the angular velocity $\delta\boldsymbol{\omega}$ of the precessing perihelion, then

$$\left(\frac{d\mathbf{A}}{dt}\right)_K = \left(\frac{d\mathbf{A}}{dt}\right)_{\text{inertial}} - \delta\boldsymbol{\omega} \times \mathbf{A}.$$

Since

$$\left(\frac{dr}{dt}\right)_{\text{inertial}} = \frac{\mathbf{L} \times \hat{\mathbf{r}}}{m\gamma r^2},$$

$$\left(\frac{d\mathbf{A}}{dt}\right)_K = \frac{\mathbf{L} \times \hat{\mathbf{r}}}{mr^2}\left(\frac{1}{\gamma} - 1\right) - \delta\boldsymbol{\omega} \times \hat{\mathbf{r}} - \frac{\delta\boldsymbol{\omega} \times (\mathbf{L} \times \mathbf{p})}{km},$$

or

$$\frac{\mathbf{L} \times \hat{\mathbf{r}}}{2mr^2}\beta^2 = \delta\boldsymbol{\omega} \times \hat{\mathbf{r}} - \mathbf{p}\frac{(\delta\boldsymbol{\omega} \cdot \mathbf{L})}{km} + \left(\frac{d\mathbf{A}}{dt}\right)_K.$$

We have chosen $\delta\boldsymbol{\omega}$ so as to make the average of $(d\mathbf{A}/dt)_K$ equal to zero. When averaged over the nonrelativistic orbit, the term in $\mathbf{p}$ vanishes. Hence

$$\langle\delta\boldsymbol{\omega}\rangle = \frac{\mathbf{L}\langle\beta^2/r^2\rangle}{2m}.$$

In the limit that the nonrelativistic motion is circular, of radius R,

$$\delta\omega = \frac{L\beta^2}{2mR^2} = \frac{\omega^3\beta^2}{2},$$

and the precession is given by $\Delta\theta = 2\pi\delta\omega/\omega = \pi\beta^2$. But $v^2 = GM/R$; thus $\Delta\theta = \pi GM/RC^2$.

45. The balloon moves in the direction of the acceleration. The acceleration of the frame, by the equivalence principle, has the same physical consequences as a uniform gravitational field in the opposite direction. Hence we may imagine the container stationary on the earth's surface. Then the balloon rises.

---

* W. Lenz, *Z. Physik* **24**, 197 (1924); W. Pauli, *Z. Physik* **36**, 336 (1926); D. F. Greenberg, *Am. J. Phys.* **34**, 1101 (1966).

3

ELECTROMAGNETISM

1. From the symmetry of the cube the currents in the resistors must be as shown in the figure. Conservation of current at the corners requires:

$$I = 2x + y, \quad \text{and} \quad y = 2z,$$

where I is the input current. The requirement that the voltage between A and B be independent of path yields the additional equation $2xR = 2(y + z)R$. These three equations have the solution $x = 3I/8$; $y = I/4$; and $z = I/8$. The resistance between A and B is $(2xR/I)$; therefore $R_{AB} = 3R/4$.

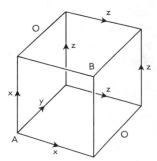

2. If 1 A is fed into A and taken out at infinity, then from symmetry, $\frac{1}{4}$ A will flow in AC. Likewise if 1 A is taken from C and fed in at infinity, $\frac{1}{4}$ A will flow in AC. By the superposition of the two solutions, we obtain the solution to the given problem with the current in AC being $\frac{1}{2}$ A.

3. Place the bars as shown at the right: If 1 is magnetized and 2 is not, there will be no net attraction between the magnets, by symmetry. If 2 is magnetized and 1 is not, the bars will attract one another, because of the poles induced in 1 by the field of 2.

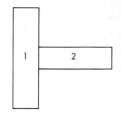

4. Due to the linearity of the equations of electrostatics, one has $q_c = C_1 V$, and $q_p = C_2 V$ for the charges on the conductor and plate respectively when they are in contact, and V is their common voltage. Thus

$$q_c/q_p = C_1/C_2 = \text{const.}$$

After the first contact $q_c = q$ and $q_p = Q - q$, and ultimately $q_p \to Q$; hence $q_c \to Qq/(Q - q)$.

5. The battery supplies the constant power, $P = EI$. The electrostatic energy of the capacitor, $U = qE/2$, is changing at the rate

$$\frac{dU}{dt} = \frac{E}{2}\frac{dq}{dt} = \frac{EI}{2}.$$

Thus the battery is doing twice as much work as is being stored in the capacitor. The difference appears as work done by the capacitor on the external agent that is causing the capacitance to change.

6. Let Q be the charge on one plate; let σ be the charge density on this plate surface; let I, J be the interplate current and current density respectively; let V be the voltage between the plates.

$$\frac{V}{R} = I = \int \mathbf{J}\cdot d\mathbf{A} = g \int \mathbf{E}\cdot d\mathbf{A} = \frac{g}{\epsilon} \int \sigma dA \qquad \text{(Gauss' Law)}$$

$$= \frac{gQ}{\epsilon} = \left(\frac{g}{\epsilon}\right) CV.$$

Hence $RC = \epsilon/g$.

7. The linearity of Maxwell's equations allows us to think of the magnetic field as arising from two current densities:

(1) A current density $j = I/\pi(b^2 - a^2)$, carried by the cylinder of radius b, and

(2) a current density $-j$ carried by a cylinder of radius a.

The sum of the current densities (1) and (2) is the current distribution of the bored-out cylinder. From Ampère's circuital law $\oint \mathbf{H}\cdot d\mathbf{l} = (4\pi/c) \int \mathbf{j}\cdot d\mathbf{A}$, one finds that (1) produces a magnetic field $H = 2Id/c(b^2 - a^2)$ at the center of the hole, while (2) produces no magnetic field at the center of the hole. The resultant magnetic field, H, is thus given by $H = 2Id/c(b^2 - a^2)$.

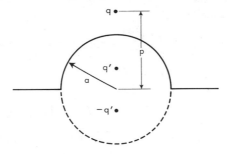

8. The solution is by the method of images. We choose image charges inside the conductor, with location and strength so as to make the conductor an equipotential. To make the surface of the boss equipotential, we must place an image $q' = -qa/p$ at distance a^2/p from the origin, on the line joining the origin with q. In order to keep the plane at constant potential, it is necessary to add additional images $-q'$, $-q$ at distances $-a^2/p$ and $-p$ inside the conductor. The force on q is then

$$F = q^2 \left[\frac{-ap}{(p - a^2/p)^2} + \frac{a/p}{(p + a^2/p)^2} - \frac{1}{4p^2} \right] = -q^2 \left[\frac{4a^3 p^3}{(p^4 - a^4)^2} + \frac{1}{4p^2} \right].$$

9. At the origin, the magnetic field is

$$\mathbf{H} = \int \frac{\hat{\mathbf{r}} \rho_M d^3 x}{r^2} + \int \frac{d\sigma_M \hat{\mathbf{r}}}{r^2}, \tag{1}$$

where $\rho_M = -\nabla \cdot \mathbf{M} = 0$ and $d\sigma_M = \mathbf{M} \cdot d\mathbf{A}$ act as magnetic volume and surface densities. On each surface, $|\mathbf{r}|$ is constant, and

$$\frac{-d\sigma_M}{r^2}\bigg|_b = \frac{d\sigma_M}{r^2}\bigg|_a = -M \cos\theta \sin\theta\, d\theta\, d\phi.$$

Upon substituting in (1), $\mathbf{H} = 0$.

10. From Gauss' Law, $E = 4\pi\sigma$, where σ is the surface charge density. $\mathbf{E}$ is normal to the plane. When the disk rises an infinitesimal amount dx, the energy of the field is decreased by an amount equal to the volume of the excluded space times the electrostatic energy density: $dU = -(E^2/8\pi)$ $(A dx)$, where A is the area of the disk. There is, therefore, a repulsive force

$$F = -\frac{dU}{dx} = \frac{AE^2}{8\pi} = \frac{A(4\pi\sigma)^2}{8\pi} = 2\pi\sigma^2 A.$$

The disk rises when this exceeds the weight of the disk.

Alternate solution: The field has two sources—the charge on the disk and the charge on the plane. The former cannot give rise to a net force on the disk. We calculate the field due to the charge on the disk, and subtract from the total field, as given by Gauss' Law. The potential at a distance x above the plane, on the disk axis, is

$$V(x) = \int_0^R \frac{2\pi\, \sigma\, r\, dr}{(r^2 + x^2)^{1/2}} = 2\pi\, \sigma[(R^2 + x^2)^{1/2} - x],$$

where R is the disk radius. Then

$$E(0) = -\frac{dV}{dx}\bigg|_{x=0} = -2\pi\sigma \left[\frac{x}{\sqrt{R^2 + x^2}} - 1 \right]_{x=0} = 2\pi\sigma.$$

The E which acts on the disk is therefore $2\pi\sigma$, and the force on it is $F = \sigma \int \mathbf{E} \cdot d\mathbf{A} = 2\pi\, \sigma^2 A$.

11. The electrostatic potential $\phi(r)$ is a function of the radius alone. This is because $\phi(R_1, \theta, \phi) = V_1$ independent of θ and ϕ, and the continuation of this function into the dielectric must also be independent of θ and ϕ.

Thus on a surface of radius R the electric field is radial and constant in magnitude over the entire surface.

Applying Gauss' Law to a sphere with radius R such that $R_1 < R < R_2$, one obtains

$$\int \mathbf{D} \cdot d\mathbf{A} = 4\pi Q$$

$$= -2\pi R^2 E \int_{+1}^{-1} (\epsilon_0 + \epsilon_1 \cos^2 \theta) d(\cos \theta)$$

$$= \frac{4\pi R^2}{3} E(3\epsilon_0 + \epsilon_1),$$

from which $E = 3Q/(3\epsilon_0 + \epsilon_1)R^2$. The potential difference between the plates is

$$V = \int_{R_1}^{R_2} \mathbf{E} \cdot d\mathbf{R} = \frac{3Q(R_2 - R_1)}{(3\epsilon_0 + \epsilon_1)R_1 R_2} \equiv \frac{Q}{C}.$$

Thus

$$C = \frac{R_1 R_2 (3\epsilon_0 + \epsilon_1)}{3(R_2 - R_1)}.$$

12. We write $\mathbf{H} = -\nabla\phi$. Then, since an infinite wire in empty space with current I has field $B_\theta = 2I/r$ (Ampère's Law), we find that

$$\phi = -2I\{\text{Im part of } \ln (x + iy)\}$$

to within a constant. Putting an image current I_2 inside the magnetic medium at distance a from the interface, we get

$$\phi = -2I \ln [(x - a) + iy] - 2I_2 \ln [(x + a) + iy]$$
$$\phi_2 = -2I_3 \ln [(x - a) + iy],$$

where ϕ and ϕ_2 are the potentials in vacuum and in the medium respectively.

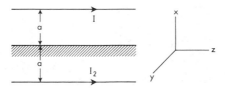

The boundary conditions are

  (1) The tangential component of $\mathbf{H}$ is continuous.

  (2) The normal component of $\mathbf{B} = \mu\mathbf{H}$ is continuous. This gives two equations, from which

$$I_2 = I\frac{(\mu - 1)}{(\mu + 1)} \quad \text{and} \quad I_3 = \frac{2I}{\mu + 1}.$$

The force on I is just that due to the image current I_2. The field due to I_2

at the wire which carries current I is

$$B = \frac{2I_2}{r} = \frac{(\mu - 1)I}{(\mu + 1)a}.$$

This results in a force per unit length of $F/L = IB = [(\mu - 1)/(\mu + 1)]$ (I^2/a) on the wire. The force is attractive.

13. We use $+$ to designate the medium where $\mu = 1$, and $-$ to designate the medium where $\mu = 2$. The insertion of the loop on the interface evokes image currents only at the position of the true current, but these change only the strength, not the character of the **B** field. This is because the sources of **B** are the true current density **j** and $\nabla \times \mathbf{M}$ where **M** is the magnetization/cm³. However, **M** is proportional to **H** and $\nabla \times \mathbf{H}$ vanishes everywhere except where $\mathbf{j} \neq 0$.

Now $\oint \mathbf{H} \cdot d\mathbf{l} = I_{\text{true}}$ around any loop, where I is the enclosed current. In case A,

$$\mathbf{B}_+^A = \mu_+ \mathbf{H}_+ = \mathbf{H}_+$$

and

$$\mathbf{B}_-^A = \mu_- \mathbf{H}_- = 2\mathbf{H}_-,$$

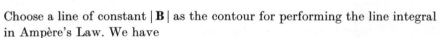

whereas in case B,

$$\mathbf{B}_+^B = \mathbf{H}_+ \quad \text{and} \quad \mathbf{B}_-^B = \mathbf{H}_-.$$

Choose a line of constant $|\mathbf{B}|$ as the contour for performing the line integral in Ampère's Law. We have

$$I = \oint \mathbf{H}^A \cdot d\mathbf{l} = \int_+ \mathbf{B}_+^A \cdot d\mathbf{l} + \frac{1}{\mu} \int_- \mathbf{B}_-^A \cdot d\mathbf{l}$$

$$= \frac{3}{2} \int_+ \mathbf{B}_+^A \cdot d\mathbf{l} = \int \mathbf{H}^B \cdot d\mathbf{l} = 2 \int_+ \mathbf{B}_+^B \cdot d\mathbf{l}.$$

Hence $\mathbf{B}_+^A = (4/3)\mathbf{B}_+^B$. If ϕ is the magnetic flux through the loop,

$$\frac{\phi_A}{\phi_B} = \frac{L_A I}{L_B I} = \frac{L_A}{L_B} = \frac{\int \mathbf{B}^A \cdot d\mathbf{A}}{\int \mathbf{B}^A \cdot d\mathbf{A}} = \frac{4}{3}.$$

14. Far away from the inclusion the electric field may be written $\mathbf{E} = \mathbf{E}_0 - \nabla V$, where $\mathbf{E}_0$ is the constant electric field. The perturbative effects of the inclusion are represented in V, where

$$V = \sum_{l, m} A_{l, m} \frac{1}{r^{l+1}} Y_l^m(\theta, \phi).$$

In this spherical harmonic expansion, the origin is taken in the interior of the inclusion. The problem now is to find the first nonzero term in this expansion. It is easily shown that the term with $l = 0$ vanishes. This is because the integral of $\int \mathbf{E} \cdot d\mathbf{A}$ over a surface enclosing the inclusion is equal to $4\pi A_0$. If this did not vanish, then charge would be leaving the inclusion at the rate $\int \mathbf{j} \cdot d\mathbf{A} = \sigma_0 \int \mathbf{E} \cdot d\mathbf{A} = 4\pi\sigma_0 A_0$. This violates the steady-state nature of the problem. Thus $A_0 \equiv 0$, and the leading term in the perturbation is

$$V \sim \frac{1}{r^2} \sum_m \alpha_m Y_1^m(\theta, \phi).$$

Thus

$$|\mathbf{E} - \mathbf{E}_0| \propto \frac{1}{r^3}.$$

15. Begin with the equation

$$\nabla^2 V = 0 = \frac{\partial^2 V}{\partial r^2} + \frac{1}{r} \frac{\partial V}{\partial r} + \frac{1}{r^2} \frac{\partial^2 V}{\partial \theta^2}.$$

We separate variables and find

$$V = \sum_{m=0}^{\infty} \left(\frac{r}{a}\right)^m [A_m \cos(m\theta) + B_m \sin(m\theta)].$$

Now

$$V(r = a, \theta) = \sum_{m=0}^{\infty} [A_m \cos(m\theta) + B_m \sin(m\theta)].$$

$B_m = 0$ by symmetry since $V(\theta) = V(-\theta)$. A_m may be calculated when $V(a, \theta)$ is multiplied by $\cos(n\theta)$, and integrated over θ from zero to 2π; then

$$\int_{-\pi/2}^{\pi/2} V_1 \cos n\theta\, d\theta + \int_{\pi/2}^{3\pi/2} V_2 \cos n\theta\, d\theta = \frac{2(V_1 - V_2)}{n} \sin\left(\frac{n\pi}{2}\right) = \pi A_n,$$

and $\pi(V_2 + V_1) = 2\pi A_0$. Thus

$$V = \frac{(V_1 + V_2)}{2} + \frac{2(V_1 - V_2)}{\pi} \sum_{n=1}^{\infty} \frac{(-1)^{n-1}}{(2n-1)} \left(\frac{r}{a}\right)^{2n-1} \cos[(2n-1)\theta].$$

Finally, it is worthwhile to note that the series expansion may be summed. This is accomplished by writing

$$\sum_{n=1}^{\infty} (-1)^{n-1} \frac{x^{2n-1}}{(2n-1)} \cos[(2n-1)\theta] = \mathrm{Re} \int_0^x dy \left(-\frac{i}{y}\right) \sum_{n=1}^{\infty} (iye^{i\theta})^{2n-1},$$

which, after making use of the expansion

$$\frac{1}{(1-x)} = \sum_{n=0}^{\infty} x^n \qquad \text{for} \qquad |x| < 1,$$

becomes

$$\mathrm{Re} \int_0^x dy \left(\frac{-i}{2y}\right) \left\{\frac{1}{1 - iye^{i\theta}} - \frac{1}{1 + iye^{i\theta}}\right\}.$$

Upon completing the integration, we obtain

$$\sum_{n=1}^{\infty} \frac{(-1)^{n-1}}{(2n-1)} x^{2n-1} \cos\left[(2n-1)\theta\right] = \frac{1}{2} \operatorname{Im} \log\left\{\frac{1+ixe^{i\theta}}{1-ixe^{i\theta}}\right\}$$

$$= \frac{1}{2} \tan^{-1}\left\{\frac{2x\cos\theta}{1-x^2}\right\}.$$

Thus

$$V(r,\theta) = \frac{(V_1+V_2)}{2} + \frac{(V_1-V_2)}{\pi} \tan^{-1}\left\{\frac{2ar\cos\theta}{a^2-r^2}\right\}.$$

This closed-form expression can also be obtained through the use of Green's-function techniques.

16. Combining Gauss' Law, div $\mathbf{E} = 4\pi\rho$; Ohm's Law, $\mathbf{J} = \sigma\mathbf{E}$; and the continuity equation $\partial\rho/\partial t + \operatorname{div} \mathbf{J} = 0$, we find an equation for the time dependence of ρ:

$$\rho = \rho_0 e^{-4\pi\sigma t}.$$

The characteristic time for a charge to disappear is therefore $(4\pi\sigma)^{-1}$; for copper this is 1.5×10^{-19} sec. This time is so short that a charge carrier would have to exceed the speed of light to travel even a very short distance. This indicates that the solution must be invalid. This may be attributed to a failure of Ohm's Law. The charge cannot remain inside the conductor; if the conductor is completely insulated, the charge appears on the interface as a surface charge density.

17. The conducting sphere is the source of an electric field $E = bV/r^2$, which polarizes the dielectric. An induced dipole $\mathbf{p}$ has energy $U = -\frac{1}{2}\mathbf{p}\cdot\mathbf{E}$ in an applied field $\mathbf{E}$. However, $\mathbf{p} = \alpha\mathbf{E}$ and the force on the dielectric is given by $\mathbf{F} = -\boldsymbol{\nabla}U$. Thus

$$F_r = -\frac{\partial U}{\partial r} = -\frac{2\alpha b^2 V^2}{r^5}.$$

18. $\mathbf{P} = N \alpha \mathbf{E}_{\text{local}} = N\alpha\left[\mathbf{E}_{\text{mac}} + 4\pi \mathbf{P}/3\right]$, where $\mathbf{E}_{\text{mac}}$ is the macroscopic electric field.

Now $\epsilon \mathbf{E}_{\text{mac}} \equiv \mathbf{E}_{\text{mac}} + 4\pi\mathbf{P}$, so

$$\mathbf{P} = 4\pi N\alpha\left[\frac{(\epsilon+2)}{(\epsilon-1)}\right]\frac{\mathbf{P}}{3}.$$

Consistency requires that $(\epsilon - 1)/(\epsilon + 2) = 4\pi N \alpha/3$.

19. The field of the atoms outside the cavity is calculated from the average macroscopic polarization density $\mathbf{P}$ by assigning surface charge densities $\sigma = \mathbf{P}\cdot\hat{\mathbf{n}}$ and a volume charge density $\rho = -\boldsymbol{\nabla}\cdot\mathbf{P}$ to the medium. Thus

$$\mathbf{E}' = \int_{\substack{\text{Cavity}\\\text{wall}}} \frac{(-\mathbf{P}\cdot\hat{\mathbf{n}})(-\hat{\mathbf{n}})dA}{r^2} + \mathbf{E}_{\text{s}} + \mathbf{E}_{\text{v}}.$$

The first term is the contribution of the surface charge density $(-\mathbf{P} \cdot \hat{n})$ on the cavity wall ($\hat{n}$ is taken outward from the cavity). The term $\mathbf{E}_s$ represents the contribution from the other surface boundaries of the dielectric, and $\mathbf{E}_v$ is the contribution from the volume density $\rho = -\nabla \cdot \mathbf{P}$. The essential point now is to consider $\mathbf{P}$ as constant in the immediate neighborhood of the cavity. Then one may identify $\mathbf{E} + \mathbf{E}_s + \mathbf{E}_v$ with the average macrosopic field $\mathbf{E}_{mac}$; in addition,

$$\int_{\text{Cavity}} \frac{(-\mathbf{P} \cdot \hat{n})(-\hat{n})dA}{r^2} = \frac{4\pi \mathbf{P}}{3}.$$

Thus $\mathbf{E} + \mathbf{E}' = \mathbf{E}_{mac} + (4\pi/3)\,\mathbf{P}$. The electric field due to nearest neighbors located at $\mathbf{x}_i$, with electric dipole moment $\mathbf{p}_i$, is

$$\mathbf{E}'' = \sum_{i=1}^{6} \frac{3(\mathbf{p}_i \cdot \mathbf{x}_i)\mathbf{x}_i - x_i^2 \mathbf{p}_i}{x_i^5}.$$

In case (a), two atoms have $|\mathbf{x}_i| = x_L$, while the other four have $|\mathbf{x}_i| = x_s$. All dipoles have the same magnitude and direction along $\mathbf{E}$; thus

$$\mathbf{E}'' = \frac{12\mathbf{p}(x_s - x_2)}{x_s^4} \qquad \text{for} \qquad (x_L - x_s) \ll x_s.$$

In addition, $\mathbf{p} = \alpha \mathbf{E}_l$ and $\mathbf{P} = N\alpha \mathbf{E}_l$; hence

$$\mathbf{E}_l = \frac{\mathbf{E}_{mac}}{\left(1 - \dfrac{4\pi N\alpha}{3} - 12\dfrac{(x_s - x_L)\alpha}{x_s^4}\right)}.$$

The index of refraction n is found from

$$\mathbf{P} = N\alpha \mathbf{E}_l \equiv \left[\frac{(\epsilon - 1)}{4\pi}\right]\mathbf{E}_{mac} \qquad \text{with} \qquad \epsilon = n^2.$$

From the given lattice constant, one finds $N = 1.25 \times 10^{23}/\text{cm}^3$, and from the value of n for the undistorted lattice, $\alpha = 0.83 \times 10^{-24}$ cm^3. Then in case (a) one has $n = 2$, which is lower than the undistorted value. In case (b) $\mathbf{E}'' = 6\mathbf{p}(x_L - x_s)/x_s^4$, and is of half the magnitude of and opposite in sign to the contribution in (a); thus $n = 2.1$.

20. The critical angle is determined from Snell's Law. This requires the calculation of the index of refraction $n(\lambda)$. This proceeds as follows:

$$m\ddot{x} = +e\mathbf{E} = e\mathbf{E}_0\, e^{-i\omega t} = -m x_0 \omega^2 e^{-i\omega t},$$

where $\omega = 2\pi c/\lambda$, and the electron is assumed to oscillate with the same frequency as the x-ray, and with amplitude x_0 about a given ion. The maximum dipole moment induced, assuming stationary ions, is

$$e x_0 = -\frac{e^2 \mathbf{E}_0}{m\omega^2} \tag{1}$$

for a single ion-electron pair. In the presence of binding, the denominator of (1) would contain an additional term $m\omega_0^2$, representing the characteristic frequency of the electron bound to the ion.

The polarization of the metal is $\mathbf{P} = -Ne^2\mathbf{E}_0/m\omega^2$, and the polarizability $\alpha = \mathbf{P}/\mathbf{E} = -Ne^2/m\omega^2$. Now $\mathbf{D} = \epsilon\mathbf{E} = \mathbf{E} + 4\pi\mathbf{P} = \mathbf{E}(1 + 4\pi\alpha)$, from which

$$n^2 \equiv \epsilon = 1 - \frac{4\pi Ne^2}{m\omega^2} \qquad \text{(less than 1!)}.$$

From Snell's Law $n_1 \cos\theta_1 = n_2 \cos\theta_2$ (angles measured with respect to the surface), the critical angle θ_1, for which $\theta_2 = 0$, is found to be given by

$$\cos^2\theta_c = n^2 = 1 - \frac{4\pi Ne^2}{m\omega^2}$$

or

$$\sin\theta_c = \left(\frac{4\pi Ne^2}{m\omega^2}\right)^{1/2} \equiv \frac{\omega_p}{\omega},$$

where ω_p is the plasma frequency. For $\omega < \omega_p$ the index of refraction is pure imaginary, and one obtains total reflection at all angles.

21. We choose coordinates as in the figure. We ignore the magnetic field of the traveling wave, in comparison with that of the earth's field. The ionospheric electrons have the equation of motion

$$m\left(\frac{d\mathbf{v}}{dt}\right) = e\mathbf{E}e^{-i\omega t} + e\mathbf{v} \times \frac{\mathbf{H}}{c}.$$

We regard $\mathbf{E}$ as a superposition of right- and left-hand polarized beams, $E_0(\hat{\mathbf{x}} \pm i\hat{\mathbf{y}})e^{-i\omega t}$. Motion of the electrons is in the $z = 0$-plane, and must have the same time dependence as $\mathbf{E}$. This motivates putting $\mathbf{v}$ equal to $v_0(\hat{\mathbf{x}} \pm i\hat{\mathbf{y}})e^{-i\omega t}$. Then

$$\mathbf{v} \times \mathbf{H} = \pm iHv_0(\hat{\mathbf{x}} \pm i\hat{\mathbf{y}})e^{-i\omega t}$$

and

$$v_0(-im\omega \mp ieH/c) = eE_0,$$

hence

$$v_0 = \frac{ieE_0}{m(\omega \pm \omega_0)}, \qquad \text{where} \qquad \omega_0 = \frac{eH}{mc}.$$

The current density is $J_0 = Nev_0 = iNe^2E_0/m(\omega \pm \omega_0)$. But

$$\text{curl } \mathbf{H} = \frac{1}{c}\frac{\partial\mathbf{E}}{\partial t} + \frac{4\pi\mathbf{J}}{c} = -\frac{i\omega}{c}\left[1 - \frac{\omega_p^2}{\omega(\omega \pm \omega_0)}\right]\mathbf{E},$$

where $\omega_p^2 = 4\pi Ne^2/m$ is the square of the plasma frequency. On the other hand, in the absence of a current, but in a dielectric medium,

$$\mathbf{curl\ H} = \frac{\epsilon}{c}\frac{\partial \mathbf{E}}{\partial t} = -\left(\frac{i\omega\epsilon}{c}\right)\mathbf{E}.$$

By comparison, $\epsilon_{\pm} = n_{\pm}^2 = 1 - \omega_p^2/\omega(\omega \pm \omega_0)$. Right- and left-hand po-larized beams travel with different phase velocities c/n_+ and c/n_-, rotating $\mathbf{E}$. If at $z = 0$, $\mathbf{E} = \mathbf{E}_+ + \mathbf{E}_-$ is in the x-direction, then after propagating a distance z,

$$\mathbf{E}_+ + \mathbf{E}_- = E_0\{\hat{\mathbf{x}}[e^{i\omega((n_+z/c)-t)} + e^{i\omega((n_-z/c)-t)}] \\ + i\hat{\mathbf{y}}[e^{i\omega((n_+z/c)-t)} - e^{i\omega((n_-z/c)-t)}]\},$$

which implies a rotation through an angle θ such that

$$\tan \theta = \frac{i(e^{i\omega n_+z/c} - e^{i\omega n_-z/c})}{(e^{i\omega n_+z/c} + e^{i\omega n_-z/c})}.$$

Putting $n_+ - n_- = \delta n$, one finds $\tan \theta = -\tan (\omega\delta nz/2c)$ or $\theta = -\omega z\delta n/2c$.

22. Consider a solution of a Maxwell's equation inside the cavity of the form (known as transverse electric)

$$E_z = 0, E_x = E_1(x, y)e^{i(kz - \omega t)}$$

$$E_y = E_2(x, y)e^{i(kz - \omega t)}.$$

From the wave equation

$$\nabla^2 \mathbf{E} - \frac{1}{c^2}\frac{\partial^2}{\partial t^2}\mathbf{E} = 0,$$

one obtains the equation

$$\left(\frac{\partial^2}{\partial x^2} + \frac{\partial^2}{\partial y^2}\right)E_i - (k^2 - \omega^2/c^2)E_i = 0. \quad (1)$$

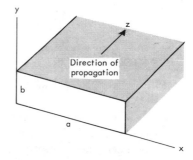

The solution satisfying the boundary conditions that the tangential com-ponent of E and normal component of H vanish on the boundaries is:

$$E_1 = E_{01} \cos\left(\frac{n\pi x}{a}\right)\sin\left(\frac{m\pi y}{b}\right), \qquad E_2 = E_{02}\sin\left(\frac{n\pi x}{a}\right)\cos\left(\frac{m\pi y}{b}\right),$$

with $nE_{01}/a + mE_{02}/b = 0$ in order to guarantee $\nabla\cdot\mathbf{E} = 0$. Equation (1) then becomes

$$\frac{\omega^2}{c^2} = k^2 + \pi^2\left(\frac{n^2}{a^2} + \frac{m^2}{b^2}\right) \qquad (n, m \text{ are integers, and we take } a > b).$$

Since $k^2 \geq 0$ for transmission, there is a minimum frequency given by $\omega_0 = c\pi/a$. The phase velocity is given by

$$v_p = \frac{\omega}{k} = c\left\{1 + \left(\frac{\pi^2}{k^2}\right)\left(\frac{n^2}{a^2} + \frac{m^2}{b^2}\right)\right\}^{1/2},$$

while the group velocity is given by

$$v_g = \frac{d\omega}{dk} = c\left\{1 + \left(\frac{\pi^2}{k^2}\right)\left(\frac{n^2}{a^2} + \frac{m^2}{b^2}\right)\right\}^{-1/2}$$

Similar results are obtained for transverse magnetic modes; i.e., $H_z = 0$; however, the cutoff frequency is higher. One notices also that $v_p v_g = c^2$.

23. $\mathbf{J}$ is constant in time, since $\mathbf{E} = 0$ everywhere. Because the slab is infinite, $\mathbf{H}$ and $\mathbf{J}$ can be functions of z only. From Maxwell's equation

$$\mathbf{curl\ B} = \frac{4\pi\mathbf{J}}{c},$$

one obtains

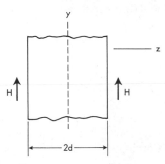

$$\nabla^2\mathbf{B} - \frac{4\pi\mathbf{B}}{\lambda}c^2 = 0,$$

where we have used the identity $\mathbf{curl\ curl} = \mathbf{grad}\ \mathrm{div} - \nabla^2$. A solution must be found obeying the boundary condition $\mathbf{B}(\pm d) = \mathbf{H}_0$. The solution desired is

$$\mathbf{B}(z) = \mathbf{H}_0\frac{(e^{kz} + e^{-kz})}{(e^{kd} + e^{-kd})} = \mathbf{H}_0\frac{\cosh{(kz)}}{\cosh{(kd)}},$$

where $k^2 = 4\pi/\lambda c^2$. The current density is determined from

$$\frac{4\pi\mathbf{J}}{c} = \mathbf{curl}\ B = -\hat{\mathbf{x}}\frac{\partial B}{\partial z} = -\hat{\mathbf{x}}kH_0\frac{\sinh{(kz)}}{\cosh{(kd)}}.$$

The field $\mathbf{H}$ has only the external currents as its sources and hence $\mathbf{H} = \mathbf{H}_0$ everywhere. This is no contradiction since $\mathbf{B} = \mathbf{H} + 4\pi\mathbf{M}$, where $\mathbf{M}$ is the the magnetization per unit volume, and should satisfy $c\nabla \times \mathbf{M} = \mathbf{J}$. This is easily checked in this problem since $\nabla \times \mathbf{H} = 0$. Then $\nabla \times \mathbf{M} = (1/4\pi)$ $\nabla \times \mathbf{B} = (1/4\pi)\ (4\pi\ \mathbf{J}/c) = \mathbf{J}/c$, and everything is consistent. One sees that $\mathbf{B}$ is confined to a region of the surface of skin depth $1/k$ and that a superconductor does not allow $\mathbf{B}$ to penetrate the interior.

24. Consider Maxwell's equations

$$\nabla \cdot \mathbf{D} = 0, \tag{1}$$

$$\nabla \cdot \mathbf{B} = 0, \tag{2}$$

$$\nabla \times \mathbf{E} = \frac{1}{c}\frac{\partial \mathbf{B}}{\partial t}, \tag{3}$$

$$\nabla \times \mathbf{H} = \frac{1}{c}\frac{\partial \mathbf{D}}{\partial t}. \tag{4}$$

If $\mathbf{P}$ is the polarization in the rest frame of the cylinder, then to first order in v/c, one has

$$\mathbf{P} = \alpha\left(\mathbf{E} + \frac{\mathbf{v} \times \mathbf{B}}{c}\right) = \left(\frac{\epsilon - 1}{4\pi}\right)\left(\mathbf{E} + \frac{\mathbf{v} \times \mathbf{B}}{c}\right) \quad \text{and} \quad \mathbf{M} = \frac{\mathbf{P} \times \mathbf{v}}{c},$$

with

$$\mathbf{D} = \mathbf{E} + 4\pi\mathbf{P} \quad \text{and} \quad \mathbf{H} = \mathbf{B} - 4\pi\mathbf{M}.$$

Thus to first order in v/c, one may write

(a) $\mathbf{D} = \epsilon\mathbf{E} + (\epsilon - 1)\dfrac{\mathbf{v} \times \mathbf{B}}{c},$ (b) $\mathbf{H} = \mathbf{B} + (\epsilon - 1)\dfrac{\mathbf{v} \times \mathbf{E}}{c},$

which may be rewritten in a form more convenient for our purposes:

(a') $\mathbf{E} = \dfrac{1}{\epsilon}\left\{\mathbf{D} - (\epsilon - 1)\dfrac{\mathbf{v} \times \mathbf{B}}{c}\right\},$ (b') $\mathbf{H} = \mathbf{B} + \dfrac{(\epsilon - 1)}{\epsilon}\dfrac{\mathbf{v} \times \mathbf{D}}{c}.$

Equations (a) and (b) are the nonrelativistic limit of Minkowski's equations for moving media (see Landau and Lifshitz, *Electrodynamics of Continuous Media*. Reading, Mass: Addison–Wesley, 1960).

We now look for circularly polarized solutions of the type

$$\begin{aligned} \mathbf{D} &= D_0(\hat{\mathbf{x}} \pm i\hat{\mathbf{y}})e^{i(k_{\pm}z - \omega t)} \quad \text{with} \quad D_z = 0, \\ \mathbf{B} &= B_0(\hat{\mathbf{x}} \pm i\hat{\mathbf{y}})e^{i(k_{\pm}z - \omega t)} \quad \text{with} \quad B_z = 0. \end{aligned} \tag{5}$$

One sees that Eqs. (1) and (2) are satisfied. Substituting Eqs. (a') and (b') into Eqs. (3) and (4), we make use of the property that

$$\nabla \times (\mathbf{v} \times \mathbf{A}) = \boldsymbol{\Omega} \times \mathbf{A}$$

when $\mathbf{A} = \mathbf{A}(z, t)$, with $A_z = 0$ and $\mathbf{v} = \boldsymbol{\Omega} \times \mathbf{r}$, with $\boldsymbol{\Omega}$ constant along the z-direction, and obtain:

$$\frac{1}{\epsilon}\nabla \times \mathbf{D} - \frac{(\epsilon - 1)}{\epsilon c}\boldsymbol{\Omega} \times \mathbf{B} = -\frac{1}{c}\frac{\partial \mathbf{B}}{\partial t}, \tag{3'}$$

and

$$\nabla \times \mathbf{B} + \frac{(\epsilon - 1)}{\epsilon c}\boldsymbol{\Omega} \times \mathbf{D} = \frac{1}{c}\frac{\partial \mathbf{D}}{\partial t}. \tag{4'}$$

For circularly polarized solutions of the type in Eq. (5), these become

$$\pm \frac{k_{\pm}}{\epsilon}D_0 \pm \frac{i\Omega(\epsilon - 1)}{\epsilon c}B_0 = \frac{i\omega}{c}B_0, \tag{3''}$$

and

$$\pm k_{\pm}B_0 \mp \frac{i\Omega(\epsilon - 1)}{\epsilon c}D_0 = -\frac{i\omega}{c}D_0. \tag{4''}$$

If this system of homogeneous equations for D_0 and B_0 is to have a solution, the determinant of the coefficients of D_0 and B_0 must vanish. Thus

$$k_{\pm}^2 = \frac{\epsilon\omega^2}{c^2}\left\{1 \mp \frac{(\epsilon - 1)\Omega}{\epsilon\omega}\right\}^2. \tag{6}$$

A wave, which at $z = 0$ is polarized along the x-axis (i.e. $E(0) = E_0\hat{x}$), becomes at $z = L$

$$\mathbf{E}(L) = \frac{E_0}{2}[\hat{x}(e^{ik_+L} + e^{ik_-L}) + i\hat{y}(e^{ik_+L} - e^{ik_-L})].$$

Hence the plane of polarization is rotated through an angle $\theta = (k_- - k_+)L/2$ in the same sense of rotation as the cylinder. Using Eq. (6), we find

$$\theta = \frac{(n^2 - 1)}{n}\left(\frac{L\Omega}{c}\right),$$

where $n = \sqrt{\epsilon}$ is the index of refraction of the dielectric.

This problem was first treated (incorrectly) by J. J. Thomson; then (correctly) by E. Fermi in *Rend. Lincei.* **32** (I) (1923). Professor Fermi put the problem on the 1948 "Basic," as a library problem. One student located Fermi's original paper, translated it, and was the only student to give a correct solution.

25. Since $V_0 \gg E_1x_1$ or E_2x_2, it is a good approximation to take the trajectory of a particle to be a straight line except in the immediate vicinity of the slit. Here the particle is acted upon by a transverse electric field which bends the trajectory. This transverse electric field may be found by use of the equation $\nabla \cdot \mathbf{E} = 0$ at the center of the slit. There, $(\partial E_x/\partial x)_0 \simeq (E_2 - E_1)/t$, where t is the thickness of the lens. Expanding E_y in the neighborhood of $y = 0$, we write $E_y \simeq ay$. Thus $\nabla \cdot \mathbf{E} = 0$ implies $a \simeq -(E_2 - E_1)/t$.

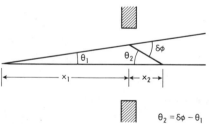

$$\theta_2 = \delta\phi - \theta_1$$

A particle of charge e entering the slit at height y is therefore acted upon by a force $-e(E_2 - E_1)y/t$. This force acts for a time t/v, where v is the velocity of the particle. The net impulse given is then $\Delta p = -e(E_2 - E_1)y/v$. This deflects the trajectory by the angle $\delta\phi = -\Delta p/p$. The particle will therefore intersect the x-axis at a point

$$x_2 = \frac{y}{\theta_2} \qquad \text{or} \qquad \frac{1}{x_2} = \frac{\theta_2}{y}.$$

Substituting $\theta_2 = \delta\phi - \theta_1$ and $\theta_1 = y/x_1$, we obtain

$$\frac{1}{x_1} + \frac{1}{x_2} = \frac{e(E_2 - E_1)}{pv} = \frac{(E_2 - E_1)}{2V_0}.$$

26. Since the magnetic intensity changes gradually, it will be assumed that the motion in the plane perpendicular to the axis is approximately circular. The angular frequency of revolution is the cyclotron frequency,

$$\omega = -\frac{eB}{mc}. \tag{1}$$

In addition, the cylindrical symmetry of the solenoid guarantees that the component of angular momentum along the axis (taken to be the z-axis) is conserved. Thus

$$\hat{\mathbf{z}} \cdot (\mathbf{r} \times \mathbf{p}) = \hat{\mathbf{z}} \cdot \left[\mathbf{r} \times \left(m\mathbf{v} + \frac{e}{c}\mathbf{A} \right) \right] = mr^2\omega + \frac{er^2 B}{2c} = \text{const},$$

and using Eq. (1) we obtain

$$mr^2\omega = mr_1^2\omega_1, \tag{2}$$

where $\frac{1}{2}mr_1^2\omega_1$ is the angular momentum for $B = B_1$. In addition, for motion in a magnetic field, the kinetic energy is constant; thus

$$\tfrac{1}{2}mv_l^2 + \tfrac{1}{2}mr^2\omega^2 = \tfrac{1}{2}mv_{1l}^2 + \tfrac{1}{2}mr_1^2\omega_1^2. \tag{3}$$

If the point of reflection (where $v_l = 0$) occurs at field strength B, then Eqs. (1) through (3) yield

$$v_{1l}^2 = v_{1t}^2\left\{ \frac{B}{B_1} - 1 \right\}, \qquad \text{where} \qquad v_{1t} \equiv r_1\omega_1.$$

Since $B \leq B_2$, the condition that reflection occur somewhere is

$$v_{1l} \leq v_{1t}\left(\frac{B_2}{B_1} - 1 \right)^{1/2}$$

27. Choosing a cylindrical coordinate system, the θ-component of the force equation reads

$$\frac{d}{dt}(mr^2\dot{\theta}) = \frac{e}{c}r\dot{r}H,$$

and conservation of energy requires that

$$\tfrac{1}{2}m(\dot{r}^2 + r^2\dot{\theta}^2) - eV(r) = 0.$$

The z-coordinate does not change.

Equation (1) has the solution $mr^2\dot{\theta} = (eH/2c)(r^2 - a^2)$. When this expression for $\dot{\theta}$ is substituted in the energy equation, we have

$$\frac{1}{2}m\left\{ \dot{r}^2 + \left(\frac{eH}{2mcr} \right)^2 (r^2 - a^2)^2 \right\} = eV(r).$$

Threshold is obtained when $\dot{r} = 0$ at $r = b$; therefore

$$V_T = eH^2(b^2 - a^2)^2/8\, mc^2 b^2.$$

28. Consider first the quadrupole lens. The magnetic field **H** is derivable from a potential; $\mathbf{H} = -\mathbf{grad}\ U$. Since

$$\operatorname{div} \mathbf{H} = 0, \qquad (1)$$

U must be a harmonic function, i.e. a function of $z = x + iy$ but not of $z^* = x - iy$ or vice versa. Now any polynomial in z is a solution of Laplace's equation in two dimensions. The correct exponent is chosen on the basis of symmetry. The quadrupole lens has symmetry under rotation by π rad; the potential chosen must reflect this symmetry. Hence we choose $U = kz^2$. Both the real and imaginary parts of U are solutions of (1), but only the imaginary part embodies the boundary conditions given in the figure, namely, that the lines ($xy = $ const) represent equipotentials.

The forces are determined from the Lorentz equation $\mathbf{F} = q(\mathbf{v}/c) \times \mathbf{B}$. Positively charged particles are focused in the yz-plane, and defocused in the x-direction; the opposite applies for negative particles. On the other hand, a particle with no charge but magnetic moment $\boldsymbol{\mu}$ in a nonuniform magnetic field experiences a force given by $\mathbf{F} = -\boldsymbol{\nabla}(-\boldsymbol{\mu}\cdot\mathbf{B}) = (\boldsymbol{\mu}\cdot\boldsymbol{\nabla})\mathbf{B}$. To maintain focusing analogous to that for charged particles in a quadrupole lens, one must have

$$F_x = -\mu_x \frac{\partial^2 U}{\partial x^2} \sim -x \qquad \text{or} \qquad \frac{\partial^2 U}{\partial x^2} \sim y,$$

when $\boldsymbol{\mu} = \mu_x\,\hat{\mathbf{x}}$. This suggests taking $U = kz^3$, and choosing the real part. This potential is symmetric under rotations of $2\pi/3$ rad, as may be seen by writing in polar coordinates, $z = re^{i\theta}$. The sextupole lens, shown in the figure, embodies this symmetry. (Such lenses are actually used for focusing neutrons and neutral atoms. A beam composed of some particles with magnetic moments parallel to the x-axis, and of some others with antiparallel magnetic moments, may be polarized by such a lens, because one of the components is focused while the other is defocused.)

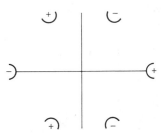

29. For a system of charges all moving with velocity **v** and generating an electric field **E**, the magnetic field is given by $\mathbf{B} = \mathbf{v} \times \mathbf{E}/c$. The force on a given charge is computed from the Lorentz force equation

$$\mathbf{F} = e\left\{\mathbf{E} + \frac{\mathbf{v}}{c} \times \mathbf{B}\right\} = e\left\{1 - \left(\frac{v}{c}\right)^2\right\}\mathbf{E} = e\mathbf{E}.$$

The electric field **E** may be found from Gauss' Law

$$\int \mathbf{E}\cdot d\mathbf{A} = 4\pi \text{ (charge enclosed)},$$

yielding $\mathbf{E} = 2\pi\rho e\mathbf{r}$. Thus there is a net outward force, making the beam unstable in the absence of focusing devices.

30. Magnetic monopoles are introduced into electromagnetic theory by modifying one of the Maxwell equations to read div $\mathbf{B} = 4\pi\rho_M$. An isolated monopole will produce a $\mathbf{B}$ field which can be written down in analogy with electrostatics:

$$\mathbf{B} = \frac{\Gamma}{r^2}\hat{\mathbf{r}}.$$

An electrically charged particle of mass m and charge q then has the equation of motion

$$m\frac{d\mathbf{v}}{dt} = q\mathbf{v} \times \mathbf{B} = \frac{q\Gamma}{r^2}\mathbf{v} \times \hat{\mathbf{r}}.$$

The kinetic energy is conserved, as are the components of the Fierz vector

$$\mathbf{Z} = \left[\mathbf{r} \times \mathbf{p} + \frac{q\Gamma\mathbf{r}}{r}\right].$$

31. (a) The equation of motion of a charge is $dp/dt = (e/c)\,\mathbf{v} \times \mathbf{B}$. Since $\mathbf{v}\cdot\mathbf{F} = 0$, $|\mathbf{v}|$ is constant, we shall write

$$\mathbf{p} = p\frac{d\mathbf{r}}{ds}; \qquad \mathbf{v} = v\frac{d\mathbf{r}}{ds}; \qquad \text{and} \qquad \frac{d}{dt} = v\frac{d}{ds},$$

where ds is an element of the path. Under these conditions the force equation becomes

$$\frac{d^2\mathbf{r}}{ds^2} = \frac{e}{pc}\frac{d\mathbf{r}}{ds} \times \mathbf{B}.$$

(b) For the equilibrium of a current-carrying wire we must have:

$$\frac{I}{c}\mathbf{dl} \times \mathbf{B} + T\left(\frac{d\mathbf{r}}{ds}\right)_2 - T\left(\frac{d\mathbf{r}}{ds}\right)_1 = 0.$$

Upon expanding $(d^2\mathbf{r}/ds^2)_2 = (d\mathbf{r}/ds)_1 + (d^2\mathbf{r}/ds^2)_1\,ds + \cdots$, we obtain:

$$\frac{d^2\mathbf{r}}{ds^2} = -\frac{I}{Tc}\frac{d\mathbf{r}}{ds} \times \mathbf{B}.$$

Thus the current must be chosen according to:

$$I = -\frac{Te}{p}.$$

32. In a weak magnetic field the system of electrons has an additional motion over that in no magnetic field. This motion is a rotation about the nucleus of angular velocity $\mathbf{\Omega} = -e\mathbf{H}/2mc$ (known as Larmor precession). $\mathbf{\Omega}$ is determined by transforming to a rotating frame where the magnetic force is canceled by a Coriolis force.

The equations of transformation from an inertial to a rotating frame are

$$\left(\frac{d\mathbf{r}}{dt}\right)_{\text{inertial}} = \left(\frac{d\mathbf{r}}{dt}\right)_{\text{rot.}} + \mathbf{\Omega} \times \mathbf{r}$$

and

$$\left(\frac{d^2\mathbf{r}}{dt^2}\right)_{\text{inertial}} = \left(\frac{d^2\mathbf{r}}{dt^2}\right)_{\text{rot.}} + 2\boldsymbol{\Omega} \times \left(\frac{d\mathbf{r}}{dt}\right)_{\text{inertial}} + \boldsymbol{\Omega} \times (\boldsymbol{\Omega} \times \mathbf{r}).$$

The force $(e\mathbf{v}/c) \times \mathbf{H}$ may be canceled by the Coriolis force if one chooses $\boldsymbol{\Omega} = -e\mathbf{H}/2mc$. Thus the electron cloud rotates. (Terms of magnitude Ω^2 are neglected, i.e. weak field limit.)

These rotating electrons produce a magnetic field at the origin given by:

$$\Delta\mathbf{H} = \frac{1}{c} \int \frac{\mathbf{j} \times (-\mathbf{r})}{r^3} \, dV \qquad \text{where} \qquad \mathbf{j} = \rho\mathbf{v} = \rho\boldsymbol{\Omega} \times \mathbf{r};$$

hence

$$\Delta\mathbf{H} = -\frac{1}{c} \int \frac{\rho(\boldsymbol{\Omega} \times \mathbf{r}) \times \mathbf{r}}{r^3} \, dV = \frac{1}{c} \int \frac{\{\boldsymbol{\Omega} r^2 - \mathbf{r}(\mathbf{r}\cdot\boldsymbol{\Omega})\}}{r^3} \rho \, dV.$$

But $\rho(\mathbf{r})$ is spherically symmetric, which means that the integral of $\rho\mathbf{r}(\mathbf{r}\cdot\boldsymbol{\Omega})/r^3$ equals that of $\frac{1}{3}(\boldsymbol{\Omega}\rho r^2)/r^3$. Finally

$$\Delta\mathbf{H} = -\left(\frac{e\mathbf{H}}{3mc^2}\right) \int \frac{\rho \, dV}{r} = -\left(\frac{e\mathbf{H}}{3mc^2}\right)\phi(0).$$

Letting $\phi(0) \approx Ze/R$, we have $\Delta H/H = -(Ze^2/mc^2 R) = -Zr_0/R$, where r_0 is the classical electron radius, e^2/mc^2. For $Z = 50$ and $R \sim 10^{-8}$ cm,

$$\frac{\Delta H}{H} \sim -10^{-3}.$$

33. Outside the charge distribution, the electric field is longitudinal (by symmetry) and constant (Gauss' Law). Hence no experiment performed outside the charge radius will detect the pulsations. The apparatus must probe inside the charge distribution. The apparatus required will depend on the extent of the distribution, and the frequency ω. For a very large radius and small ω, one might use an electroscope. For a microscopic system, it might be necessary to use a charged orbiting particle, whose interactions with the electric structure of the nucleus could be made to act as a pulse detector.

34. There is no radiation. This is obvious, because the charge and current distributions are constant in time.

35. Electric-dipole radiation is proportional to $|d\mathbf{D}/dt|^2$, where $\mathbf{D}$ is the electric dipole moment. But $\mathbf{D} = e(\mathbf{r}_1 + \mathbf{r}_2)$, and this has zero time derivative because it is proportional to the center-of-mass vector. The magnetic moment is

$$\mathbf{M} = \frac{1}{2c} \sum_{i=1}^{2} (\mathbf{r}_i \times \mathbf{J}_i) = \frac{e}{2c} \sum_{i=1}^{2} (\mathbf{r}_i \times \mathbf{v}_i) = \frac{e}{2mc}\mathbf{L}.$$

Thus $\mathbf{M}$ is proportional to the angular momentum of the system, which is constant. Magnetic-dipole radiation is proportional to $|d\mathbf{M}/dt|^2$, and so is equal to zero.

36. Let the incident radiation be monochromatic: $\mathbf{E} = \mathbf{E}_0\, e^{-i\omega t}$. Then the induced dipole moment is $\mathbf{P} = \alpha\,\mathbf{E}_0\, e^{-i\omega t}$. If $\hat{\mathbf{n}}$ is a unit vector in the direction of the observer at large distance R, then the electric field is

$$\mathbf{E}(R, \hat{\mathbf{n}}) = \frac{(\ddot{\mathbf{P}} \times \hat{\mathbf{n}}) \times \hat{\mathbf{n}}}{c^2 R} = -\frac{\alpha\omega^2(\mathbf{E}_0 \times \hat{\mathbf{n}}) \times \hat{\mathbf{n}}}{c^2 R}.$$

The total time-averaged power radiated by the scatterer is $|\ddot{\mathbf{P}}|^2/3c^3 = \alpha^2\omega^4\, E_0^2/3c^3$.

Let the ratio (energy scattered per unit time)/(incident energy flux) be σ:

$$\sigma = \frac{\left(\dfrac{\alpha^2\omega^4 E_0^2}{3c^3}\right)}{\left(\dfrac{E_0^2 c}{8\pi}\right)} = \frac{8\pi\alpha^2}{3}\left(\frac{\omega}{c}\right)^4.$$

Note that for free electrons, $\alpha = e^2/m\omega^2$, and one obtains the Thomson cross section

$$\sigma_0 = \frac{8\pi}{3}\, r_0^2,$$

where $r_0 = e^2/mc^2$ is the classical electron radius.

37. The voltage developed in the wire while rotating with angular frequency ω is

$$\mathscr{E} = -\frac{1}{c}\frac{d}{dt}\,(\text{magnetic flux}).$$

Thus

$$\mathscr{E} = -\frac{H_0\pi a^2\omega}{c}\sin{(\omega t)} = IR,$$

where R is the electrical resistance and I the current in the ring. The average power loss to Joule heating per cycle is

$$P = \langle I^2 R\rangle = \frac{H_0^2\pi^2 a^4\omega^2}{2c^2 R}.$$

The only source for this heating is the kinetic energy of the ring, $I\omega^2/2$, where I is the moment of inertia about the axis of rotation, $I = ma^2/2$.

Conservation of energy requires that

$$\frac{d}{dt}\frac{ma^2\omega^2}{4} = -\frac{H_0^2\pi^2 a^4\omega^2}{2c^2 R}.$$

The solution to this equation is $\omega = \omega_0 e^{-t/\tau}$, with $\tau = mRc^2/(\pi H_0 a)^2$.
But $m = 2\pi a A\,\rho$ and $R = 2\pi a/\sigma A$; hence τ becomes

$$\tau = 4\rho c^2/\sigma H_0^2 = 1.6 \text{ sec.}$$

38. The potential energy of the charge and its image is

$$V = -\frac{e^2}{4h} \sim -\frac{e^2}{4(d-L)} + \frac{e^2 L\theta^2}{8(d-L)^2}$$

for small oscillations. The kinetic energy is $K = mL^2\dot{\theta}^2/2$. Thus the frequency is given by $\omega^2 = e^2/4mL(d-L)^2$.

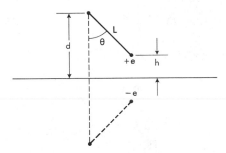

When calculating the energy loss due to radiation, one must include the radiation field of the image so that the electric field will be normal to the conducting plane. Choosing the origin of a spherical coordinate system to be on the conducting plane midway between the source and its image, we find that the radiated electric field in the upper half-plane, in the direction $\hat{\mathbf{n}}$, is proportional to

$$\mathbf{E} \sim \sum e^{-i\mathbf{k}\cdot\mathbf{x}_i}\{\hat{\mathbf{n}} \times (\mathbf{p}_i \times \hat{\mathbf{n}})\}\frac{e^{ikR}}{R}, \tag{1}$$

the sum being over both the source and its image (note $\mathbf{p}$(source) = $-\mathbf{p}$(image) = $e\mathbf{a}$). After squaring and integrating over ϕ, we find that the average power is proportional to

$$|e\mathbf{a}|^2 \sin\theta d\theta \left\{\cos^2\theta + \frac{\sin^2\theta}{2}\right\} \sin^2(kh\cos\theta).$$

Upon change of variables $x = \cos\theta$ and use of the identity

$$\sin^2 x = (1 - \cos 2x)/2,$$

the power is given by

$$P = k|e\mathbf{a}|^2 \int_0^1 dx(1 + x^2)[1 - \cos(2khx)].$$

From this expression we may now determine the constant of proportionality K, since if the interference term proportional to $\cos(2khx)$ is omitted, this expression must reduce to the power radiated by a single dipole, i.e.

$$P \longrightarrow \frac{\omega^4 e^2 a^2}{3c^3}.$$

Hence $K = \omega^4/4c^3$. Of course one could have used Eq. (1) with the multiplicative factors given by radiation theory, and the result would have been the same. Although the integral may now be completed, we shall be content with the approximation $1 - \cos(2khx) \sim 2k^2h^2x^2$, valid for $kh \ll 1$. When use is made of the expression for ω, the condition $kh \ll 1$ becomes $e^2/L \ll 4mc^2$. The power is then

$$P = \frac{4e^2\omega^6a^2h^2}{15c^5} = \frac{e^8a^2}{240m^3c^5L^3h^4}.$$

39. For a single antenna at $\mathbf{r}_i'$,

$$\mathbf{E} = -k^2\hat{\mathbf{n}} \times (\hat{\mathbf{n}} \times \mathbf{p}_i)\frac{e^{i\mathbf{k}\cdot(\mathbf{R}-\mathbf{r}_i')}}{|\mathbf{R} - \mathbf{r}_i'|} = \mathbf{B} \times \hat{\mathbf{n}}.$$

Here $\hat{\mathbf{n}}$ is a unit vector from $\mathbf{r}_i'$ to the field point; $\mathbf{p}_i$ is the electric dipole moment; $k = 2\pi/\lambda$ and R is the position vector to the field point. The fields from the antenna add coherently.

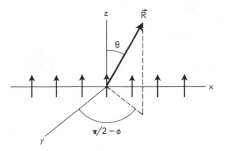
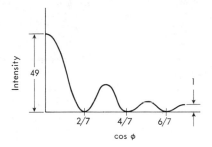

In the radiation zone,

$$-\mathbf{E} = \frac{k^2\hat{\mathbf{n}} \times (\hat{\mathbf{n}} \times \mathbf{p}_0)}{R}e^{ikR}\sum_{i=1}^{7} e^{-ik\hat{\mathbf{n}}\cdot\mathbf{r}_i}.$$

The sum of the phases is

$$e^{3\pi i n_z}\sum_{j=0}^{6}(e^{-in_z\pi})^j = e^{3\pi i n_z}\left(\frac{1 - e^{-i7\pi n_z}}{1 - e^{-i\pi n_z}}\right)$$

and

$$|\Sigma\, e^{-ik\hat{\mathbf{n}}\cdot\mathbf{r}_i}|^2 = \frac{\sin^2\left(\dfrac{7\pi}{2}\sin\theta\cos\phi\right)}{\sin^2\left(\dfrac{\pi}{2}\sin\theta\cos\phi\right)}.$$

a) The average power radiated into solid angle $d\Omega$ is

$$dP = \frac{c\bar{k}^4p_0^4}{8\pi}\sin^2\theta\,\frac{\sin^2\left(\dfrac{7\pi}{2}\sin\theta\cos\phi\right)d\Omega}{\sin^2\left(\dfrac{\pi}{2}\sin\theta\cos\phi\right)}.$$

b) In the xy-plane,

$$\frac{dP}{d\Omega} \sim \frac{\sin^2\left(\dfrac{7\pi}{2}\cos\phi\right)}{\sin^2\left(\dfrac{\pi}{2}\cos\phi\right)}.$$

c) The ratio is 49 : 1. For the array of 7 antennas, the greatest power is radiated along the y-axis; for a single antenna, any direction in the xy-plane.

40. In order to calculate the power radiated by a relativistic particle, one may make use of the fact that the power radiated is an invariant; $P = dE/dt$, and after transforming from rest frame to a moving frame, $dE' = \gamma dE$ and $dt' = \gamma dt$. Hence $dE'/dt' = dE/dt$. Nonrelativistically, the rate of radiation is

$$P = \frac{2e^2}{3m^2}\left(\frac{d\mathbf{p}}{dt}\right)^2,$$

where $\mathbf{p}$ represents the momentum. The Lorentz-invariant generalization reducing to this in the nonrelativistic limit is

$$P = \frac{2e^2}{3m^2}\left(\frac{dp_\mu}{d\tau}\right)\left(\frac{dp_\mu}{d\tau}\right),$$

where $d\tau = dt/\gamma$ is the proper time, and p_μ is the four-momentum. Since $dE = \beta dp$, this is

$$\frac{2e^2}{3m^2}\left[\left(\frac{d\mathbf{p}}{d\tau}\right)^2 - \beta^2\left(\frac{dp}{d\tau}\right)^2\right].$$

For circular motion $dp/d\tau = 0$, while

$$\frac{d\mathbf{p}}{d\tau} = \gamma\frac{d\mathbf{p}}{dt} = \gamma\boldsymbol{\omega}\times\mathbf{p};$$

so

$$P = \frac{2e^2\gamma^2\omega^2 p^2}{3m^2} = \frac{2e^2\gamma^4\beta^4}{3R^2}.$$

The energy radiated per revolution is

$$P\cdot\left(\frac{2\pi}{\omega}\right) = \frac{4\pi}{3}\beta^3\gamma^4\left(\frac{r_0}{R}\right) \qquad \text{in units of } m_0c^2.$$

41. The mirror sees a smaller frequency (longer wavelength) given by $\nu' = \gamma\nu_0(1 - v)$. There is no change of frequency upon reflection. The observer sees a still smaller frequency

$$\nu'' = \gamma\nu'(1 - v) = \nu_0\frac{(1 - v)}{(1 + v)}.$$

The intensity of the beam is the energy crossing unit area in unit time.

Photons   Mirror

This may be computed in two ways.

Method A views the light beam as a free electromagnetic wave, with $|\mathbf{E}| = |\mathbf{H}|$. If $\mathbf{E}$ is the electric field in the source frame, then the mirror sees $\mathbf{E}' = \gamma(\mathbf{E} + \mathbf{v} \times \mathbf{B}) = \sqrt{(1-v)/(1+v)}\ \mathbf{E}$. But $\mathbf{E}'$ is unchanged on reflection, whereas $\mathbf{B}'$ is reversed in direction. The observer sees

$$\mathbf{E}'' = \sqrt{\frac{1-v}{1+v}}\mathbf{E}' = \frac{(1-v)}{(1+v)}\mathbf{E}.$$

The energy density is proportional to the intensity; it depends quadratically on $\mathbf{E}$. Hence $I = I_0[(1-v)/(1+v)]^2$.

Method B regards the light beam as a beam of photons, each having energy proportional to ν, and separated by a distance z_0 from one another along the length of the beam. If at time $t = 0$, a photon is reflected from the mirror, the next photon reaches the mirror at a time t determined by $t = z_0 + vt$, or $t = z_0/(1-v)$. The distance separating the two photons is now $t + vt = z_0$ $(1+v)/(1-v)$. The rate at which photons reach the observer is inversely proportional to their spacing, and therefore depends on the velocity according to $(1-v)/(1+v)$. The frequency (and hence energy) of the photons transforms in the same way, so that the intensity is reduced according to

$$I = I_0[(1-v)^2/(1+v)^2].$$

42. In the rest frame the force per unit length F is given by $\mathbf{F} = \lambda\mathbf{E}$, where $\mathbf{E}$ is the electric field at one wire produced by the other. This is easily found from Gauss' Law,

$$\int \mathbf{E}\cdot d\mathbf{A} = 4\pi \text{ (charge enclosed)}.$$

Thus $E = 2\lambda/a$, and the force $F = 2\lambda^2/a$ (repulsive).

In a frame in which the rods are seen to move with velocity v, there is a magnetic field $\mathbf{B} = v \times \mathbf{E}'/c$, in addition to the electric field $\mathbf{E}'$. The total force per unit length $\mathbf{F}'$ is then

$$\mathbf{F}' = \lambda'\left(\mathbf{E}' + \frac{\mathbf{v} \times \mathbf{B}'}{c}\right) = \lambda'\left(1 - \frac{v^2}{c^2}\right)\mathbf{E}'.$$

However, $\mathbf{E}' = 2\lambda'/a$ where λ' is the charge as seen in the new frame ($\lambda' = \gamma\lambda$ because of the Lorentz contraction of lengths). Thus

$$F' = \frac{2(\lambda')^2(1 - v^2/c^2)}{a} = \frac{2\lambda^2}{a} = F.$$

The fact that $F' = F$ may be seen easily by an alternate argument. If in its rest frame, one of the rods is allowed to move under the action of the force FL on it, it would gain momentum $dp = FL\,dt$, while in the frame in which the rods move, the gain is $dp' = F'L'dt'$. But $dp = dp'$ because momenta normal to the direction of a Lorentz transformation are invariant under such a transformation, and $dt' = \gamma dt$. Hence $LF = \gamma F'L'$. In addition $L' = L/\gamma$, due to Lorentz contraction; hence $F = F'$.

43. If we choose a cylindrical coordinate system, the Lagrangian equations of motion for r and θ are

$$\frac{d}{dt}\left\{m\dot{r}\left(1 - \frac{v^2}{c^2}\right)^{-1/2}\right\} = mr\dot{\theta}^2\left(1 - \frac{v^2}{c^2}\right)^{-1/2}\frac{e\dot{\theta}}{c}\frac{\partial}{\partial r}(rA_\theta),\tag{1}$$

$$\frac{d}{dt}\left\{mr^2\dot{\theta}\left(1 - \frac{v^2}{c^2}\right)^{-1/2} + \frac{erA_\theta}{c}\right\} = 0.\tag{2}$$

Since $B_\theta \equiv 0$, only A_θ has been taken nonzero with

$$B_z = \frac{1}{r}\frac{\partial}{\partial r}(rA_\theta) \qquad \text{and} \qquad B_r = -\frac{\partial A_\theta}{\partial z}.$$

If $\dot{r} = 0$, then Eq. (1) reduces to

$$mr\dot{\theta}\left(1 - \frac{r^2\dot{\theta}^2}{c^2}\right)^{-1/2} + \frac{e}{c}\frac{\partial}{\partial r}(rA_\theta) = 0,\tag{1'}$$

while Eq. (2) has the solution

$$mr\dot{\theta}\left(1 - \frac{r^2\dot{\theta}^2}{c^2}\right)^{-1/2} = -\frac{e}{c}A_\theta.\tag{3}$$

Elimination of the term proportional to m between Eqs. (1') and (3) yields

$$\frac{\partial}{\partial r}(rA_\theta) = A_\theta.\tag{4}$$

From $\oint \mathbf{A}\cdot d\mathbf{l} = \int \mathbf{B}\cdot d\mathbf{A}$, it follows that

$$A_\theta = \frac{1}{2\pi R}\int_0^R 2\pi r B_z(r)dr$$

and Eq. (4) becomes

$$B_z(r_0) = \frac{\langle B_z\rangle}{2},\tag{5}$$

where $\langle B_z\rangle$ is the average of B_z over the area of the orbit, and r_0 is the orbit radius. In addition to this condition, one must also have $B_r = 0$ in the plane of orbit to guarantee $\dot{z} = 0$.

The angular frequency $\dot{\theta}$ is found from Eq. (1') to be

$$\dot{\theta} = -\omega\left(1 + \frac{r_0^2\omega^2}{c^2}\right)^{-1/2},\tag{6}$$

where ω is the cyclotron frequency, $\omega = eB_0/mc$. The kinetic energy of the particle is given by

$$E = mc^2\left\{\left(1 - \frac{r_0^2\dot{\theta}^2}{c^2}\right)^{-1/2} - 1\right\} = mc^2\left\{\left(1 + \frac{r_0^2\omega^2}{c^2}\right)^{1/2} - 1\right\}.$$

In order to calculate the radial oscillations, assume a solution for $r(t)$ defined by

$$r(t) = r_0[1 + \epsilon(t)] \qquad \text{with} \qquad |\epsilon| \ll 1.$$

We shall keep only terms of lowest order in ϵ; the left-hand side of Eq. (1) is then

$$mr_0\ddot{\epsilon}\left(1 + \frac{r_0^2\omega^2}{c}\right)^{1/2},$$

while the right-hand side becomes

$$\frac{e}{c}\dot{\theta}\left(\frac{\partial}{\partial r}(rA_\theta) - A_\theta\right),$$

when Eq. (2) is used. This is to be expanded in ϵ. Now $A_\theta(r)$ is calculated from

$$B_z = \frac{1}{r}\frac{\partial}{\partial r}(rA_\theta) = B_0\left(\frac{r_0}{r}\right)^n;$$

hence

$$A_\theta = -\frac{B_0 r_0^n}{(n-2)r^{(n-1)}} + \frac{K}{r}, \qquad \text{where } K = \text{const.}$$

The constant K is found from the requirement that

$$\left[\frac{\partial}{\partial r}(rA_\theta) - A_\theta\right]_{r=r_0} = 0.$$

Thus $K = B_0 r_0^2 (n-1)/(n-2)$. Upon expanding in ϵ,

$$\frac{\partial}{\partial r}(r\overset{*}{A}_\theta) - A_\theta = -B_0 r_0(n-1)\epsilon.$$

With these small oscillation approximations, Eq. (1) becomes

$$mr_0\left(1 + \frac{r_0^2\omega^2}{c^2}\right)^{1/2}\ddot{\epsilon} = -\frac{eB_0 r_0\dot{\theta}}{c}(n-1)\epsilon.$$

Using Eq. (6) for $\dot{\theta}$ and the definition of the cyclotron frequency, this equation becomes

$$\ddot{\epsilon} + \frac{\omega^2(1-n)\epsilon}{(1 + r_0^2\omega^2/c^2)} = 0,$$

which has bounded sinusoidal solutions only if $n < 1$. For this case the radial frequency is given by ω_r where

$$\omega_r^2 = \frac{\omega^2(1-n)}{(1 + r_0^2\omega^2/c^2)}. \tag{7}$$

For small oscillations in the z-direction, we have the equation

$$m\ddot{z}\left(1 + \frac{r_0^2\omega^2}{c^2}\right)^{1/2} = -\frac{e}{c}B_r(z)r_0\dot{\theta}, \tag{8}$$

obtained from the force equation

$$\frac{dp_z}{dt} = \frac{e}{c}(\mathbf{v} \times \mathbf{B})_z.$$

To find $B_r(z)$ we note that $\nabla \times \mathbf{B} = 0$ in the vicinity of the orbit and thus

$$(\nabla \times \mathbf{B})_\theta = \frac{\partial B_r}{\partial z} - \frac{\partial B_z}{\partial r} = 0.$$

Hence

$$\frac{\partial B_r}{\partial z} = \frac{\partial B_z}{\partial r} = -\frac{n B_0}{r_0}.$$

Thus $B_r = -m B_0 z/r_0$ and when this is substituted in Eq. (8), one obtains

$$\ddot{z} + \frac{\omega^2 n z}{(1 + r_0^2 \omega^2/c^2)} = 0.$$

Hence

$$\omega_z^2 = \frac{n\omega^2}{(1 + r_0^2 \omega^2/c^2)} \qquad \text{and} \qquad \omega_z^2 + \omega_r^2 = \dot{\theta}^2.$$

44. (a) Let p_μ be the four-momentum, with the property

$$p_\mu p_\mu = \mathbf{p}^2 - E^2 = -m^2.$$

The force $d\mathbf{p}/dt$ generalizes naturally to the four-vector $dp_\mu/d\tau$, where τ is the proper time. This is to be set equal to a four-vector function of the electromagnetic field tensor $\mathbf{F}_{\mu\nu} = \partial_\mu A_\nu - \partial_\nu A_\mu$ and p_ν. One also wants an expression linear in the electromagnetic field. This is guaranteed by setting

$$\frac{dp_\mu}{d\tau} = \alpha F_{\mu\nu} p_\nu + \beta \epsilon^{\mu\nu\rho\gamma} F_{\nu\rho} p_\gamma.$$

Here $\epsilon^{\mu\nu\rho\gamma}$ is the completely antisymmetric tensor with $\epsilon^{1234} = 1$. The spatial part of this equation becomes

$$\frac{d\mathbf{P}}{d\tau} = \alpha m(\mathbf{E} + \mathbf{v} \times \mathbf{B}) + 2i\beta m(\mathbf{B} - \mathbf{v} \times \mathbf{E}).$$

Thus one must choose $\alpha = e/m$ and $\beta = 0$ to obtain the Lorentz force equation. The term proportional to β would be the force on a magnetic monopole, if one existed. Since, under parity, or space inversion $\mathbf{p} \longrightarrow -\mathbf{p}$, $\mathbf{E} \longrightarrow -\mathbf{E}$, $\mathbf{B} \longrightarrow \mathbf{B}$, α must be a scalar, even under parity, and β a pseudoscalar, in a parity-preserving theory. Finally the covariant generalization of the Lorentz force equation is

$$\frac{dp_\mu}{d\tau} = \frac{e}{m} F_{\mu\nu} p_\nu.$$

(b)   We invent a four-vector S_μ, with nonrelativistic limit $(0, \mathbf{S})$. Note that $S_\mu p_\mu = 0$ in the rest frame, and therefore is zero in any frame. The covariant generalization of dS/dt is $dS_\mu/d\tau$. Thus, in analogy to the procedure of part (a), one sets

$$\frac{dS_\mu}{d\tau} = \frac{ge}{2m} F_{\mu\nu} S_\nu + b\,p_\mu.$$

The term linear in p_μ is necessary to guarantee $p_\mu S_\mu = 0$. Thus from

$$\frac{d}{d\tau}(S_\mu p_\mu) = p_\mu \frac{dS_\mu}{d\tau} + S_\mu \frac{dp_\mu}{d\tau} = 0 \quad \text{and} \quad \frac{dp_\mu}{d\tau} = \frac{e}{m} F_{\mu\nu} p_\nu,$$

one finds $b = e(g - 2)(p_\mu F_{\mu\nu} S_\nu)/2m^3$.

The covariant equation for the spin is thus

$$\frac{dS_\mu}{d\tau} = \frac{e}{2m} \left\{ g F_{\mu\nu} S_\nu + \frac{(g - 2)}{m^2} p_\alpha F_{\alpha\beta} S_\beta p_\mu \right\}.$$

It is of interest to write explicitly the equation for the component $S_4 = iS_0 = i\mathbf{S} \cdot \mathbf{p}/p_0 = i\mathbf{S} \cdot \mathbf{v}$, in the case $\mathbf{E} = 0$. Then

$$\frac{d(\mathbf{S} \cdot \mathbf{v})}{dt} = \frac{e\gamma}{2m}(g - 2)\mathbf{B} \cdot (\mathbf{v} \times \mathbf{S}).$$

The helicity $\mathbf{S} \cdot \mathbf{v}$ is conserved when $g = 2$. This is indicated in the figure, where the double arrow indicates spin direction. For both the electron and the muon, the g value differs from 2 by a small amount. As a result, the helicity is not quite conserved. This fact was used in the CERN experiment to measure accurately the muon g value.

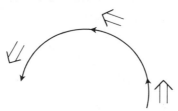

It is worth noting that, given the components of the spin in the rest frame of a particle, $(0, \mathbf{S})$, one may determine the components in a frame moving with velocity $\mathbf{v} = \mathbf{p}/E$ by using the two Lorentz-invariant relations $p_\mu S_\mu = 0$ and $S_\mu S_\mu = -\mathbf{S}^2$. Note that in any frame, the number of independent components of the spin is three, due to the subsidiary condition $p_\mu S_\mu = 0$, which imposes a constraint on the components of the four-vector. This is necessary if the spin is to be described by a three-vector in the rest frame, since the number of independent components should not depend on the frame of the observer.

4

ELECTRONICS

1. The circuit will fail to operate as a generator when ω is so great that an electron cannot respond to the changes in grid voltage, i.e. $\omega\tau > 1$ where τ is the electron's transit time in the tube. This corresponds to $\omega > 3000$ Mc.

2. We solve the more general problem, in which L and C are replaced by Z_1 and Z_2, arbitrary impedances. Let Z be the impedance of the network. Then, using the translational symmetry of the network, we have

$$Z_1 + \frac{Z_2 Z}{Z_2 + Z} = Z,$$

with solution

$$Z = \frac{Z_1 + \sqrt{Z_1^2 + 4Z_1 Z_2}}{2}.$$

The plus sign is used for the square root, since $Z_2 = 0$ implies $Z = Z_1$. In the problem given, $Z_1 = j\omega L$, $Z_2 = -j/\omega C$. Then

$$Z = \frac{j\omega L}{2} + \frac{1}{2}\sqrt{-\omega^2 L^2 + \frac{4L}{C}}.$$

After passing through n stages, an input voltage is reduced to $V_n = V(1 - Z_1/Z)^n \equiv \alpha^n V$, where

$$\alpha \equiv \left(1 - \frac{Z_1}{Z}\right) = \left(\frac{-Z_1 + \sqrt{Z_1^2 + 4Z_1 Z_2}}{Z_1 + \sqrt{Z_1^2 + 4Z_1 Z_2}}\right) = \frac{-j\omega L + \sqrt{4L/C - \omega^2 L^2}}{j\omega L + \sqrt{4L/C - \omega^2 L^2}}.$$

It is easily seen that $|\alpha| = 1$ if $\omega < 2\omega_0$ but $|\alpha| < 1$ if $\omega > 2\omega_0$. Thus the system acts as a low pass filter without attenuation for $\omega < 2\omega_0 = \omega_c$.

3. The presence of C_0 introduces frequency dependence in the amplification ($= V_{\text{out}}/V_{\text{in}}$); this frequency dependence is to be eliminated. This can be done by placing capacitors C_1, C_2, and C_3 across resistors R_1, R_2, and R_3, provided the values of the capacitances are chosen properly. The impedance Z of a parallel combination of R and C is

$$\frac{1}{Z} = \frac{1}{R} + j\omega C - \frac{1}{R}(1 + j\omega RC).$$

It is evident that if the capacitors are such that $R_1C_1 = R_2C_2 = R_3C_3 = R_0C_0$, each resistor-capacitor combination will have the same frequency dependence for its impedance. The amplification, a ratio of impedances, will then be independent of frequencies, and equal to the amplification when the circuit is purely resistive. In other words, the effect of C_0 on the amplification is nullified by the introduction of the other capacitors.

4. The equivalent circuit, using a potential-source representation for the triode, is shown in the figure. Because of the hint, the grid resistor and grid capacitor have been omitted.

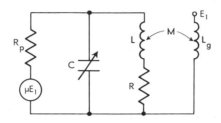

Now $I_L Z_L = I_C Z_C$ and $I_L + I_C = I$, from which $I = I_L(1 + Z_L/Z_C)$. Because of the mutual inductance, $E_1 = I_L Z_M = j\omega M I_L$, and from Ohm's Law,

$$\mu(I_L Z_M) \equiv \mu E_1 = I R_P + I_L Z_L = I_L\left[R_P\left(1 + \frac{Z_L}{Z_C}\right) + Z_L\right].$$

Hence the impedances are related by

$$\mu Z_M = R_P\left(1 + \frac{Z_L}{Z_C}\right) + Z_L. \tag{1}$$

But $Z_C = -j/\omega C$ and $Z_L = R + j\omega L$. Equating real and imaginary parts yields two equations:

$$\omega^2 = \frac{1}{LC}\left(1 + \frac{R}{R_P}\right) \qquad \text{and} \qquad M = \frac{(RR_P C + L)}{\mu}.$$

The latter condition on M is actually a condition for the onset of instability. For $M \geq (RR_P C + L)/\mu$, oscillation occurs. This can be seen by solving Eq. (1) with $\omega = \omega_R + j\omega_I$. Since the time dependence is taken to be $\exp(j\omega t)$, one obtains exploding solutions (i.e., oscillations are possible) if $\omega_I < 0$. Substituting this expression for ω in Eq. (1), one solves for ω_I,

$$-\omega_I = \frac{\mu M - RR_P C - L}{2R_P LC}.$$

Thus $\omega_I \leq 0$ if $M \geq (RR_P C + L)/\mu$. The circuit will fail to oscillate when

$$M < \frac{(RR_P C + L)}{\mu}.$$

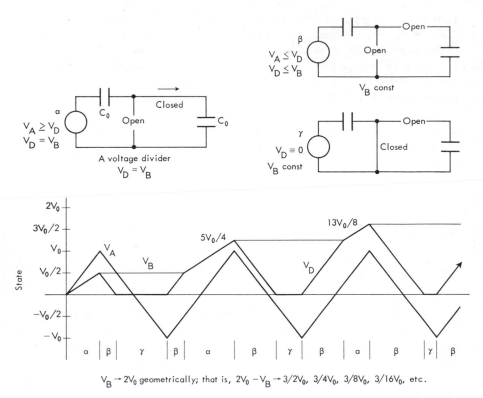

$V_B \to 2V_0$ geometrically; that is, $2V_0 - V_B \to 3/2V_0,\ 3/4V_0,\ 3/8V_0,\ 3/16V_0,$ etc.

5. The operation falls into threé phases, which we label $\alpha,\ \beta,\ \gamma$. During the first quarter-cycle, V_A increases. $V_A > V_D > 0$. The diode D_2 is hence open, and $V_{DB} \geq 0$ allows the diode D_1 to remain closed. In this phase $V_D = V_B$. Then V_A begins to decrease. $V_D > 0$ still; hence D_2 is still open. But V is decreasing, so D_1 opens. Since there is nowhere for the charge on the capacitor to go, V_B remains constant, and V_{AD} remains constant at $V_0/2$, for the same reason. This phase continues until V_D reaches 0. But $V_D < 0$ is not possible, so $V_D = 0$. The first half-cycle has been completed. Now V_A again increases from $-V_0$, and again V_{AD} is constant at $-V_0$ (we are again in the β-phase), and $V_B > V_D$ stays at $V_0/2$ until V_D rises to meet V_B. When $V_D = V_B$, the β-phase has concluded, and a new α-phase begins.

Each cycle goes through the phases $\alpha - \beta - \gamma - \beta$, in that order; the circuit is not periodic in time, however, as may be seen from inspecting the graph. V_B approaches $2V_0$ geometrically; the circuit is known as a voltage doubler.

5

1. Call the object distance for the first lens surface S_1. Then the object distance for the second lens surface is $-S_1'$, where S_1' is given by Snell's law at small angles:

$$\frac{n_1}{S_1} + \frac{n}{S_1'} = \frac{n - n_1}{R_1}.$$

Also,

$$\frac{-n}{S_1'} + \frac{n_2}{S_2} = \frac{n_2 - n}{R_2}.$$

Adding these two equations gives

$$\frac{n_1}{S_1} + \frac{n_2}{S_2} = \frac{n - n_1}{R_1} + \frac{n_2 - n}{R_2}.$$

The focal lengths are obtained by setting $S_1 = \infty$ and $S_2 = \infty$, respectively. Thus

$$f_1 = \frac{n_1}{(n - n_1)/R_1 + (n_2 - n)/R_2},$$

and

$$f_2 = \frac{n_2}{(n - n_1)/R_1 + (n_2 - n)/R_2}.$$

It follows that $f_1/S_1 + f_2/S_2 = 1$.

Special case: When $n_1 = n_2$, then $f_1 = f_2 = f$, and $1/S_1 + 1/S_2 = 1/f$.

2. If the pattern of incident, reflected, and transmitted waves is as is shown in Fig. 1, then time reversal invariance requires that the pattern in Fig. 2 is also possible. In addition, the superposition principle allows Fig. 2 to

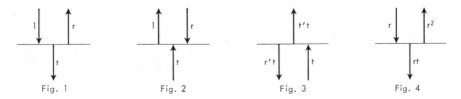

Fig. 1   Fig. 2   Fig. 3   Fig. 4

be considered as the sum of Figs. 3 and 4. Equating the incoming and out-going waves of this sum to those of Fig. 2 yields the Stokes' relations

$$r^2 + tt' = 1 \qquad \text{and} \qquad r + r' = 0.$$

3. If the amplitude of the incident light is denoted by E, the amplitude of the light upon the first reflection from A is defined to be Er_A, and that of the transmitted light, Et_A. The transmitted light undergoes multiple reflections, and some fraction emerges again into the vacuum, the amount being governed by the reflection and transmission coefficients $r_A, r_A', r_B, t_A, t_A'$ defined in the figure. The total amplitude reflected into the vacuum is

$$E_r = Er_A + Et_A t_A' r_B e^{2ikd}[1 + (r_A' r_B e^{2ikd}) + (r_A' r_B e^{2ikd})^2 + \cdots]$$

$$= Er_A + \frac{Et_A t_A' r_B e^{2ikd}}{1 - r_A' r_B e^{2ikd}}$$

$$= E \frac{[r_A + r_B e^{2ikd}(t_A t_A' - r_A' r_A)]}{1 - r_A' r_B e^{2ikd}}.$$

In this expression for E_r, $k = 2\pi n_1/\lambda$ and λ is the vacuum wavelength.

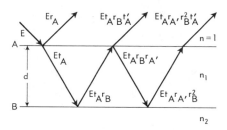

The factor e^{2ikd} is the change in phase for the light traveling a distance $2d$ in the dielectric n_1. Using the Stokes' relations derived in the previous problem, $r_A' = -r_A$ and $t_A t_A' + r_A^2 = 1$, one finds

$$\frac{E_r}{E} = \frac{[r_A + r_B e^{2ikd}]}{[1 + r_A r_B e^{2ikd}]}.$$

We wish now to relate the coefficients r_A and r_B to the indices of refraction, n_1 and n_2. For this it is necessary to use the boundary conditions required by Maxwell's equations. Thus at the interface A the tangential components of the electric and magnetic fields must be continuous across the interface.

$$E + Er_A = Et_A \qquad \text{or} \qquad 1 + r_A = t_A, \tag{1}$$

$$H + H_r = H_t. \tag{2}$$

Now for a plane wave $H = E$, $H_r = -E_r$ (the change of sign is due to a reversal of the direction of motion for the reflected wave), and $H_t = n_1 E_t$. Thus Eq. (2) becomes $1 - r_A = n_1 t_A$ and one finds

$$r_A = (1 - n_1)/(1 + n_1).$$

Similarly

$$r_B = (n_1 - n_2)/(n_1 + n_2).$$

The condition that the reflected wave vanish is

(1) $\begin{Bmatrix} r_A = -r_B \\ e^{2ikd} = +1 \end{Bmatrix}$ or   (2) $\begin{Bmatrix} r_A = r_B \\ e^{2ikd} = -1 \end{Bmatrix}$.

In terms of the indices of refraction, these conditions become

(1) $\begin{Bmatrix} n_2 = 1 \\ n_1 d/\lambda = p/2 \end{Bmatrix}$ or   (2) $\begin{Bmatrix} n_2 = n_1^2 \\ n_1 d/\lambda = (2p+1)/4 \end{Bmatrix}$,

where p is an integer. It should be mentioned that E_r/E may also be calculated by postulating a standing wave electromagnetic field in the dielectric n_1, and solving the system of equations required by matching boundary conditions at the interfaces A and B. Thus for the electric field these equations are

at A, $E_r + E = E_1 + E_2$;   at B, $E_1 e^{ikd} + E_2 e^{-ikd} = E_t$.

The continuity of the magnetic field requires (when one substitutes $H = nE$),

at A, $E - E_r = n_1 E_1 - n_1 E_2$;   at B, $n_1 E_1 e^{ikd} - n_1 E_2 e^{-ikd} = n_2 E_t$.

When E_1, E_2 and E_t are eliminated between these four equations, one finds the above expression for E_r/E.

4. Standing waves of light are formed in the emulsion. Each wavelength λ will therefore develop the film at depths $\lambda(2n+1)/4$ ($n = 1,2\ldots$) in the emulsion (where the intensity has maxima). When white light is incident on a particular layer of silver, a small amount is reflected; most is transmitted. Light of a given wavelength λ will be reflected coherently only from the layers with spacing $\lambda/2$, incoherently from all other layers, thus reproducing the initial colors. A serious limitation to the method is the color distortion caused by developing the emulsion and by temperature effects. A quantum field theoretic treatment of the Lippmann process may be found in E. Fermi, *Revs. Mod. Phys.* **4**, 87 (1932).

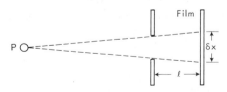

5. Consider a source of light P on the sun. The image of this source is spread over a region of linear dimension $\delta x = D + l(\lambda/D)$, where D is the pin-hole diameter. The first term in δx is due to the spread of rays over the pin-hole opening. In addition, due to diffraction, there is a spread in angle of approximately λ/D upon passing through the opening, and hence a further increase in δx of about $(\lambda/D)l$. The total δx has a minimum for $D^2 = \lambda l$ which for $\lambda = 5000$ Å and $l = 10$ cm gives $D \approx .22$ mm.

6. The Rayleigh criterion asserts that two lines are barely resolvable if the maximum intensity of one coincides with the minimum of the other. Principle maxima in the mth order occur for $d \sin \theta = m\lambda$. Here, $m\lambda$ represents the difference in path lengths for light traveling to a given maximum from two adjacent slits. If the grating has N slits, the difference in path length is $Nm\lambda$ between rays from opposite ends of the grating. Suppose the angle θ is increased slightly, until the path lengths from the opposite ends of the grating differ by $Nm\lambda + \lambda$. The pattern then has zero intensity, as may be seen by a simple argument. Rays from the top slit and the center slit differ in path length by $\lambda/2$, and hence they cancel; similarly amplitudes from other pairs of slits cancel, giving zero intensity.

To satisfy Rayleigh's criterion, these minima must occur for the same angle as the maxima in that order, for light of wavelength $\lambda + \Delta\lambda$. Hence

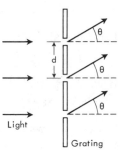

$$mN\lambda + \lambda = mN(\lambda + \Delta\lambda) \quad \text{or} \quad N = \frac{\lambda}{m(\Delta\lambda)}.$$

A grating requires at least 982 slits to separate the sodium doublet in first order ($m = 1$). Note that the result is independent of the grating spacing. The corresponding angular separation is 10^{-5} rad.

7. Ray 1 changes its phase when it reflects from the surface of the lake. Therefore the difference in phases between the two rays arriving at the detector is

$$\phi_1 - \phi_2 = \pi + \frac{2\pi d(1 - \cos 2\theta)}{\lambda \sin \theta}.$$

When the star first appears over the horizon, the difference is π; the waves are out of phase, and the intensity at the detector is minimum. As θ increases, the intensity increases; a maximum is reached when $\sin \theta = \lambda/4d$; this happens for $\theta = 6°$ above the horizon.

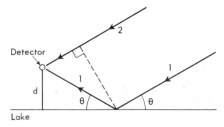

8. Each atom of the ribbon radiates independently of the other atoms of the ribbon. Hence each point produces its own diffraction pattern, and these overlap on the screen. The interference pattern is expected to disappear when the first maximum from the pattern due to A coincides with the first minimum from the pattern due to B (the Rayleigh criterion).

Light from A has interference pattern centered on the observation screen at $\theta_A = -W/2L$; the first minimum is at $\theta_A \pm \lambda/2d$. Light from B has pattern centered at $\theta_B = W/2L$. The pattern vanishes when $\theta_A + \lambda/2d < \theta_B$, or $W > L\lambda/2d$. If, as d is increased, the pattern vanishes at $d = d_0$, then

$$W \simeq \frac{L\lambda}{2d_0}.$$

9. Let a be the amplitude of light coming from a single slit, and let $\delta = (2\pi/\lambda)d \sin \theta$ be the phase difference between rays of light from adjacent slits. The total amplitude is

$$A = a\left(1 + \frac{1}{2} e^{i\delta} + \frac{1}{4} e^{2i\delta} + \frac{1}{2^{(N-1)}} e^{i(N-1)\delta}\right)$$

$$= a\frac{(1 - 1/2^N e^{iN\delta})}{(1 - 1/2\, e^{i\delta})}.$$

The intensity is

$$|A|^2 = a^2 \frac{(1 + 1/4^N - \cos N\delta/2^{(N-1)})}{(\frac{5}{4} - \cos \delta)}.$$

16. $A = a[(1 + \alpha) + (1 - \alpha)e^{i\delta} + (1 + \alpha)e^{2i\delta} + (1 - \alpha)e^{3i\delta}$
$$+ \cdots + (1 - \alpha)e^{i(N-1)\delta}],$$

where the notation is that of the preceding problem, and we have taken N even. The series is finite; we may rearrange terms:

$$A = a[\sum_{n=0}^{N-1} e^{in\delta} + \alpha \sum_{n=0}^{N-1} (-1)^n e^{in\delta}]$$

$$= ae^{i(N-1)\delta/2}\left[\frac{\sin (N\delta/2)}{\sin (\delta/2)} - i\alpha \frac{\sin (N\delta/2)}{\cos (\delta/2)}\right].$$

The intensity is proportional to

$$AA^* = a^2\left[\frac{\sin^2 (N\delta/2)}{\sin^2 (\delta/2)} + \alpha^2 \frac{\sin^2 (N\delta/2)}{\cos^2 (\delta/2)}\right],$$

which may be written as

$$I = I_0 \frac{\sin^2 (N\delta/2)}{N^2 \sin^2 (\delta/2)}\left(1 + \alpha^2 \tan^2 \frac{\delta}{2}\right).$$

The effect of α is most detectable when $\delta \approx (2k + 1)\pi$, when the intensity is $I = \alpha^2 I_0$.

11. The amplitude of the wave on the positive z-axis which is transmitted through the opening is given by

$$\psi = A_0 \int \frac{d\sigma \, e^{ikr}}{r},$$

where $d\sigma$ is an element of surface area of the opening and r is the distance from $d\sigma$ to the observation point. This expression follows from the simple version of Huygen-Fresnel theory where the obliquity factor is taken constant (valid for $(a/z) \ll 1$). Using $d\sigma = 2\pi\rho \, d\rho = 2\pi r \, dr$, we find that the amplitude integrates to

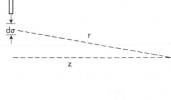

$$\psi = -iA_0\lambda\{\exp(ik\sqrt{a^2 + z^2}) - \exp(ikz)\},$$

which has zeros for $\sqrt{a^2 + z^2} = z + n\lambda$; n is a positive integer, thus $z = (a^2 - n^2\lambda^2)/2n\lambda$.

12. After passing through a polarizer whose relative angle of deviation from the preceding polarizer is θ, the intensity is decreased by the factor $\cos^2\theta$. The attenuation of this polarizer is thus $(1 - \cos^2\theta)$. The probability that the relative angle is between θ and $\theta + d\theta$ is $Be^{-a\theta^2}d\theta$, where B is chosen so that the total probability is unity. The average attenuation per polarizer is thus

$$A = \int_{-\infty}^{\infty} (1 - \cos^2\theta) Be^{-a\theta^2} \, d\theta.$$

Actually the limits of integration are from $-\pi$ to $+\pi$, but $a \gg 1$ and one may extend the limits to infinity to a good approximation. In addition, the integrand is large only in the neighborhood of $\theta = 0$, allowing the approximation $\cos^2\theta \approx 1 - \theta^2$. The average attenuation then becomes

$$A = \int_{-\infty}^{+\infty} d\theta \, \theta^2 Be^{-a\theta^2} = -B\frac{d}{da}\int_{-\infty}^{\infty} e^{-a\theta^2} \, d\theta.$$

As a function of a, $\int_{-\infty}^{\infty} e^{-a\theta^2} \, d\theta = Ca^{-1/2}$, where C is some constant, ($C = \int_{-\infty}^{\infty} d\theta \, e^{-\theta^2}$). Thus $B = a^{1/2}/C$, and

$$A = -\left(\frac{a^{1/2}}{C}\right)\frac{d}{da}(Ca^{-1/2}) = \frac{1}{2a}.$$

The average attenuation per polarizer is thus $1/2a$.

13. For a perfect grating with grating spacing a, the amplitude of the diffracted wave is

$$A = \alpha[1 + e^{i\delta} + e^{2i\delta} + \cdots + e^{i(N-1)\delta}],$$

where α is the amplitude of the wave diffracted by a single slit opening, and

$\delta = (2\pi d/\lambda) \sin \theta = k\, d \sin \theta$. The intensity is

$$|A|^2 = \alpha^2 \frac{\sin^2 (N\delta/2)}{\sin^2 (\delta/2)} = |A_0|^2 \frac{\sin^2 \beta}{\beta^2} \frac{\sin^2 (N\delta/2)}{\sin^2 (\delta/2)},$$

where $\beta = (\pi a/\lambda) \sin \theta$, and A_0 is the amplitude from a single slit, in the direction $\theta = 0$.

When a given slit is displaced by an amount x from equilibrium, the amplitude from that slit is

$$\int_{-a/2+x}^{a/2+x} A_0 e^{iky \sin \theta}\, dy = A_0 \left(\frac{\sin \beta}{\beta}\right) e^{ikx \sin \theta}.$$

Then

$$A = \sum_{m=0}^{N-1} A_0 e^{ikx_m \sin \theta} e^{im\delta} \frac{\sin \beta}{\beta}.$$

The average intensity observed over a time long compared to the period of oscillations is

$$I = \langle |A|^2 \rangle = |A_0|^2 \frac{\sin^2 \beta}{\beta^2} \sum_{m,\,n=0}^{N-1} e^{i(m-n)\delta} \left\langle e^{ik \sin \theta (x_m - x_n)} \right\rangle.$$

Assuming that the displacements of each slit are uncorrelated, with a distribution $e^{-x^2/2\sigma^2}$, where σ is the root-mean-square displacement, one has

$$\left\langle e^{ik(x_m - x_n) \sin \theta} \right\rangle = \delta_{mn} + \frac{(1 - \delta_{mn})}{2\pi\sigma^2} \int_{-\infty}^{+\infty}\int_{-\infty}^{+\infty} dx_m dx_n\, e^{-x_m^2/2\sigma^2} e^{-x_n^2/2\sigma^2} e^{ik \sin \theta (x_m - x_n)}$$

$$= \delta_{mn} + (1 - \delta_{mn}) e^{-k^2\sigma^2 \sin^2 \theta}.$$

Therefore, $I = \phi I_0 + N(1 - \phi)i_0$, where

$$\phi = e^{-k^2\sigma^2 \sin^2 \theta}; \qquad i_0 = |A_0|^2 \frac{\sin^2 \beta}{\beta^2}; \qquad \text{and} \qquad I_0 = i_0 \frac{\sin^2 (N\delta/2)}{\sin^2 (\delta/2)}.$$

14. The velocity of light in the lower conduit is c/n. As seen in the laboratory, the velocity in the upper conduit is (using the relativistic addition formula for velocities)

$$\frac{c/n + u}{1 + [(c/n)/c](u/c)} \approx \left[\frac{c}{n} + u\right]\left[1 - \frac{u}{nc}\right] \approx \frac{c}{n} + u\left(1 - \frac{1}{n^2}\right).$$

Then $\lambda_1 f = c/n$ and $\lambda_2 f = c/n + u(1 - 1/n^2)$. Upon emerging from the conduits, the two rays differ in phase by

$$\phi = 2\pi L\left(\frac{1}{\lambda_1} - \frac{1}{\lambda_2}\right) = 2\pi\left(\frac{Lf}{c}\right)\left(\frac{u}{c}\right)(n^2 - 1).$$

The first minimum occurs when $\phi = \pi$, or

$$\frac{u}{c} = \frac{1}{2}\left(\frac{c}{Lf}\right)\frac{1}{(n^2 - 1)}.$$

15. In a birefringent crystal, the electromagnetic waves polarized parallel and perpendicular to the optical axis travel with different phase velocities. Thus a linear polarized wave $E_0(\hat{\mathbf{x}} + \hat{\mathbf{y}})/\sqrt{2}$, after leaving one polarizer and traveling a distance z, becomes

$$\mathbf{E} = \frac{E_0\{\hat{\mathbf{x}}e^{in_+kz} + \hat{\mathbf{y}}^{in_-kz}\}e^{-i\omega t}}{\sqrt{2}}.$$

Here $\hat{\mathbf{y}}$ has been chosen parallel to the optic axis, $\hat{\mathbf{x}}$ perpendicular to $\hat{\mathbf{y}}$ and the direction of propagation. A wave polarized along $\hat{\mathbf{x}}$ has been taken to have phase velocity c/n_+ while along $\hat{\mathbf{y}}$, c/n_-. The plane of the polarizers is parallel to $\hat{\mathbf{n}} = (\hat{\mathbf{x}} + \hat{\mathbf{y}})/\sqrt{2}$. After traveling the complete length, $2^n d$, of a crystal, the amplitude which is transmitted through a polarizer is $\mathbf{E} \cdot \hat{\mathbf{n}}$, where

$$\mathbf{E} \cdot \hat{\mathbf{n}} = \frac{E_0}{2} [\exp(in_+k2^n d) + \exp(in_-k2^n d)]$$

$$= E_0 \exp[ikd2^{n-1}(n_+ + n_-)] \cos(2^n\phi),$$

and $\phi \equiv (n_+ - n_-)kd/2$. Therefore, after passing through S elements, the intensity is reduced by the factor

$$T = [\cos(\phi) \cdot \cos(2\phi) \ldots \cos(2^{S-1}\phi)]^2.$$

Using the identity $\cos\theta = (\sin 2\theta/2 \sin\theta)$ repeatedly, we find the transmission factor to be

$$T = \left(\frac{\sin(2^S\phi)}{2^S \sin\phi}\right)^2.$$

Examination shows that $T(\phi) = T(\phi + \pi)$, and that transmission occurs primarily for $\phi = p\pi$, where $p = $ integer. The width $\delta\phi \approx 2\pi/2^S$. In terms of the wavelength, we see that transmission occurs in bands located at $\lambda = (n_+ - n_-)d/p$ and of width $\delta\lambda/\lambda \approx 2/p2^S$.

6

QUANTUM MECHANICS

1. If the operators B and C anticommute, the corresponding eigenvalues b and c must have zero as their product: $bc = 0$. Since $C^2 = 1$, we have $c = \pm 1$; hence $b = 0$. Thus a state may be an eigenstate of charge conjugation only if the baryon number is zero.

2. The matrix $\mathbf{M} = (M_x, M_y, M_z)$ represents an angular momentum matrix, because of the commutation rules. It is evident that the matrices do not represent irreducible representations; rather they represent *several* irreducible representations.

If a state with spin J is represented, M_x has $(2J + 1)$ eigenvalues, ranging in integral steps from $+J$ to $-J$, each appearing once. Hence there are no states of spin greater than 2, only one of spin 2, and eight of spin $\frac{3}{2}$. One of the 28 entries of ± 1 is accounted for by the $J = 2$ state; there are, therefore 27 representations of $J = 1$. Similarly, there are $(56 - 8) = 48$ representations of $J = \frac{1}{2}$ and 42 of $J = 0$.

Each eigenvalue of M^2 corresponding to spin J has value $J(J + 1)$; to each representation there are $(2J + 1)$ such values. We then construct the following table:

J	$J(J + 1)$	$(2J + 1)$	Number of representations	Number of entries in M^2
2	6	5	1	5
$\frac{3}{2}$	$\frac{15}{4}$	4	8	32
1	2	3	27	81
$\frac{1}{2}$	$\frac{3}{4}$	2	48	96
0	0	1	42	42

3. Repeated use of the commutator $[L_r, x_s] = i\epsilon_{rst}x_t$ and the identities $[AB, C] = A[B, C] + [A, C]B$, and $\epsilon_{ijk}\epsilon_{imn} = \delta_{jm}\delta_{kn} - \delta_{jn}\delta_{km}$ yields

$$\lambda_{ik} = 2\{r^2\delta_{ik} - x_ix_k\}, \qquad \text{where } r^2 = \sum_i x_ix_i.$$

If ϵ is an eigenvector of this matrix with eigenvalue λ, then $2r^2\epsilon - 2r(r\cdot\epsilon) = \lambda\epsilon$. Thus one may choose two degenerate eigenvectors perpendicular to r and with eigenvalues $2r^2$. The eigenvector parallel to r has eigenvalue $\lambda = 0$.

4. Taking matrix elements of Eq. (1) between states of different m, one obtains

$$\langle jm\,|\,U\,|\,j'm'\rangle(m' - m - \tfrac{1}{2}) = 0,$$

or $m' = m + \tfrac{1}{2}$ if $\langle jm\,|\,U\,|\,j'm'\rangle$ is to be nonzero.

Likewise from Eq. (2), $\langle jm\,|\,U\,|\,j'm'\rangle$ is nonzero only if $(X_j - X_{j'})^2 = \tfrac{1}{2}(X_j + X_{j'}) + \tfrac{3}{16}$. This is equivalent to

$$[j(j + 1) - j'(j' + 1)]^2 = \tfrac{1}{2}[j(j + 1) + j'(j' + 1) + \tfrac{3}{8}],$$

which has as solution $j = j' \pm \tfrac{1}{2}$.

5. All $(2S_{\max} + 1)$ of the wave functions belonging to $S_{\max}$ have the same symmetry, since the raising and lowering operators, $S_x \pm iS_y$, of S_z are symmetric functions of the individual particle operators (*e.g.* $S_x = S_{1x} + \cdots + S_{Nx}$, which is symmetric under particle exchange).

Now S_z attains its maximum value $N/2$ only when each S_{iz} is oriented along the z-direction; i.e., if

$$\psi = \psi_1\,(\tfrac{1}{2}, \tfrac{1}{2})\,\psi_2\,(\tfrac{1}{2}, \tfrac{1}{2}) \ldots \psi_N\,(\tfrac{1}{2}, \tfrac{1}{2}),$$

where the argument of the single-electron wave function gives the spin and its projection on the z-axis. Now ψ is completely symmetric under interchange of spinors; therefore, by the argument of the first paragraph, all wave functions in question are symmetric.

6. From the operator identities

$$\dot{x} = \frac{p_x}{m}; \quad \text{and} \quad \dot{x} = \left(\frac{i}{\hbar}\right)[H, x], \quad \text{and} \quad [x, p_x] = i\hbar,$$

where $[A, B] \equiv AB - BA$, we obtain

$$[x, [H, x]] = \hbar^2/m.$$

Expanding the commutators and taking the expectation value for the state ψ_0 we obtain

$$\langle 0\,|\,2xHx - x^2H - Hx^2\,|\,0\rangle = \frac{\hbar^2}{m}.$$

We now introduce a complete set of intermediate states:

$$\langle 0\,|\,xHx\,|\,0\rangle = \sum_n \langle 0\,|\,x\,|\,n\rangle\langle n\,|\,x\,|\,0\rangle E_n = \sum |\,x_{n0}\,|^2 E_n,$$

$$\langle 0\,|\,Hx^2\,|\,0\rangle = \langle 0\,|\,x^2H\,|\,0\rangle = \sum_n \langle 0\,|\,x\,|\,n\rangle\langle n\,|\,x\,|\,0\rangle E_0 = \sum |\,x_{n0}\,|^2 E_0,$$

and the identity

$$\sum_n |\,x_{n0}\,|^2 (E_n - E_0) = \frac{\hbar^2}{2m}$$

is deduced.

7. In the absence of sources, the Maxwell equations take the form:

$$\nabla \times \mathbf{H} = \frac{1}{c}\frac{\partial \mathbf{E}}{\partial t}, \qquad \nabla \times \mathbf{E} = -\frac{1}{c}\frac{\partial \mathbf{H}}{\partial t}, \qquad \nabla \cdot \mathbf{E} = 0 = \nabla \cdot \mathbf{B}.$$

In terms of $\mathbf{F}$ and $\mathbf{F}^*$ (which are linearly independent), these become

$$\nabla \times \mathbf{F} = \frac{i}{c}\frac{\partial \mathbf{F}}{\partial t}, \qquad \text{and} \qquad \nabla \times \mathbf{F}^* = -\frac{i}{c}\frac{\partial \mathbf{F}^*}{\partial t},$$

and div $\mathbf{F} = 0 =$ div $\mathbf{F}^*$. Noting that

$$(\nabla \times F)_\alpha = \epsilon_{\alpha\beta\gamma}\partial_\beta F_\gamma = -\partial_\beta \epsilon_{\beta\alpha\gamma}F_\gamma,$$

where we have used the notation $\partial_\beta = \partial/\partial x_\beta$, the equation for $\mathbf{F}$ becomes

$$(-i\partial_\beta)(-i\epsilon_{\beta\alpha\gamma})F_\gamma = \frac{i\dot{F}_\alpha}{c}.$$

The momentum operator is $p_\beta = -i\partial_\beta$; this therefore becomes

$$-p_\beta i\epsilon_{\beta\alpha\gamma}F_\gamma = \frac{i}{c}\frac{\partial F_\gamma}{\partial t}.$$

Now for fixed β, $-i\epsilon_{\beta\alpha\gamma}$ is a 3×3 matrix, $S_{\beta(\alpha,\gamma)}$. The equation for $\mathbf{F}$ then takes the form

$$p_\beta S_\beta \mathbf{F} = (\mathbf{p} \cdot \mathbf{S})\mathbf{F} = \frac{i}{c}\frac{\partial}{\partial t}\mathbf{F}.$$

This is called the Dirac form of the Maxwell equations because of similarity with the Dirac equation for leptons:

$$(\mathbf{p} \cdot \boldsymbol{\alpha} + \beta m)\psi = i\frac{\partial \psi}{\partial t}.$$

When the Dirac field is massless (*i.e.* $m = 0$), the two equations have the same form. Using the standard definition of ϵ_{ijk}, in which

$$\epsilon_{ijk} = \pm 1 \qquad \text{if } ijk \text{ are an } \left(\begin{matrix}\text{even}\\\text{odd}\end{matrix}\right) \text{ permutation of the integers 123,}$$

$$= 0 \qquad \text{otherwise,}$$

we obtain a representation for the matrices,

$$S_1 = i \begin{pmatrix} 0 & 0 & 0 \\ 0 & 0 & -1 \\ 0 & 1 & 0 \end{pmatrix}, \quad S_2 = i \begin{pmatrix} 0 & 0 & 1 \\ 0 & 0 & 0 \\ -1 & 0 & 0 \end{pmatrix}, \quad \text{and} \quad S_3 = i \begin{pmatrix} 0 & -1 & 0 \\ 1 & 0 & 0 \\ 0 & 0 & 0 \end{pmatrix}.$$

We can verify easily that the S's obey the angular-momentum commutation relations $\mathbf{S} \times \mathbf{S} = i\mathbf{S}$, and furthermore, that

$$S^2 = S_1^2 + S_2^2 + S_3^2 = 2 \times \text{identity matrix}.$$

This shows that the equation describes a particle of spin 1, for which

$$S_{\text{op}}^2 = S(S + 1)I.$$

8. From the rule $\oint p\,dx = nh$ and the energy equation $(p^2/2m) + mgx = E$, we obtain

$$2 \int_0^{E/mg} dx\,[2m(E - mgx)]^{1/2} = nh.$$

Thus $E = (9mg^2 h^2 n^2/32)^{1/3}$, where n is an integer.

9. A state of the three-dimensional oscillator is uniquely specified by the set of three numbers (n_1, n_2, n_3) with $n_1, n_2, n_3 \geq 0$ and $n_1 + n_2 + n_3 = n$. The numbers n_1, n_2, n_3 are the harmonic-oscillator quantum numbers for excitations along the x-, y-, and z-axes respectively. For fixed n_3 and n, the number of pairs (n_1, n_2) for which $n_1 + n_2 = n - n_3$, is

$$\sum_{n_1=0}^{n-n_3} 1 = (n - n_3 + 1).$$

Finally summing over n_3 we obtain the total degeneracy, d_n:

$$d_n = \sum_{n_3=0}^{n} (n + 1 - n_3) = (n + 1)^2 - \frac{n(n + 1)}{2} = \frac{(n + 2)(n + 1)}{2}.$$

10. The Hamiltonian of this system is given by $H = L_z^2/2I$, where $I = 3mr^2$ and L_z is the angular momentum perpendicular to the plane of the three masses. The eigenfunctions are $e^{il\phi}$ with l an integer; thus the energy levels are $E = \hbar^2 l^2/2I$. However, the three particles are bosons, and the wave function must be invariant under rotations of $120°$. Hence $l = 3n$ with n an integer, and $E_n = 9n^2\hbar^2/2I$.

11. The proton and antiproton magnetic moments are given by $\boldsymbol{\mu}_1 = 2\mu_0\mathbf{S}_1$ and $\boldsymbol{\mu}_2 = -2\mu_0\mathbf{S}_2$. Choosing the axis joining the two particles for the z-axis, one has

$$V = \frac{\mu_0^2}{a^3}[-4\mathbf{S}_1 \cdot \mathbf{S}_2 + 12S_{1z}S_{2z}].$$

Now

$$2S_{1z}S_{2z} = (S_z^2 - \tfrac{1}{2}) \atop 2S_1 \cdot S_2 = S(S+1) - \tfrac{3}{2}\}} \quad \text{for spin } \tfrac{1}{2} \text{ particles.}$$

The eigenstates are $S = 0, S_z = 0$ and $S = 1, S_z = 1, 0, -1$. The corresponding energies are

$$V_{0,0} = 0, \qquad V_{1,0} = \frac{-4\mu_0^2}{a^3}, \qquad V_{1,\pm1} = \frac{2\mu_0^2}{a^3}.$$

12. The total spin operators $S_z = (\sigma_z^{(1)} + \sigma_z^{(2)})/2$ and $S^2 = (\boldsymbol{\sigma}_1 + \boldsymbol{\sigma}_2)^2/4$ commute with the Hamiltonian, as is easily checked. In terms of these operators, the Hamiltonian may be rewritten

$$H = 2AS_z + B(2S^2 - 3),$$

where $\sigma_x^2 = \sigma_y^2 = \sigma_z^2 = 1$ has been used. Thus for a singlet state

$$S^2 = S_z = 0 \qquad \text{and} \qquad E_1 = -3B,$$

while for the triplet states $S^2 = 2$ and $E_3 = 2AS_z + B$, with $S_z = \pm1, 0$.

13. Choosing the magnetic field B along the z-axis, we may write the Hamiltonian

$$H = -\frac{(\alpha + \beta)}{2}(\sigma_{1z} + \sigma_{2z})B - \frac{(\alpha - \beta)}{2}(\sigma_{1z} - \sigma_{2z})B + J\boldsymbol{\sigma}_1 \cdot \boldsymbol{\sigma}_2.$$

Upon going to the representation where S_z and S^2 are diagonal with $\mathbf{S} = \tfrac{1}{2}(\boldsymbol{\sigma}_1 + \boldsymbol{\sigma}_2)$, the Hamiltonian becomes

$$H = -(\alpha + \beta)BS_z + \frac{J}{2}(4S^2 - 6) - \frac{(\alpha - \beta)}{2}B(\sigma_{1z} - \sigma_{2z}).$$

The first two terms are diagonal in this basis, and in addition the triplet, $S = 1$, and singlet, $S = 0$, both have definite parity under exchange of particles (1) and (2) (even and odd respectively). Thus the only nonzero matrix element of the last term (which is odd under exchange) is between the singlet and triplet states. One calculates directly

$$(\sigma_{1z} - \sigma_{2z})\left(\frac{\alpha(1)\beta(2) - \beta(1)\alpha(2)}{\sqrt{2}}\right) = 2\left(\frac{\alpha(1)\beta(2) + \beta(1)\alpha(2)}{\sqrt{2}}\right).$$

The only nonzero matrix element of the last term in H is thus

$$\langle 10 | H | 00 \rangle = -(\alpha - \beta)B.$$

Hence

$$S = 1, \ S_z = \pm1$$

are eigenstates with eigenvalues

$$E_\pm = \mp(\alpha + \beta)B + J,$$

while for those states with $S_z = 0$, the Hamiltonian may be represented by a matrix

$$H(S_z = 0) = \begin{pmatrix} J & -(\alpha - \beta)B \\ -(\alpha - \beta)B & -3J \end{pmatrix},$$

where the off-diagonal elements are due to triplet-singlet mixing. The eigenvalues of this matrix are

$$E_\pm(S_z = 0) = -J \pm [4J^2 + (\alpha - \beta)^2 B^2]^{1/2}.$$

14. By solving Schrödinger's equation, one finds the energy $E(R)$. Now suppose the sphere expands uniformly by a small amount. The work done is

$$dW = P\,dV = 4\pi R^2 P\,dR = -dE(R) = -\frac{dE(R)}{dR}\,dR.$$

Hence $P = -(dE/dR)/4\pi R^2$. For the lowest s-state,

$$\psi \sim \frac{\sin(kr)}{kr}.$$

The boundary condition $\psi(R) = 0$ implies $kR = \pi$. Then

$$E = \frac{k^2\hbar^2}{2m} = \frac{\pi^2\hbar^2}{2mR^2},$$

from which $P = \pi\hbar^2/4mR^5$. The lowest p-state is

$$\psi_1 \sim \frac{\cos(kr)}{kr} - \frac{\sin(kr)}{(kr)^2} = \frac{d\psi}{d(kr)}.$$

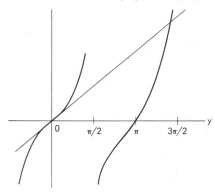

Here, the condition $\psi(R) = 0$ implies $kR \cot(kR) = 1$. It is easy to solve this numerically, obtaining $kR \approx 4.5$, and $P = (4.5)^2\hbar^2/4\pi mR^5$. The numerical solution is managed by determining the smallest nonzero, positive value of y for which $y = \tan y$. This is the intersection of the two graphs, as seen in the figure. We first guess $y = 3\pi/2 = 4.71$ and then "zero in" on the intersection by solving $\tan y = 3\pi/2$ for y and iterating. One or two iterations are sufficient if the initial guess is good, and the functions vary smoothly near the intersection.

15. Since there is no angular momentum, the Schrödinger equation reduces to

$$-\frac{\hbar^2}{2m}\left\{\frac{\partial^2}{\partial r^2}+\frac{2}{r}\frac{\partial}{\partial r}\right\}\psi = E\psi \qquad \text{(outside the well)}$$

and

$$-\frac{\hbar^2}{2m}\left\{\frac{\partial^2}{\partial r^2}+\frac{2}{r}\frac{\partial}{\partial r}\right\}\psi = (V_0 + E)\psi \qquad \text{(inside the well)}.$$

With the introduction of $U = r\psi$, the wave equation is

$$U'' - \alpha^2 U = 0 \qquad (r > a),$$

and

$$U'' + \beta^2 U = 0 \qquad (0 \le r \le a),$$

where

$$\alpha = \left[\frac{-2mE}{\hbar^2}\right]^{1/2} \quad \text{and} \quad \beta = \left[\frac{2m(V_0 + E)}{\hbar^2}\right]^{1/2}$$

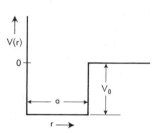

We are interested in the limit $E \to 0^-$. Solutions are

$$\psi = A\frac{e^{-\alpha r}}{r} \quad (r > a), \qquad \psi = B\frac{\sin\beta r}{r} \quad (r < a),$$

where we have eliminated the solution singular at the origin. Continuity of ψ and its derivative at $r = a$ (equivalently, and easier, continuity of U and its derivative) requires that $\beta \cot\beta a = -\alpha$. As $E \to 0$, $\alpha \to 0$; hence $\cot(\beta a) \to 0$. This happens when $\beta a = \pi/2$, or $V_0 = \pi^2\hbar^2/8ma^2$.

16. In the region $x > 0$, ψ obeys the same differential equation as the two-sided harmonic oscillator; however, the only acceptable solutions are those that vanish at the origin. Therefore, the eigenvalues are those of the ordinary harmonic oscillator belonging to wave functions of odd parity. Now the parity of the S.H.O. wave functions alternates with increasing n, starting with an even-parity ground state. Hence,

$$E = \frac{(4n + 3)\hbar\omega}{2} \qquad \text{with } n = 0, 1, \ldots$$

17. Choosing the z-axis along the magnetic field **B**, the Hamiltonian is

$$H = \frac{1}{2m}\left\{\left(P_x - \frac{e}{c}A_x\right)^2 + \left(P_y - \frac{e}{c}A_y\right)^2 + P_z^2\right\} \qquad \text{with } A_z = 0.$$

Defining new variables

$$Q = \left(\frac{c}{eB}\right)^{1/2}\left(P_x - \frac{e}{c}A_x\right), \qquad P = \left(\frac{c}{eB}\right)^{1/2}\left(P_y - \frac{e}{c}A_y\right),$$

one finds $[Q, P] = i\hbar$; i.e. P and Q are canonically conjugate variables. In terms of P and Q, the Hamiltonian becomes

$$H = \frac{eB}{2mc}[P^2 + Q^2] + \frac{P_z^2}{2m}.$$

The term in brackets represents a harmonic oscillator in QP-space. Motion in the z-direction is not quantized. Thus the energy levels are

$$E = \frac{eB\hbar}{mc}\left(n + \frac{1}{2}\right) + \frac{P_z^2}{2m}.$$

If the orbit is large, we may use the semiclassical approach. We assume a closed orbit, and the Bohr-Sommerfeld quantization rule gives (with $n = $ integer)

$$nh = \oint \mathbf{p} \cdot d\mathbf{r} = \oint \left(m\mathbf{v} + \frac{e\mathbf{A}}{c}\right) \cdot d\mathbf{r}.$$

Now the integral $\oint \mathbf{A} \cdot d\mathbf{r} = \int \mathbf{B} \cdot d\mathbf{A} = \Phi$, represents the magnetic flux enclosed by the orbit. The integral

$$\oint m\mathbf{v} \cdot d\mathbf{r} = \oint m\mathbf{v}^2\, dt = -\oint m\mathbf{r} \cdot \frac{d\mathbf{v}}{dt}\, dt,$$

because of the equation of motion $m(d\mathbf{v}/dt) = (e\mathbf{v}/c) \times \mathbf{B}$, becomes

$$\oint m\mathbf{v} \cdot d\mathbf{r} = -\oint \mathbf{r} \cdot \left(\frac{e}{c}\, d\mathbf{r} \times \mathbf{B}\right) = -\oint \frac{e}{c}(\mathbf{B} \times \mathbf{r}) \cdot d\mathbf{r}$$

$$= -\frac{e}{c} \int \boldsymbol{\nabla} \times (\mathbf{B} \times \mathbf{r}) \cdot d\mathbf{A}.$$

For a constant $\mathbf{B}$, one has $\boldsymbol{\nabla} \times (\mathbf{B} \times \mathbf{r}) = 2\mathbf{B}$. Finally, then, $\oint \mathbf{p} \cdot d\mathbf{r} = -e\Phi/c$, and one finds that the flux is quantized in units of hc/e.

The prediction of flux quantization by F. London received confirmation through experiments in superconductors; however the unit of flux quantization was found to be $hc/2e$, half the unit predicted by London. This is explained on the basis of two electron correlations in spin space and momentum space, for the superconducting state.

Whereas in nonrelativistic quantum mechanics the energies are given by

$$\frac{(p_x - eA_x/c)^2}{2m} + \frac{(p_y - eA_y)^2}{2m} + \frac{p_z^2}{2m} = E_{NR},$$

they are given in relativistic quantum mechanics by

$$E_R^2 = \left(p_x - \frac{eA_x}{c}\right)^2 + \left(p_y - \frac{eA_y}{c}\right)^2 + p_z^2 + m^2c^4 = (2mE_{NR}) + m^2c^4,$$

from which

$$E_R = mc^2\left(1 + \frac{2E_{NR}}{mc^2}\right)^{1/2}.$$

18. Let the amplitudes for states 1 and 2 be C_1 and C_2.
Then

$$i\frac{dC_1}{dt} = H_{11}C_1 + H_{12}C_2 = E_1C_1 + V_{12}C_2,$$

$$i\frac{dC_2}{dt} = H_{21}C_1 + H_{22}C_2 = V_{12}^*C_1 + E_2C_2.$$

If $C_1 = A_1 e^{-iwt}$ and $C_2 = A_2 e^{-iwt}$, one has

$$A_1(W - E_1) - V_{12}A_2 = 0, \qquad A_1 V_{12}^* + (E_2 - W)A_2 = 0.$$

Self-consistency requires

$$\frac{V_{12}}{W - E_1} = \frac{W - E_2}{V_{12}^*}, \quad \text{or} \quad W_\pm = \frac{(E_1 + E_2)}{2} \pm \frac{[(E_1 - E_2)^2 + 4\,|\,V_{12}\,|^2]^{1/2}}{2}.$$

Then, one may write

$$C_1 = A_1 e^{-iw_+t} + B_1 e^{-iw_-t}, \qquad C_2 = A_2 e^{-iw_+t} + B_2 e^{-iw_-t}.$$

The coefficients obey the constraints (not all of them independent):

$$\frac{A_1}{A_2} = \frac{2V_{12}}{(E_2 - E_1) + \sqrt{(E_1 - E_2)^2 + 4\,|\,V_{12}\,|^2}},$$

$$\frac{B_1}{B_2} = \frac{2V_{12}}{(E_2 - E_1) - \sqrt{(E_1 - E_2)^2 + 4\,|\,V_{12}\,|^2}},$$

and

$$A_1 + B_1 = 1, \qquad A_2 + B_2 = 0, \qquad A_1^2 + B_1^2 + A_2^2 + B_2^2 = 1,$$

$$A_1 B_1 + A_2 B_2 = 0.$$

The solutions are

$$A_1 = \frac{2\,|\,V_{12}\,|^2}{(E_1 - E_2)^2 + 4\,|\,V_{12}\,|^2 + (E_2 - E_1)\sqrt{(E_1 - E_2)^2 + 4\,|\,V_{12}\,|^2}},$$

$$B_1 = \frac{2\,|\,V_{12}\,|^2}{(E_1 - E_2)^2 + 4\,|\,V_{12}\,|^2 - (E_2 - E_1)\sqrt{(E_1 - E_2)^2 + 4\,|\,V_{12}\,|^2}},$$

$$A_2 = \frac{V_{12}^*}{\sqrt{(E_1 - E_2)^2 + 4\,|\,V_{12}\,|^2}} = -B_2.$$

19. For a trial function of the type $\psi = 0$ for $x < 0$ and $\psi = x e^{-ax}$ for $x > 0$, the expectation value for the energy is given by

$$\langle E \rangle = \int_0^\infty dx\, x e^{-ax} \left\{ -\frac{\hbar^2}{2m}\frac{d^2}{dx^2} + cx \right\} x e^{-ax} \bigg/ \int_0^\infty dx\, x^2 e^{-2ax}$$

$$= \frac{3c}{2a} + \frac{\hbar^2 a^2}{2m}.$$

This expectation value has a minimum for $a = (3cm/2\hbar^2)^{1/3}$. Thus the ground-state energy is greater than or equal to $\frac{9}{4}(2\hbar^2 c^2/3m)^{1/3}$.

20. The unperturbed eigenfunctions and energies are given by $\psi_n = (2\pi)^{-1/2} e^{in\theta}$ with $E_n = \hbar^2 n^2 / 2mr^2 = n^2/a$, where n is an integer. Note that states with $+n$ and $-n$ are degenerate. The perturbing interaction

$$V = eFr \cos \theta$$

has matrix elements

$$\langle k \,|\, V \,|\, n \rangle = \frac{eFr}{2\pi} \int_0^{2\pi} d\theta \, \cos \theta \, e^{i(k-n)\theta} = \frac{eFr}{2}(\delta_{k,(n+1)} + \delta_{(k+1),n}).$$

Since $\langle k \,|\, V \,|\, k \rangle = 0$ and V has no matrix elements connecting degenerate states, the first-order perturbation correction to the energy levels vanishes.

The second-order shift in energy is given by

$$\Delta E_n = \sum_{n \neq l} \frac{\langle n \,|\, V \,|\, l \rangle \langle l \,|\, V \,|\, n \rangle}{E_n - E_l}, \quad \text{if } E_m \text{ is nondegenerate.}$$

However, from the rules of degenerate perturbation theory, one must diagonalize

$$H_{nn'} = \sum_l \frac{\langle n \,|\, V \,|\, l \rangle \langle l \,|\, V \,|\, n' \rangle}{E_n - E_l}$$

with respect to *degenerate* states n and n'. For the problem discussed here, the only nonzero $H_{nn'}$ with $E_n = E_{n'}$ and $n \neq n'$ occurs for $n = +1$, and $n' = -1$, or $n = -1$ and $n' = +1$. Therefore to (correctly) obtain the shift in the energy of the states with $n = \pm 1$, one must diagonalize the matrix

$$\begin{pmatrix} V_{-1,-1} & V_{-1,1} \\ V_{1,-1} & V_{1,1} \end{pmatrix} = \frac{\alpha e^2 F^2 r^2}{6} \begin{pmatrix} 1 & \frac{3}{2} \\ \frac{3}{2} & 1 \end{pmatrix}.$$

The eigenvalues are $5\alpha e^2 F^2 r^2 / 12$ and $-\alpha e^2 F^2 r^2 / 12$.

For those states $|n\rangle$ for which $|n| \neq 1$, the energy shift is given by

$$\Delta E_n = \frac{\langle n \,|\, V \,|\, n+1 \rangle \langle n+1 \,|\, V \,|\, n \rangle}{E_n - E_{n+1}} + \frac{\langle n \,|\, V \,|\, n-1 \rangle \langle n-1 \,|\, V \,|\, n \rangle}{E_n - E_{n-1}}$$

$$= \frac{\alpha e^2 F^2 r^2}{2(4n^2 - 1)}.$$

21. Taking the electric field to be along the z-axis, the perturbing interaction V is given by $V = eFz$. The first-order correction to the energy vanishes, i.e., $\langle 1S \,|\, z \,|\, 1S \rangle = 0$, and one must go to second order;

$$\Delta E = -\frac{1}{2} \alpha F^2 = -\sum_{n>1} \frac{|\langle 1S \,|\, V \,|\, nlm \rangle|^2}{(E_n - E_1)},$$

where $|nlm\rangle$ is an eigenstate of the hydrogen atom with principal quantum n, angular momentum l, and $l_z = m$. The energy levels of the hydrogen atom, E_n, are given by $E_n = -me^4 / 2\hbar^2 n^2$. Since $|\langle 1S \,|\, V \,|\, nlm \rangle|^2$ is always positive, and E_n is monotonically increasing, it follows that

$$-\frac{e^2}{E_1} \sum |\langle 1S|z|nlm\rangle|^2 < \frac{\alpha}{2} < \frac{e^2}{(E_2 - E_1)} \sum |\langle 1S|z|nlm\rangle|^2.$$

In setting the lower limit, the contribution from the continuum states is assumed negligible. Since the states $|nlm\rangle$ form a complete set (neglecting the continuum),

$$\sum \langle 1S|z|nlm\rangle\langle nlm|z|1S\rangle = \langle 1S|z^2|1S\rangle.$$

Expressing the state $|1S\rangle$ in spherical coordinates it follows that:

$$\langle 1S|z^2|1S\rangle = \frac{1}{\pi a^3} \int \frac{4\pi}{3} r^4 e^{-2r/a}\, dr = a^2.$$

Writing $E_1 = -e^2/2a$ and $E_2 - E_1 = 3e^2/8a$, one finally obtains $4a^3 < \alpha < (16/3)a^3$. The experimental value is $\alpha = 4.5a^3$.

22. The total spin of the system commutes with the Hamiltonian of the system, thus allowing the total wave function, ψ, to be written

$$\psi = \phi(r_1, r_2)\chi(1, 2),$$

where ϕ is a function of the space variables r_1, r_2, and χ is the spin wave function of the two particles. However, the total wave function, ψ, must be antisymmetric; thus by choosing the total spin to be $S = 1$, $\chi(1, 2)$ is totally symmetric, and hence $\phi_A(r_1, r_2)$ must be totally antisymmetric. Similarly for $S = 0$, $\chi(1, 2)$ is antisymmetric, and $\phi_S(r_1, r_2)$ must be symmetric. Therefore, for the purposes of energy calculation, there are two types of spatial wave functions: ϕ_S and ϕ_A.

In the absence of interaction between the particles, the wave function $\phi(r_1, r_2)$ is given by symmetrized products, $\phi_m(r_1)\phi_n(r_2) \pm \phi_n(r_1)\phi_m(r_2)$, where

$$\phi_n(\mathbf{r}) = \left(\frac{2}{d}\right)^{3/2} \sin\left(\frac{n_1\pi x}{d}\right) \sin\left(\frac{n_2\pi y}{d}\right) \sin\left(\frac{n_3\pi z}{d}\right),$$

with energy equal to $\pi^2\hbar^2(n^2 + m^2)/2Md^2$ where $d = 10^{-8}$ cm and m_i, n_i are integers, $\mathbf{n} = (n_1, n_2, n_3)$, etc. The interaction between two particles is

$$V(\mathbf{r}_1 - \mathbf{r}_2) = \begin{cases} 0 \text{ for } |\mathbf{r}_1 - \mathbf{r}_2| > a \\ -V_0 \text{ for } |\mathbf{r}_1 - \mathbf{r}_2| < a \end{cases} \quad \text{and} \quad \begin{array}{l} a = 10^{-10} \text{ cm} \\ V_0 = 10^{-3} \text{ eV}. \end{array}$$

Hence

$$\int V(\mathbf{r}_1 - \mathbf{r}_2)d^3\mathbf{r}_2 = \frac{-4\pi a^3 V_0}{3}.$$

This inspires the useful approximation that

$$V(\mathbf{r}_1 - \mathbf{r}_2) = -(4\pi a^3 V_0/3)\, \delta^{(3)}(\mathbf{r}_1 - \mathbf{r}_2).$$

The first-order perturbation correction to the energy is given by

$$\Delta E = \int d^3r_1 d^3r_2 \phi^*(\mathbf{r}_1, \mathbf{r}_2) V(\mathbf{r}_1 - \mathbf{r}_2) \phi(\mathbf{r}_1, \mathbf{r}_2).$$

Since the interaction is nonzero only when $\mathbf{r}_1 = \mathbf{r}_2$, and an antisymmetrical function of $\mathbf{r}_1, \mathbf{r}_2$ vanishes when $\mathbf{r}_1 = \mathbf{r}_2$, the energies of the $S = 1$ states are unshifted. However, the symmetrical states have their energy shifted by the amount

$$\Delta E_S = -\left(\frac{4\pi a^3 V_0}{3}\right) \int d^3r_1 \, |\phi(\mathbf{r}_1, \mathbf{r}_1)|^2.$$

For any set of quantum numbers $(\mathbf{n}, \mathbf{m})$ the $S = 0$ state will always have a smaller energy than the $S = 1$ state. The energy of the ground state $(n_i = m_i = 1)$ is then $E_0 = 3\pi^2\hbar^2/Md^2 - (9\pi a^3/2d^3)V_0$.

23.   The perturbation V may be rewritten in decreasing order of symmetry,

$$V = \frac{(A + B)r^2}{2} - \frac{3(A + B)z^2}{2} + \frac{(A - B)}{2}(x^2 - y^2).$$

For $V = 0$ one has three degenerate levels with $L = 1$, $L_z = 0, \pm 1$. In order to show that the expectation value of L_z is zero when V is turned on, we will show that the eigenfunctions are of the form:

$$\psi_0 = |\, L = 1, \, L_z = 0\rangle$$
$$\psi_\pm = |\, L = 1, \, L_z = +1\rangle \pm |\, L = 1, \, L_z = -1\rangle,$$

and that they are *nondegenerate*. (States with principal quantum number $n \neq 2$ are neglected). Since $L_z = +1$ and -1 occur with equal probability for these states, the expectation value of L_z is zero. It is important to show that ψ_+ and ψ_- are nondegenerate since otherwise linear combinations (such as $\psi_+ + \psi_-$) are eigenfunctions which have $\langle L_z \rangle \neq 0$ (for $\psi_+ + \psi_-$, $L_z = +1$).

The first term in V commutes with both L^2 and L_z and thus adds the *same* energy to the three states $L_z = 0, \pm 1$. The next term commutes with L_z; thus the energy levels are still eigenfunctions of L_z, but $L_z = 0$ is split from $L_z = \pm 1$. Note $L_z = +1$ and $L_z = -1$ are degenerate because V is an even function of z. Finally, the last term, having the form $(x^2 - y^2) = r^2 \sin^2\theta \cos 2\phi$, in spherical coordinates, is seen to have nonvanishing matrix elements only between $L_z = +1$ and $L_z = -1$. (This is because of the $\cos 2\phi$ dependence and $|\, L_z = m\rangle \sim e^{im\phi}$.) Thus the last term in V has eigenfunctions $\psi_\pm = |\, L_z = +1\rangle \pm |\, L_z = -1\rangle$. The two eigenfunctions $\psi_\pm$ are separated by an energy difference proportional to $(A - B)$. Hence, in the general case, i.e. $A \neq B$, the functions ψ_+ and ψ_- are nondegenerate and one may conclude that $\langle L_z \rangle = 0$.

24. Defining new variables $x = \theta_1 + \theta_2$ and $y = \theta_1 - \theta_2$, one finds that the Hamiltonian separates:

$$H = -2A\hbar^2 \left(\frac{\partial^2}{\partial x^2} + \frac{\partial^2}{\partial y^2} \right) - B \cos y.$$

Unperturbed eigenfunctions (corresponding to $B = 0$) are

$$\psi_{lm} = \frac{1}{2\pi} \exp \left(\frac{ilx}{2} \right) \exp \left(\frac{imy}{2} \right), \tag{1}$$

with energy eigenvalues

$$E_{lm} = \frac{A\hbar^2}{2} (l^2 + m^2),$$

where l and m are integers which are either both even or both odd to guarantee the singlevaluedness of ψ; that is

$$\psi(\theta_1 + j2\pi, \theta_2 + k2\pi) = \psi(\theta_1, \theta_2),$$

with j and k arbitrary integers.

(a) Here we treat $V = -B \cos y$ as a perturbation, the matrix elements of which are

$$\langle l'm' | V | lm \rangle = \frac{-B}{(2\pi)^2} \int_0^{2\pi} d\theta_1 \int_0^{2\pi} d\theta_2 \cos (\theta_1 - \theta_2)$$

$$\exp \left\{ \frac{i\Delta l}{2} (\theta_1 + \theta_2) + \frac{i\Delta m}{2} (\theta_1 - \theta_2) \right\},$$

where

$$\Delta m = m - m' \quad \text{and} \quad \Delta l = l - l'.$$

These matrix elements are nonzero only if $\Delta l = 0$, in which case

$$\langle lm' | V | lm \rangle = \frac{-B}{2} \{ \delta(m' - m - 2) + \delta(m' - m + 2) \}. \tag{2}$$

To proceed further with first-order perturbation theory, one must diagonalize $\langle lm' | V | lm \rangle$ with respect to degenerate states, with $m' \neq m$ but $E_{lm'} = E_{lm}$. From Eq. (1) one finds that $E_{lm} = E_{lm'}$ with $m' \neq m$ implies $m' = -m$, while $\langle l, -m | V | l, m \rangle$ is nonzero only if $m = \pm 1$ as seen from Eq. (2). Thus for $m \neq \pm 1$, the shift $\Delta E_{lm} = \langle lm | V | lm \rangle = 0$, while for $|m| = 1$ the basis $[| l, 1 \rangle \pm | l, -1 \rangle]/\sqrt{2}$ diagonalizes V with respect to degenerate levels, with $\Delta E_{l\pm} = \mp B/2$. Finally then, to first order in B, the eigenfunctions and energy levels are given by

$$\psi_{lm} = \frac{1}{(2\pi)^2} \exp \left\{ \frac{il}{2} (\theta_1 + \theta_2) + \frac{im}{2} (\theta_1 - \theta_2) \right\}$$

and

$$E_{lm} = \frac{A\hbar^2}{2} (l^2 + m^2),$$

for $|m| \neq 1$, while for $|m| = 1$ we have

$$\psi_{l\pm} = \frac{\sqrt{2}}{(2\pi)}\left\{\begin{matrix} \cos\tfrac{1}{2}(\theta_1 - \theta_2) \\ i\sin\tfrac{1}{2}(\theta_1 - \theta_2) \end{matrix}\right\} \exp\left\{\frac{il}{2}(\theta_1 + \theta_2)\right\},$$

with energies $E_{l\pm} = \mp B/2 + (A\hbar^2/2)(l^2 + 1)$.

(b) For small oscillations $|\theta_1 - \theta_2| \ll 1$, we expand $\cos y \simeq 1 - y^2/2$. The Schrödinger equation was shown to be separable; thus the solution is

$$\psi(x, y) = \exp\left(\frac{ilx}{2}\right)f(y),$$

where l is an integer. Note that since the oscillations are small (i.e. $\theta_1 \sim \theta_2$), it is required that the wave function be singlevalued and take the form

$$\exp\left\{\frac{il}{2}(x + 4\pi)\right\} = \exp\left\{\frac{ilx}{2}\right\},$$

so l must be an integer. The Schrödinger equation for $f(y)$ is

$$-2A\hbar^2\frac{\partial^2 f}{\partial y^2} + \frac{B}{2}y^2 f = \left(E + B - \frac{A\hbar^2 l^2}{2}\right)f,$$

which is the equation for a harmonic oscillator of frequency $\omega = (4AB)^{1/2}$, of which the energy eigenvalues are $\hbar\omega(n + \tfrac{1}{2})$, with $n = 0, 1, \dots$ Thus

$$E_{l,n} = -B + \frac{A\hbar^2 l^2}{2} + \hbar(4AB)^{1/2}\left(n + \frac{1}{2}\right),$$

with normalized eigenfunctions

$$\psi_{l,n} = \frac{1}{\sqrt{2\pi}}\exp\left\{\frac{il}{2}(\theta_1 + \theta_2)\right\}H_n(\theta_1 - \theta_2),$$

where $H_n(y)$ are the normalized harmonic-oscillator wave functions.

25. To first order in the F's the ground-state wave function is

$$\psi' = \psi_0 + \sum_{k \neq 0}\psi_k\frac{V_{k0}}{E_0 - E_k},$$

where the ψ_k are the unperturbed wave functions of a particle in a box. Here $V_{k0} = F_1 x_{k0} + F_2 y_{k0}$ is the matrix element of V between ψ_k and ψ_0. Thus the expectation value of x is given by

$$\langle\psi'|x|\psi'\rangle = 2\sum_{k \neq 0}\frac{F_1 x_{k0}x_{0k} + F_2 y_{k0}x_{0k}}{E_0 - E_k}$$

to first order in F's. In deriving this expression the following properties are used:

(1) $\langle\psi_0|x|\psi_0\rangle = 0$, which follows from the fact that the ground state has definite parity, i.e. $\psi_0 = (1/a)\cos(\pi x/2a)\cos(\pi y/2a)$.

(2) Matrix elements $x_{k0} = \langle \psi_k | x | \psi_0 \rangle$ are real. This follows from choosing real eigenfunctions.

Now if $x_{k0} \neq 0$, then $y_{k0} = 0$, and vice versa. This must be true, for if $x_{k0} \neq 0$, then the y-dependence of ψ_k must be $\cos(\pi y/2a)$, and hence $y_{k0} = 0$. Finally then $\langle \psi' | x | \psi' \rangle = CF_1$ and $\langle \psi' | y | \psi' \rangle = CF_2$ where

$$C = 2 \sum_{n \neq 0} \frac{x_{0n}^2}{E_0 - E_n}.$$

Here

$$x_{n0} = x_{0n} = \frac{1}{a} \int_{-a}^{a} x \sin\left(\frac{\pi n x}{a}\right) \cos\left(\frac{\pi x}{2a}\right) dx$$

and

$$E_0 - E_n = \frac{\pi^2}{8ma^2}(1 - 4n^2).$$

26. (a) The total Hamiltonian is

$$H = -\frac{\hbar^2}{2m}\left(\frac{\partial^2}{\partial x_1^2} + \frac{\partial^2}{\partial x_2^2}\right) + \frac{k}{2}(x_1^2 + x_2^2) + cx_1 x_2.$$

Defining new variables ξ, η by

$$x_1 = \frac{(\xi + \eta)}{\sqrt{2}}, \qquad x_2 = \frac{(\xi - \eta)}{\sqrt{2}},$$

we express the Hamiltonian as

$$H = -\frac{\hbar^2}{2m}\left(\frac{\partial^2}{\partial \xi^2} + \frac{\partial^2}{\partial \eta^2}\right) + \frac{1}{2}(k + c)\xi^2 + \frac{1}{2}(k - c)\eta^2.$$

The exact energy levels are thus given by

$$E = n_1 \hbar \omega_1 + n_2 \hbar \omega_2 + \left(\frac{\hbar}{2}\right)(\omega_1 + \omega_2)$$

where n_1 and n_2 are positive integers, and

$$\omega_1^2 = \frac{(k - c)}{m}; \qquad \omega_2^2 = \frac{(k + c)}{m}.$$

(b) Treating H' as a perturbation, we write the unperturbed eigenfunctions as

$$|n_1 n_2\rangle = U_{n_1}(x_1) U_{n_2}(x_2),$$

where the U's are simple harmonic-oscillator wave functions, the energy levels of which are

$$E_{n_1 n_2} = \hbar \omega (n_1 + n_2 + 1), \qquad \text{where} \qquad \omega^2 = \frac{k}{m}.$$

The first excited state of energy $2\hbar\omega$ is two-fold degenerate, the states corresponding to $(n_1, n_2) = (1, 0)$ or $(0, 1)$. The Hamiltonian matrix of the perturbation, taken with respect to the unperturbed states, is

$$H' = \begin{pmatrix} 0 & a \\ a & 0 \end{pmatrix}, \qquad \text{where } a = \langle 10 | H' | 01 \rangle = c\langle 0 | x_1 | 1 \rangle \langle 1 | x_2 | 0 \rangle.$$

Rather than calculate a by doing an integral ($\int x U_1 U_0 dx$), we may write

$$a = c\langle 0 | x | 1 \rangle \langle 1 | x | 0 \rangle = c\langle 0 | x^2 | 0 \rangle.$$

This is so since $\langle n | x | 0 \rangle \neq 0$ only if $n = 1$. Therefore the completeness relation $\sum |n\rangle\langle n| = 1$ yields

$$\langle 0 | x^2 | 0 \rangle = \sum_n \langle 0 | x | n \rangle \langle n | x | 0 \rangle = \langle 0 | x | 1 \rangle \langle 1 | x | 0 \rangle.$$

The matrix element of x^2 may be deduced from the quantum-mechanical version of the virial theorem, which in this case yields that the expectation value of the potential energy is one-half the total energy. Thus

$$\langle 0 | x^2 | 0 \rangle = \frac{2}{k} \left(\frac{1}{4} \hbar\omega \right).$$

Finally the energies of the two states corresponding to

$$\psi_\pm = \frac{|10\rangle \pm |01\rangle}{\sqrt{2}}$$

are $2\hbar\omega(1 \pm c/4k)$ to first order in c. Notice that these energies are equal to the leading terms of the expansion in c, of the exact result.

27. The three lowest eigenstates, labeled by quantum numbers n and m, are

$$\psi_{0,0} = \frac{1}{\sqrt{\pi}} \exp\left[-\frac{1}{2}(x^2 + y^2) \right],$$

$$\psi_{1,0} = \sqrt{\frac{2}{\pi}} \, y \exp\left[-\frac{1}{2}(x^2 + y^2) \right],$$

$$\psi_{0,1} = \sqrt{\frac{2}{\pi}} \, x \exp\left[-\frac{1}{2}(x^2 + y^2) \right],$$

with corresponding energies $E_{0,0} = 1$, $E_{1,0} = E_{0,1} = 2$. The perturbation Hamiltonian is $H' = (\delta xy/2)(x^2 + y^2)$. The (0, 0) level is shifted by an amount $\langle 0, 0 | H' | 0, 0 \rangle = 0$ by symmetry under reflection. The (10) and (01) levels are degenerate; therefore the 2×2 matrix

$$\begin{pmatrix} \langle 10 | H' | 10 \rangle & \langle 10 | H' | 01 \rangle \\ \langle 01 | H' | 10 \rangle & \langle 01 | H' | 01 \rangle \end{pmatrix}$$

must be diagonalized. The diagonal elements vanish by symmetry under reflection; the off-diagonal elements are each equal to $3\delta/4$. After diagonalization, H' has as diagonal elements, $3\delta/4$ and $-3\delta/4$. The perturbation, therefore, splits the degenerate levels, whose energies are now

$$E_\pm = 2 \pm \frac{3\delta}{4},$$

with corresponding wave functions $\psi_\pm = (1/\sqrt{2})(\psi_{0,1} \pm \psi_{1,0})$.

28. We take as the Hamiltonian of the electron, $H = \mu_0 B\sigma_z$, where μ_0 represents the magnetic moment of the electron, and B the magnetic field strength. In addition, we choose the representation

$$\sigma_x = \begin{pmatrix} 0 & 1 \\ 1 & 0 \end{pmatrix}, \qquad \sigma_y = \begin{pmatrix} 0 & -i \\ i & 0 \end{pmatrix}, \qquad \sigma_z = \begin{pmatrix} 1 & 0 \\ 0 & -1 \end{pmatrix}.$$

The spin wave function satisfies the equation,

$$i\frac{\partial \psi}{\partial t} = H\psi \qquad \text{where } \psi = A(t)\frac{1}{\sqrt{2}}\begin{pmatrix} 1 \\ 1 \end{pmatrix} + C(t)\frac{1}{\sqrt{2}}\begin{pmatrix} 1 \\ -1 \end{pmatrix}.$$

A and C are to be determined subject to the constraint $A(0) = e^{i\delta}$, $C(0) = 0$. This guarantees that for $t = 0$ the spin is in the eigenstate $S_x = \frac{1}{2}$. The equations of motion give $i\dot{A}(t) = \mu_0 BC$ and $i\dot{C}(t) = \mu_0 BA$, and the appropriate solution is

$$A(t) = e^{i\delta} \cos(\mu_0 Bt)$$

$$C(t) = -ie^{i\delta} \sin(\mu_0 Bt).$$

(a) (Probability that $S_x = +\frac{1}{2}$) $= |A(t)|^2 = \cos^2(\mu_0 Bt) \equiv P_+$,

(b) (Probability that $S_x = -\frac{1}{2}$) $= |C(t)|^2 = \sin^2(\mu_0 Bt) \equiv P_-$,

(c) (Probability that $S_z = +\frac{1}{2}$) $= \frac{1}{2}$.

Alternate solution: We know that expectation values of quantum-mechanical operators obey their classical equations of motion (Ehrenfest's theorem). We therefore have the equation $d\langle \mathbf{S}\rangle/dt = \langle \boldsymbol{\mu}\rangle \times \mathbf{B}$, where $\boldsymbol{\mu} = 2\mu_0 \mathbf{S}$ is the electron magnetic moment. Subject to the initial conditions that $\langle S_x\rangle = \frac{1}{2}$, $\langle S_y\rangle = 0$, and $\langle S_z\rangle = 0$ at time $t = 0$, this equation has the solution $\langle S_x\rangle = \frac{1}{2}\cos(2\mu_0 Bt)$; $\langle S_y\rangle = -\frac{1}{2}\sin(2\mu_0 Bt)$; $\langle S_z\rangle = 0$. To find the probabilities, we need only use the definition $\langle S_x\rangle = (P_+ - P_-)/2$ and $P_+ + P_- = 1$ with similar expressions for $\langle S_y\rangle$ and $\langle S_z\rangle$. The solution obtained in this manner agrees with the previous result.

29. An electron with energy in the kilovolt region moves away from the daughter ion so quickly that its influence on the atomic wave function may be neglected. If we assume the decay takes place in negligible time, then the "sudden approximation" is valid, and the wave function of the bound electron remains unchanged during the transformation. If the initial state is $n = 1$, $Z = 1$, the amplitude for finding the electron in the state $n = 2$, $Z = 2$ immediately after decay is $A = \langle n = 2, Z = 2 \mid n = 1, Z = 1\rangle$.

The normalized states are

$$|n = 1, Z = 1\rangle = (\pi a^3)^{-1/2} \exp\left(\frac{-r}{a}\right),$$

and

$$|n = 1, Z = 2\rangle = (\pi a^3)^{-1/2}\left(1 - \frac{r}{a}\right)e^{-r/a},$$

where a is the Bohr radius. Thus the amplitude is given by

$$A = \frac{1}{\pi a^3}\int_0^\infty 4\pi r^2\, dr\left(1 - \frac{r}{a}\right)e^{-2r/a} = -\frac{1}{2}.$$

The probability is given by $P = |A|^2 = \frac{1}{4}$.

30. Let the initial **B** field define the z-axis; the final **B** field, the z'-axis. In the approximation that the change in field occurs in negligible time, the wave functions just before and just after the change must be equal. Initially $\psi = \binom{1}{0}$, with respect to the z-axis. However in the basis with $J_{z'}$, diagonal

$$\psi = \begin{pmatrix} \cos(\phi/2) \\ -\sin(\phi/2) \end{pmatrix}.$$

This result is obtained by requiring that ψ be an eigenstate of $\boldsymbol{\sigma}\cdot\hat{\mathbf{n}}$, where $\hat{\mathbf{n}}$ points in the $\hat{\mathbf{z}}$-direction; i.e., one solves

$$\begin{pmatrix} \cos\phi & -\sin\phi \\ -\sin\phi & -\cos\phi \end{pmatrix}\psi = \psi.$$

The probabilities for finding $m'_j = \frac{1}{2}$ or $m'_j = -\frac{1}{2}$ are, respectively, $\cos^2(\phi/2)$ and $\sin^2(\phi/2)$; i.e., $\frac{3}{4}$ and $\frac{1}{4}$.

A less-formal approach is to make use of the fact that immediately after the field is applied, the expectation value of $\mathbf{J}\cdot\hat{\mathbf{z}}'$ is unchanged, and is given by

$$\langle\psi|\mathbf{J}\cdot\hat{\mathbf{z}}'|\psi\rangle = \cos\phi\,\langle\psi|J_z|\psi\rangle = \frac{1}{2}\cos\phi \equiv \frac{P_+ - P_-}{2}$$

with $P_+ + P_- = 1$. Here P_+, P_- are the probabilities of finding $m'_j = \pm\frac{1}{2}$ along the z'-axis. Thus:

$$P_+ = \frac{1 + \cos\phi}{2} = \cos^2\frac{\phi}{2},$$

$$P_- = \frac{1 - \cos\phi}{2} = \sin^2\frac{\phi}{2}.$$

31. Conservation of angular momentum would require that the emitted photon have total angular momentum zero. Now a photon has spin 1 (described by the polarization vector **e**) and therefore must be in an ($L = 1$)-state (described by its momentum vector **k**) in order to form a $J = 0$ state.

Thus the transition amplitude is proportional to the scalar combination $\mathbf{e} \cdot \mathbf{k}$, which vanishes because of the transversality condition $\nabla \cdot \mathbf{A} = 0$. Hence it is impossible to put a *single* photon in a $J = 0$ state, and the transition is forbidden. Radiative decays may therefore proceed only with the emission of at least two photons. A well-known example of a forbidden zero-zero transition occurs in the hydrogen-like atom. The transition from the $2S$-state to the ground state is forbidden via single photon emission; however, the decay may proceed via two photons. The two photons have actually been seen in coincidence in the decay of the $2S$-state of singly ionized helium [M. Lipeles, et al., *Phys. Rev. Letters* **15**, 690 (1965)].

Another example of a zero-zero transition proceeding, with the emission of two photons, is the decay $\pi^0 \longrightarrow \gamma + \gamma$.

Transitions in which J is large proceed slowly because the transition matrix element goes to zero at least as fast as $(R/\lambda)^{J-1}$ as J gets large, where λ is the wavelength of the emitted radiation and R is a typical radius of the emitting system. This can be seen from a multipole expansion of the transition matrix element, $\langle f | e^{i\mathbf{k} \cdot \mathbf{r}} \, \mathbf{j}_e(\mathbf{r}) \cdot \boldsymbol{\epsilon} | i \rangle$, where $\mathbf{j}_e(\mathbf{r})$ is the electromagnetic current of the emitting system.

32. The transition rate in perturbation theory is given by

$$\Gamma = 2\pi \, | \, H'_{fi} \, |^2 \times \text{(phase space)},$$

where H' is the perturbation Hamiltonian. For electromagnetic processes, the perturbing Hamiltonian is obtained by the principle of minimal electromagnetic coupling (Ampère's Hypothesis) $\mathbf{p} \longrightarrow \mathbf{p} - e\mathbf{A}/c$. Then the Hamiltonian is

$$H = H_0 - \frac{e}{2mc}(\mathbf{p} \cdot \mathbf{A} + \mathbf{A} \cdot \mathbf{p})$$

to first order in e. In the coordinate representation, $\mathbf{p}$ is the gradient operator. One may choose a gauge for $\mathbf{A}$ in which div $\mathbf{A} = 0$; hence the perturbing potential may be taken as $H' = -(e/mc)(\mathbf{A} \cdot \mathbf{p})$. This is taken between the initial state

$$\psi_i = \left(\frac{Z}{\pi a_0} \right)^{3/2} e^{-Zr/2a_0}$$

and the final state, assumed to be a plane wave $e^{i\mathbf{p} \cdot \mathbf{r}}$. Then exhibiting only Z-dependent factors,

$$H'_{fi} \sim Z^{5/2} \int d^3\mathbf{r} \exp \left\{ -i\mathbf{q} \cdot \mathbf{r} - \frac{Zr}{2a_0} \right\},$$

where $\mathbf{q} = \mathbf{p} - \mathbf{k}$, and $\mathbf{k}$ is the photon momentum. Completing the integral, we obtain

$$H'_{fi} \sim \frac{Z^{5/2}}{(q^2 + Z^2/4a_0^2)^2}.$$

At high photon energies one may neglect $Z^2/4a_0^2 \ll q^2$ and the Z-dependence of H'_{fi} is $H'_{fi} \sim Z^{5/2}$. In addition, when recoil and binding energy are neglected, the final density of states is Z-independent. Finally then, the cross section is proportional to Z^5.

33. The spherical square well may be treated as a one-dimensional square well, provided the potential is replaced by the effective potential $V(r) = -V_0 + (\hbar^2/2mr^2)L(L + 1)$ inside the well, and $V(r) = (\hbar^2/2mr^2)L(L + 1)$ outside. The transmission factor, the ratio of the flux outside and inside the well, is given, in WKB-approximation, as

$$T = \left(\frac{E}{E + V_0}\right)^{1/2} e^{-2\int k(x)dx}, \qquad \text{where } \frac{\hbar^2 k^2}{2m} = V - E;$$

thus,

$$T = \left(\frac{E}{E + V_0}\right)^{1/2} \exp\left\{-2\int_a^b dr\left[\frac{L(L + 1)}{r^2} - \frac{2mE}{\hbar^2}\right]^{1/2}\right\}.$$

Here, b is the point at which the radicand vanishes. Each time the particle, which may be pictured as bouncing back and forth inside the well, strikes the wall, the probability of escaping is T. The particle hits the wall ($v/2a = [(2/m)(E + V_0)]^{1/2}/2a$) times/sec, so

$$\frac{1}{\tau} = \left[\frac{(E + V_0)}{2ma^2}\right]^{1/2} T.$$

Upon change of variables the integral in the expression for T becomes

$$\sqrt{L(L + 1)} \int_\gamma^1 dx\left(\frac{1}{x^2} - 1\right)^{1/2}, \qquad \text{where } \gamma = \left[\frac{2mEa^2}{\hbar^2 L(L + 1)}\right]^{1/2}.$$

The exact integration is somewhat involved; however, in the case $\gamma \ll 1$, one has

$$\int_\gamma^1 \frac{dx}{x}(1 - x^2)^{1/2} \sim \log\left(\frac{1}{\gamma}\right)$$

and the expression for τ becomes

$$\tau = \left(\frac{2ma^2}{E}\right)^{1/2}\left\{\frac{\hbar^2 L(L + 1)}{2ma^2 E}\right\}^{\sqrt{L(L+1)}}.$$

34. The second-order perturbation amplitude for the transition is

$$M \propto \sum_{m_z} \langle f|\, \boldsymbol{\epsilon}_1 \cdot \mathbf{p}\,| L = 1, m_z\rangle\langle L = 1, m_z|\boldsymbol{\epsilon}_2 \cdot \mathbf{p}\,|i\rangle.$$

The interaction Hamiltonian is taken as $H = -(e\mathbf{A} \cdot \mathbf{p})/mc$, and the dipole approximation is used. Both initial and final states have $L = 0$, while the only intermediate state contributing has $L = 1$. As the matrix elements $\langle L|\mathbf{p}|0\rangle$ are zero unless $L = 1$, the sum may be extended over a complete set of intermediate states. That is, $M \propto \langle f|(\boldsymbol{\epsilon}_1 \cdot \mathbf{p})(\boldsymbol{\epsilon}_2 \cdot \mathbf{p})|i\rangle$. However the initial and final states $(L = 0)$ are spherically symmetric, so

$$M \propto \tfrac{1}{3}\boldsymbol{\epsilon}_1 \cdot \boldsymbol{\epsilon}_2 \langle f | p^2 | i \rangle.$$

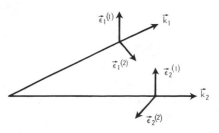

The rate Γ is obtained by summing $|M|^2$ over the final spin states of the photons:

$$\Gamma \propto \sum_{\text{Polarizations}} |\boldsymbol{\epsilon}_1 \cdot \boldsymbol{\epsilon}_2|^2.$$

By referring to the figure, one sees that $\Gamma \propto 1 + \cos^2 \phi$. Another way of obtaining the spin sum is to note that

$$\sum_{\alpha=1}^{z} \epsilon_i^{(\alpha)} \epsilon_j^{(\alpha)} = \delta_{ij} - \frac{k_i k_j}{k^2} \qquad (i, j = 1, 2, 3).$$

That this is so may be seen from the fact that the sum transforms like a second-rank tensor and is therefore equal to

$$A\delta_{ij} + Bk_i k_j.$$

The coefficients A and B are determined by the transversality condition $\mathbf{k} \cdot \boldsymbol{\epsilon}^{(\alpha)} = 0$, and the normalization of $\boldsymbol{\epsilon}$, that is $|\boldsymbol{\epsilon}|^2 = 1$. The spin sum is thus

$$\Gamma = \sum_{ij} \sum_{\alpha\beta} \epsilon_{1i}^{(\alpha)} \epsilon_{1j}^{(\alpha)} \epsilon_{2i}^{(\beta)} \epsilon_{2j}^{(\beta)} = \sum_{ij} \left(\delta_{ij} - \frac{k_{1i} k_{1j}}{k_1^2} \right) \left(\delta_{ij} - \frac{k_{2i} k_{2j}}{k_2^2} \right)$$

$$= 1 + \frac{(\mathbf{k}_1 \cdot \mathbf{k}_2)^2}{k_1^2 k_2^2} = 1 + \cos^2 \phi.$$

Hence $W(\phi) = (2/3\pi)(1 + \cos^2 \phi)$ when $W(\phi)$ is normalized so that $\int_0^\pi W(\phi)d\phi = 1$.

35. The first-order Born-approximation scattering amplitude is

$$f(\theta, \phi) = -\frac{m}{2\pi\hbar^2} \int V(\mathbf{r}) e^{i\mathbf{K}\cdot\mathbf{r}} d^3\mathbf{r},$$

where $\mathbf{K} = \mathbf{k}_i - \mathbf{k}_f$. Therefore

$$f(\theta, \phi) = a \sum_i \int \delta(\mathbf{r} - \mathbf{r}_i) e^{i\mathbf{K}\cdot\mathbf{r}} d^3\mathbf{r} = a \sum_i e^{i\mathbf{K}\cdot\mathbf{r}_i}.$$

One obtains maximum scattering when the contributions from each lattice point are in phase. We choose a lattice point $\mathbf{r}_i = d(n_1\hat{\mathbf{x}} + n_2\hat{\mathbf{y}} + n_3\hat{\mathbf{z}})$ where the n_i are integers, and this condition becomes

$$d\mathbf{K}\cdot\hat{\mathbf{x}} = 2\pi l; \qquad d\mathbf{K}\cdot\hat{\mathbf{y}} = 2\pi m; \qquad \text{and} \qquad d\mathbf{K}\cdot\hat{\mathbf{z}} = 2\pi n,$$

where (l, m, n) are integers (the so-called Miller indices). Thus $\mathbf{K}$ is normal to the set of lattice planes defined by (lmn) (see the figure on p. 174). The magnitude of K then satisfies

$$Kd = 2\pi (l^2 + m^2 + n^2)^{1/2}. \tag{1}$$

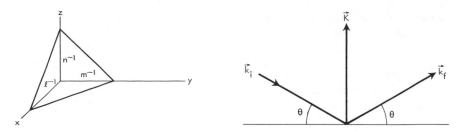

In terms of the scattering angle from the reflection planes,

$$K^2 = (\mathbf{k}_i - \mathbf{k}_f)^2 = 2k^2(1 - \cos 2\theta) = 4k^2 \sin^2 \theta,$$

Eq. (1) becomes $kd \sin \theta = \pi(l^2 + m^2 + n^2)^{1/2}$, which is the Bragg condition for reflection from a set of planes of spacing $d(l^2 + m^2 + n^2)^{-1/2}$.

36.

$$\mathbf{J} = \frac{\hbar}{2mi}[\psi^* \nabla \psi - \psi \nabla \psi^*] \tag{1}$$

and

$$\nabla \cdot \mathbf{J} = \frac{\hbar}{2mi}[\psi^* \nabla^2 \psi - \psi \nabla^2 \psi^*], \tag{2}$$

but

$$-\frac{\hbar^2}{2m} \nabla^2 \psi + (V_r - iV_i)\psi = i \frac{\partial \psi}{\partial t} \tag{3a}$$

and

$$-\frac{\hbar^2}{2m} \nabla^2 \psi^* + (V_r + iV_i)\psi^* = -i \frac{\partial \psi^*}{\partial t}. \tag{3b}$$

Multiply (3a) by ψ^* and (3b) by ψ, and subtract. Then (2) becomes

$$\frac{\partial}{\partial t}(\psi^* \psi) + \nabla \cdot \mathbf{J} = -\frac{2}{\hbar} V_i \psi^* \psi.$$

Thus particles are being absorbed at the rate $(2/\hbar)V_i\psi^*\psi$ per unit volume. By definition of the absorption cross section, this must equal $N\sigma v\psi^*\psi$, where N is the density of absorbers. Thus $\sigma = 2V_i/\hbar Nv$.

37. Semiclassical considerations show that the incident particle interacts with the sphere for $l \leq l_0 = 2\pi a/\lambda$; higher partial waves do not interact with the sphere. Thus $\delta_l = 0$ when $l > l_0$, and for these partial waves, $\eta_l = 1$; i.e. there is no scattering because there is no interaction. However, for $l < l_0$, the partial waves are simply removed from the beam. This means $\eta_l = 0$ for $l < l_0$. The scattering amplitude is then

$$f(\theta) = -\frac{1}{2ik} \sum_{l=0}^{l_0} (2l + 1)P_l(\cos \theta).$$

The cross section for elastic scattering is

$$\sigma_e = \int d\Omega \,|f(\theta)|^2 = \frac{\pi}{k^2} \sum_{l=0}^{l_0} (2l+1) \approx \frac{\pi l_0^2}{k^2} = \pi a^2$$

for large l_0. The total cross section is obtained from the optical theorem:

$$\text{Im } f(0) = \frac{k\sigma_t}{4\pi}.$$

Recalling that $P_l(\cos\theta = 1) = 1$, we obtain $\sigma_t = 2\pi a^2$. The reaction cross section is then

$$\sigma_a = \sigma_t - \sigma_e = \pi a^2.$$

38. The symmetry of the problem suggests the expansion $\psi(r, \theta) = \sum_l U_l(r) P_l(\cos\theta)$, where

$$U_l''(r) + \frac{2}{r} U_l'(r) + \left[k^2 - \frac{l(l+1)}{r^2} \right] U_l(r) = 0 \qquad (r > a),$$

and $k^2 = 2mE/\hbar^2$. The most general solution is

$$U_l(r) = A_l[\cos\delta_l\, j_l(kr) - \sin\delta_l\, \eta_l(kr)],$$

where the spherical Bessel functions are all obtained from the recursion relation and from the solutions for $l = 0$:

$$j_0(kr) = \frac{\sin(kr)}{kr} \qquad \text{and} \qquad \eta_0(kr) = -\frac{\cos(kr)}{kr}.$$

Since $U_l(a) = 0$, $\tan\delta_l = j_l(ka)/\eta_l(ka)$ gives the phase shifts. As we are told to neglect D-waves, only two phase shifts need be computed:

$$\delta_0 \approx \tan\delta_0 = -\tan(ka)$$

$$\delta_1 \approx \tan\delta_1 = \frac{ka - \tan(ka)}{1 + ka\tan(ka)} \sim -\frac{(ka)^3}{3}.$$

The differential cross section is

$$\frac{d\sigma}{d\Omega} = \frac{1}{k^2} \left| \sum_l e^{i\delta_l} \sin(\delta_l)(2l+1) P_l(\cos\theta) \right|^2$$

$$= \frac{1}{k^2} \left| e^{-ika} \sin(ka) + 3e^{-i(ka)^3/3} \sin\left[\frac{k^3 a^3}{3}\right] \cos\theta \right|^2$$

$$= a^2 \left\{ 1 - \frac{(ka)^2}{3} + 2(ka)^2 \cos\theta + \text{terms of higher order in } (ka) \right\}.$$

The total cross section is

$$\sigma = \int d\Omega \frac{d\sigma}{d\Omega} = 4\pi a^2 \left(1 - \frac{(ka)^2}{3} + \cdots \right).$$

39. In case (a) we are dealing with low-energy scattering for which the S-wave dominates. The radial wave function $U = \sin(kr + \delta_0)/r$ must vanish on the surface of the sphere; thus $\delta_0 = -ka$. The scattering amplitude $f(\theta)$ then reduces to $f(\theta) = -a$, and the total cross section $\sigma = \int d\Omega |f|^2 = 4\pi a^2$.

In case (b) we are dealing with high-energy scattering and all partial waves contribute. Since classically all particles having impact parameter greater than a are unaffected, one expects $\delta_l = 0$ for $l\hbar > \hbar ka$. However, those particles for which $l\hbar < \hbar ka$ would be eliminated from the *forward direction* (classically speaking). This intuitive classical argument suggests that one take

$$e^{2i\delta_l} = \begin{cases} 1 & \text{for } l > ka \\ 0 & \text{for } l < ka \end{cases}, \tag{1}$$

for purposes of calculation in the *forward direction* (only). Actually these arguments can be made more rigorous, and in so doing we show that for scattering on a *hard* (totally reflecting) sphere, we obtain a diffraction peak in both the forward and backward directions, while for a *black* (totally absorbing) sphere, a peak is obtained only in the forward direction.

Using the results of the last section, we calculate the phase shifts δ_l from the condition:

$$U_l(a) = \cos \delta_l j_l(ka) - \sin \delta_l \eta_l(ka) = 0. \tag{2}$$

For $ka \gg 1$ we make use of asymptotic expansions; thus
(a) for $l \gg ka \gg 1$, one has $j_l(ka) \sim 0$ and $\eta_l(ka) \gg 1$, and Eq. (2) is solved by taking $\delta_l = 0$ in this case.
(b) for $ka \gg l$, one has

$$j_l \sim \frac{1}{ka} \sin\left(ka - \frac{l\pi}{2}\right) \quad \text{and} \quad \eta_l \sim -\frac{1}{ka} \cos\left(ka - \frac{l\pi}{2}\right).$$

Eq. (2) is then solved by taking $\delta_l \sim -ka + l\pi/2$
Thus a somewhat more rigorous approach than our original one yields

$$e^{2i\delta_l} = \begin{cases} 1 & \text{for } l > ka \gg 1 \\ (-1)^l e^{-2ika} & \text{for } ka \gg l \end{cases}. \tag{3}$$

This expression for $e^{2i\delta_l}$ is to be substituted in the formula for the scattering amplitude,

$$f(\theta) = \frac{1}{2ik} \sum_{l=0}^{\infty} (2l+1)(e^{2i\delta_l} - 1)P_l(\cos\theta)$$

$$\simeq \frac{1}{2ik} \sum_{l=0}^{ka} (2l+1)(e^{2i\delta_l} - 1)P_l(\cos\theta).$$

Near the forward direction $P_l(\cos\theta) \sim 1$ for all l, and it makes little difference whether one uses $e^{2i\delta_l} = 0$ or $(-1)^l e^{-2ika}$, because the latter choice alternates in sign, thus making its contribution neglible compared to the sum $\sum_{l=0}^{ka}(2l+1)$. Thus, for either choice,

$$f(\theta) \simeq \frac{-1}{2ik} \sum_{l=0}^{ka}(2l+1)P_l(\cos\theta) \qquad \text{for small } \theta.$$

For $ka \gg 1$, one may further approximate this sum by an integral:

$$f(\theta) \sim \frac{i}{k}\int_0^{ka} x \, dx \, P_x(\cos\theta),$$

and when one uses the approximation, $P_n(\cos\theta) = J_0(n\theta)$, becomes

$$f(\theta) \simeq \frac{i}{k}\int_0^{ka} x \, dx \, J_0(x\theta) = \frac{ia}{\theta}J_1(ka\theta).$$

Thus $d\sigma/d\Omega = |f|^2$ has the approximate value a^4k^2 in the forward direction and decreases rapidly in the angular region $0 < \theta \leq 1/ka$.

Consider now the backward direction. For a *black* sphere, $e^{2i\delta_l} \sim 0$ for $l < ka$, and the scattering amplitude is

$$f_b(\theta) \simeq -\frac{1}{2ik} \sum_{l=0}^{ka}(2l+1)P_l(\cos\theta).$$

However, for $\theta \sim \pi$ one has $P_l(-1) = (-1)^l$, and alternate terms in the sum tend to cancel, giving no peak in the backward direction. However, for a *hard* sphere we have shown that $e^{2i\delta_l} \simeq (-1)^l e^{-2ika}$; thus, in the backward direction, one has

$$f_h(\theta) \simeq -\frac{e^{-2ika}}{2ik} \sum_{l=0}^{ka}(2l+1)P_l(\cos\phi), \qquad \text{where } \phi = \pi - \theta.$$

Hence for a hard sphere one obtains a diffraction pattern of the same size and shape in both the forward and backward directions.

40. The wave equation inside the well is

$$-\frac{\hbar^2}{2m}\frac{d^2\psi}{dx^2} + (V_0a)\delta(x)\psi = E\psi, \tag{1}$$

while outside the well,

$$-\frac{\hbar^2}{2m}\frac{d^2\psi}{dx^2} = E\psi. \tag{2}$$

We use the parameter (V_0a) with dimensions energy-times-length to describe the strength of the potential. Because the introduction of a delta-function in the wave equation makes a direct solution of (1) difficult, we first eliminate the delta from (1) by an integration. We integrate (1) over x, from $-\epsilon$ to $+\epsilon$, and then let $\epsilon \longrightarrow 0$. We have

$$-\frac{\hbar^2}{2m}\left[\frac{d\psi(\epsilon)}{dx} - \frac{d\psi(-\epsilon)}{dx}\right] + (V_0a)\psi(0) = E\int_{-\epsilon}^{+\epsilon}\psi(x)\,dx \longrightarrow 0. \tag{1'}$$

Outside the well, the wave function is

$$\psi(x) = \alpha^{1/2} e^{-\alpha x} \qquad \text{for } x > 0$$
$$= \alpha^{1/2} e^{+\alpha x} \qquad \text{for } x < 0,$$

normalized to $\int_{-\infty}^{+\infty} |\psi|^2 \, dx = 1$. Substituting this solution in (1') relates α to the potential strength, $\alpha = -m(V_0 a)/\hbar^2$. Then (2) gives the bound-state energy (for V_0 attractive)

$$E = -\frac{\hbar^2 \alpha}{2m} = -m \frac{(V_0 a)^2}{2\hbar^2}.$$

Thus the delta-function potential has only one bound state. It is striking that if we choose $V_0 = -e^2/a$, we obtain the energy of the ground-state hydrogen atom, and the wave function for $x > 0$ becomes that of the hydrogen atom (radial part).

To solve the scattering problem, we introduce, for $x < 0$, an incident wave and a reflected wave, while for $x > 0$ we have a transmitted wave. Thus

$$\psi(x < 0) = e^{ikx} + R e^{-ikx},$$
$$\psi(x > 0) = T e^{ikx}.$$

Substituting into (1'), with the condition $\psi(x \to 0^-) = \psi(x \to 0^+)$ which requires $1 + R = T$, we obtain $-(\hbar^2 ik R)/m + (V_0 a)(1 + R) = 0$, from which

$$R = \frac{-1}{(1 - \hbar^2 ik/mV_0 a)},$$

and

$$T = \frac{1}{(1 + imV_0 a/k\hbar^2)}.$$

The reflected intensity is therefore

$$|R|^2 = \frac{1}{1 + (\hbar^2 k/mV_0 a)^2},$$

while the transmitted intensity is

$$|T|^2 = \frac{1}{1 + (mV_0 a/\hbar^2 k)^2}.$$

As the strength of the potential well increases relative to the energy $E = \hbar^2 k^2/2m$ of the incident beam, the intensity of the reflected beam approaches unity. On the other hand, as $(V_0 a) \to 0$, the transmission coefficient approaches unity, i.e. the potential has no effect, as expected. It is interesting to note that $|R|$ and $|T|$ do not depend on the sign of V_0 and that the pole in R or T at $k = -imV_0 a/\hbar^2$ yields the bound-state energy $E = \hbar^2 k^2/2m = -m(V_0 a)^2/2\hbar^2$. Also see Problem (11–21).

An alternate solution to the bound-state problem is obtained by considering the wave equation in momentum space. The wave equation in coordinate space, for all x, is

$$-\frac{\hbar^2}{2m}\frac{d^2\psi(x)}{dx^2} + V_0 a\,\delta(x)\psi(x) = E\psi(x).$$

Upon multiplying by $e^{ipx/\hbar}$ and integrating over all x, we obtain, assuming that $\psi(x)$ and $d\psi(x)/dx \longrightarrow 0$ as $|x| \longrightarrow \infty$ in the integration by parts for the first term, the wave equation in momentum space

$$\frac{p^2}{2m}\phi(p) + V_0 a\psi(x=0) = E\phi(p),$$

where

$$\phi(p) \equiv \int_{-\infty}^{\infty} \psi(x)e^{ipx/\hbar}\,dx.$$

Thus

$$\phi(p) = -\frac{2mV_0 a\psi(x=0)}{p^2 - 2mE} \equiv -\frac{2mV_0 a\psi(x=0)}{p^2 + 2mW},$$

where $W \equiv -E = |E| > 0$. We now obtain

$$\psi(x) = \frac{1}{2\pi\hbar}\int_{-\infty}^{\infty}\phi(p)e^{-ipx/\hbar}\,dp = -\frac{m}{\pi\hbar}V_0 a\psi(x=0)\int_{-\infty}^{+\infty}\frac{dp\,e^{-ipx/\hbar}}{p^2 + 2mW}.$$

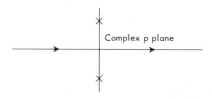

Complex p plane

The integration may be performed in the complex plane, closing the contour below when $x > 0$, and above when $x < 0$. We obtain

$$\psi(x) = -\frac{mV_0 a}{\pi\hbar}\frac{\psi(x=0)}{\sqrt{2mW}}e^{-x\sqrt{2mW}\epsilon(x)/\hbar},$$

where $\epsilon(x) = 1(-1)$ when $x > 0\ (< 0)$. The requirement that $\psi(x \to 0^+) = \psi(x \to 0^-) = \psi(x=0)$ gives a self-consistency condition

$$1 = -\frac{mV_0 a}{\hbar\sqrt{2mW}} \qquad \text{or} \qquad W = \frac{mV_0^2 a^2}{2\hbar^2} = -E.$$

Notice that a bound state is possible only when V_0 is negative.

7

THERMODYNAMICS

1. The thermodynamic potentials are defined by:

U (internal energy),

$H = U + PV$ (enthalpy),

$A = U - TS$ (Helmholtz function),*

$G = H - TS$ (Gibbs function).

The quantities

$$dU = T\,dS - P\,dV,$$
$$dH = T\,dS + V\,dP,$$
$$dA = -P\,dV - S\,dT,$$
$$dG = V\,dP - S\,dT,$$

are all exact differentials. Equations (1) follow immediately. One also obtains such identities as

$$S = -\left(\frac{\partial G}{\partial T}\right)_P \quad \text{and} \quad V = \left(\frac{\partial G}{\partial P}\right)_T.$$

To establish (2), expand

$$T\,dS = T\left(\frac{\partial S}{\partial V}\right)_T dV + T\left(\frac{\partial S}{\partial T}\right)_V dT = T\left(\frac{\partial P}{\partial T}\right)_V dV + C_V dT,$$

using (1). To establish (3),

$$T\,dS = dU + P\,dV,$$

$$T\left(\frac{\partial S}{\partial V}\right)_T = P + \left(\frac{\partial U}{\partial V}\right)_T = T\left(\frac{\partial P}{\partial T}\right)_V$$

using (1) again.

2. In a reversible, infinitesimal change of state,

$$T\,dS = dU + P\,dV,$$

---

* We will sometimes denote the Helmholtz function (free energy) by A; F is also used for this function. Both notations are common in the literature.

from which

$$\left(\frac{\partial U}{\partial V}\right)_T = T\left(\frac{\partial S}{\partial V}\right)_T - P = T\left(\frac{\partial P}{\partial T}\right)_V - P.$$

When $P = RT/(V - B) - a/V^2$, this is a/V^2.

3. Consider infinitesimal changes of state along the phase boundary.
(a) $dV_1 = dV_2$, or

$$\left[\left(\frac{\partial V}{\partial T}\right)_P^1 - \left(\frac{\partial V}{\partial T}\right)_P^2\right]dT = dP\left[\left(\frac{\partial V}{\partial P}\right)_T^2 - \left(\frac{\partial V}{\partial P}\right)_T^1\right].$$

Calling $\alpha = (1/V)(\partial V/\partial T)_P$ the coefficient of volume expansion and $\beta = -(1/V)(\partial V/\partial P)_T$ the bulk compressibility, we have

$$\frac{dP}{dT} = \frac{(\alpha_1 - \alpha_2)}{(\beta_1 - \beta_2)}.$$

(b) $dH_1 = dH_2$ implies $dS_1 = dS_2$ in a phase change at constant pressure. Then

$$\left[\left(\frac{\partial S}{\partial P}\right)_T^1 - \left(\frac{\partial S}{\partial P}\right)_T^2\right]dP = \left[\left(\frac{\partial S}{\partial T}\right)_P^2 - \left(\frac{\partial S}{\partial T}\right)_P^1\right]dT,$$

or

$$\frac{dP}{dT} = \frac{\left[\left(\frac{\partial S}{\partial T}\right)_P^2 - \left(\frac{\partial S}{\partial T}\right)_P^1\right]}{\left[\left(\frac{\partial S}{\partial P}\right)_T^1 - \left(\frac{\partial S}{\partial P}\right)_T^1\right]} = \frac{(C_P^{(2)} - C_P^{(1)})}{T\left[\left(\frac{\partial V}{\partial T}\right)_P^2 - \left(\frac{\partial V}{\partial T}\right)_P^1\right]} = \frac{C_P^{(2)} - C_P^{(1)}}{TV(\alpha^{(2)} - \alpha^{(1)})}.$$

These two expressions for dP/dT are known as Ehrenfest's equations.

4. On the assumption that the body of water is taken down with no heat exchange (i.e., $\Delta S = 0$), the change in temperature is $\Delta T = (\Delta T/\Delta P)_S \Delta P$ where

$$\Delta P = \rho g \, \Delta h = 9.8 \times 10^7 \, \text{dyne/cm}^2.$$

Using the identity

$$\left(\frac{\partial T}{\partial P}\right)_S = \frac{-\left(\frac{\partial S}{\partial P}\right)_T}{\left(\frac{\partial S}{\partial T}\right)_P} = -\frac{T\left(\frac{\partial S}{\partial P}\right)_T}{C_P}$$

and the Maxwell equation $-(\partial S/\partial P)_T = (\partial V/\partial T)_P$, one finds

$$\left(\frac{\partial T}{\partial P}\right)_S = \frac{VT\frac{1}{V}\left(\frac{\partial V}{\partial T}\right)_P}{C_P} = 300 \times .0013 \, \text{cm}^3 \, \text{deg/cal}$$

$$= 10^{-8} \, \text{cm}^2 \, \text{deg/dyne}.$$

Hence $T \approx 1°\text{C}$.

5. If the two adiabats intersect, they may be joined by an isotherm. Now consider a cycle along the closed curve. The work done is equal to the enclosed area, and heat is absorbed along the isotherm. Thus the cycle absorbs heat and does work, without giving off any heat. The efficiency of the process is one hundred percent in violation of the Second Law of Thermodynamics.

6. (a) We have

$$S = C_P \ln\left(\frac{273}{298}\right) - \frac{\lambda}{273} + C_P \ln\left(\frac{373}{298}\right) + \frac{n\lambda'}{373} = 0,$$

from which $n = 0.1$ mole.

(b) The heat removed from the first mole is

$$Q_1 = C_P \Delta T_1 + \lambda$$

$$= \left(18\frac{\text{cal}}{\text{mole} \cdot \text{deg}}\right)(25 \text{ deg}) + 1438\frac{\text{cal}}{\text{mole}} = 1888 \text{ cal},$$

while the heat absorbed by the second mole is

$$Q_2 = C_P \Delta T_2 + n\lambda' = 2320 \text{ cal}.$$

The work done by the refrigerator is $Q_2 - Q_1 = 430$ cal.

7. Since the final entropy does not depend on how the final state is reached it will be calculated as if it were reached isobarically. This is possible because the final pressure is $P_f = P$. Then, for each side separately,

$$T \, dS = C_P \, dT.$$

Hence

$$\Delta S_1 = C_P \log\frac{T_f}{T_1} \quad \text{and} \quad \Delta S_2 = C_P \log\frac{T_f}{T_2}.$$

But $T_f = (T_1 + T_2)/2$ and $C_P = (5/2) Nk$. Therefore

$$\Delta S = \Delta S_1 + \Delta S_2 = \frac{5}{2} Nk \log\left(\frac{T_f^2}{T_1 T_2}\right)$$

$$= 5Nk \log\left\{\frac{(T_1 + T_2)}{2\sqrt{T_1 T_2}}\right\},$$

which vanishes if $T_1 = T_2$ as it should.

8. (a) Mechanical equilibrium requires $V_A/V_B = 1/3$ at $t = 0$ and $V_A/V_B = 1$ at $t = \infty$.

(b) The final temperature is $T_A = T_B = 2T_0$. The process is isobaric, therefore $T \, dS = C_P \, dT$ for each compartment. Integrating, one obtains

$$\Delta S_A = C_P \log\left(\frac{2T_0}{T_0}\right) \quad \text{and} \quad \Delta S_B = C_P \log\left(\frac{2T_0}{3T_0}\right).$$

The total entropy change is thus

$$C_P \log\left(\tfrac{4}{3}\right) = \tfrac{5}{2}R \log\left(\tfrac{4}{3}\right).$$

(c) If the transfer had been reversible, we would have had

$$\Delta S = 0, \quad \text{or} \quad \int \frac{C_V dT}{T} + R\int \frac{dV}{V} = 0$$

or

$$C_V \ln\left(\frac{T_f^2}{T_A T_B}\right) + R \ln\left(\frac{V_f^2}{V_A V_B}\right) = 0.$$

Plugging in $V_f = (V_A + V_B)/2$, $V_A = V_f/2$, $V_B = 3V_f/2$, one obtains

$$T_f^2 = 3T_0^2(\tfrac{4}{3})^{-R/C_V} = 3T_0^2(\tfrac{4}{3})^{-2/3}.$$

The amount of useful work possible would have been

$$W = U_i - U_f = C_v(4T_0 - 2T_f) = 3R(2T_0 - T_f).$$

9. In a free expansion, there is no heat exchange with the outside universe, and no work is done. Hence $dU = 0$. But

$$dU = \left(\frac{\partial U}{\partial T}\right)_V dT + \left(\frac{\partial U}{\partial V}\right)_T dV = C_V dT + \frac{a dV}{V^2}$$

[see Problem (7–2)]. Hence

$$dT = -\left(\frac{a}{C_V V^2}\right), \quad (dV < 0), \quad \text{or} \quad T_2 = T_1 + \left(\frac{1}{V_2} - \frac{1}{V_1}\right)\frac{a}{C_V}.$$

A free expansion is not a reversible process; to compute the change in entropy, one must find a reversible process connecting the initial and final states. Then one computes ΔS for this process; the result is independent of the process, as dS is an exact differential. A convenient process consists of (a) expansion of the gas at constant temperature to the final volume, then (b) cooling at constant volume, to the final temperature.

Then,

$$\Delta S = \int_1^2 \frac{dQ}{T} = \int_{V_1}^{V_2} \frac{\{(\partial U/\partial V)_T + P\}dV}{T_1} + \int_{T_1}^{T_2} \frac{C_V dT}{T},$$

the first term contributing only during the expansion, the second only during the cooling.

Inserting $P = RT/(V - b) - a/V^2$, and $(\partial U/\partial V)_T = a/V^2$ in the first term, we have

$$\Delta S = C_V \ln\frac{T_2}{T_1} + R \ln\left\{\frac{V_2 - b}{V_1 - b}\right\} \approx R \ln\frac{V_2}{V_1},$$

$$H = U + PV \quad \text{and} \quad \Delta H = \Delta U + \Delta(PV).$$

However, $\Delta U = 0$; hence

$$\Delta H = P_2 V_2 - P_1 V_1 \approx R(T_2 - T_1) = \left(\frac{1}{V_2} - \frac{1}{V_1}\right)\frac{a}{C_V}.$$

10. The process described is throttling, in which the enthalpy, $H = U + PV$, remains constant. The initial enthalpy of 1 gm of water is $H_i = U_i + P_i V$. If, in the final state, a fraction, x, is converted to steam at the boiling point of water $T_f = 100°C$, then the final enthalpy is given by

$$H_f = [U_i + C(T_f - T_i) + P_f V] + xL.$$

The first term in brackets is the enthalpy of 1 gm of water at $P_f = 1$ atm and $T_f = 100°C$. The last term is the enthalpy change upon a change of phase of x gm. Here L is the latent heat of vaporization. Thus

$$C(T_f - T_i) + xL = (P_f - P_i)V,$$

where $C = 1$ cal/gm · deg; $L = 540$ cal/gm; $V = 1$ cm³. Then

$$(P_f - P_i) \approx 10^4 \text{ atm} \approx 10^{10} \text{ dyn/cm}^2; \; 1 \text{ cal} = 4.18 \times 10^7 \text{ erg}:$$

From this equation one finds $x \approx 0.3$.

11.   Assume that entropy of the gas remaining does not change. The entropy change on cooling is given by

$$T \, dS = C_V \, dT + P \, dV,$$

which, combined with the ideal gas equation, yields

$$\Delta S_C = (C_V + R) \log \frac{T_f}{T_i} - R \log \frac{P_f}{P_i}.$$

The entropy change upon liquefying is $\Delta S_l = -L/T_f$. However, $(\Delta S)_C + (\Delta S)_l = 0$ by hypothesis; thus

$$P_i = P_f \left(\frac{T_i}{T_f} \right)^{(C_V + R)/R} \exp \left(\frac{L}{RT_f} \right) = 100 \text{ atm}.$$

12.   Suppose that the center of the compression at any time has a higher temperature by ΔT than the center of dilation. These points are separated by a distance $\lambda/2$ where λ is the wavelength of the oscillation. Then in time $\tau/2$ (where τ is the period) one has approximately

$$[\text{Heat flow from crest to valley}] = K \left(\frac{\Delta T}{\lambda/2} \right) A \left(\frac{\tau}{2} \right),$$

$$[\text{Heat flow necessary to raise temperature by } \Delta T] = \rho A \left(\frac{\lambda}{2} \right) C \, \Delta T.$$

Oscillations tend to be isothermal if the heat flow is sufficient to equalize the temperature in time $\tau/2$. Thus

$$KA\tau \frac{\Delta T}{\lambda} \gg A\rho\lambda C \frac{\Delta T}{2} \qquad \text{(isothermal condition)}$$

or in terms of the frequency $f = 1/\tau$,

$$f \ll \frac{2K}{\rho C \lambda^2}.$$

However, the frequency and wavelength are related by $f\lambda = \sqrt{Y/\rho}$. Thus the above relation may be written, after eliminating λ,

$$f \gg \frac{YC}{2k} = 5 \times 10^{10} \text{ sec}^{-1}.$$

13. Let n_1 and n_2 be the concentration of CO in the first and second vessels, respectively. The current of CO from the first to the second vessel is then

$$I = \frac{D(n_1 - n_2)A}{L},$$

which must equal $-V(dn_1/dt)$. However, $n_1 + n_2 = n$ is constant, since the total amount of CO is conserved. Thus,

$$\frac{dn_1}{dt} = -\left(\frac{2DA}{LV}\right)n_1 + \frac{DAn}{LV}.$$

The solution to this equation with $n_1(0) = n$ (all CO initially in the first vessel) is

$$n_1 = \frac{n}{2}\{1 + e^{-2ADT/LV}\},$$

and since the partial pressure is proportional to the concentration,

$$P = \frac{P_0}{2}\{1 + e^{-2ADt/LV}\}.$$

14. The heat flow equation $\nabla \cdot \mathbf{H} + C\rho\,(\partial T/\partial t) = 0$, with $\mathbf{H} = -K\nabla T$, has the spherically symmetric solution for $t > 0$,

$$T(r, t) = \sum_{n=1} A_n e^{-n^2\pi^2 at}\frac{\sin\,(n\pi r/R)}{r}, \qquad \text{with} \qquad T(R, t) \equiv 0°C,$$

where $a = (K/C\rho\,R^2)$ and $R = $ radius of sphere. The coefficients A_n are found from the initial condition

$$T(r, 0) = T_0 = 100°C \qquad \text{for } r \leq R.$$

Thus $\sum_n A_n \sin\,(n\pi r/R) = T_0 r$, and from the orthogonality relations

$$\int_0^R dr \sin\frac{n\pi r}{R} \sin\frac{j\pi r}{R} = R\frac{\delta_{jn}}{2},$$

one finds

$$A_n = \frac{2T_0}{R}\int_0^R dr\, r \sin\frac{n\pi r}{R} = (-1)^{n+1}\frac{2T_0 R}{\pi n}.$$

Therefore

$$T(r, t) = \frac{2T_0 R}{\pi r}\sum_{n=1}^{\infty}\frac{(-1)^{n+1}}{n}\exp\,(-n^2\pi^2 at)\sin\frac{\pi nr}{R}$$

and $T(0, t) = 2T_0 \sum_{n=1}^{\infty} (-1)^{n+1} \exp\,(-n^2\pi^2\,at)$. When $t = 15$ min, $\pi^2 at = 2.85 \times (900)/(20)^2 = 4.16$. It is therefore a good approximation to keep only the first term in the sum

$$T(0, 15 \ m) \approx 200°C \times \exp(-4.16) = 3.1°C.$$

15. (a) The temperature satisfies the heat equation

$$\nabla^2 T = \frac{1}{a^2} \frac{\partial T}{\partial t} = 0, \qquad \text{where} \qquad a^2 = \frac{K}{C\rho}.$$

In spherical coordinates with T only a function of the radius r, one has $\nabla^2 = (1/r)(\partial^2/\partial r^2)r$, and the solution for the temperature may be written

$$T(r, t) = T_0 + \frac{R}{r} \Theta(r, t),$$

with $\Theta(r, t)$ satisfying the equation

$$\frac{\partial^2 \Theta}{\partial r^2} - \frac{1}{a^2} \frac{\partial \Theta}{\partial t} = 0. \tag{1}$$

The boundary conditions on Θ are

$$\Theta(R, t > 0) = T_1 - T_0, \qquad \Theta(r > R, t = 0) = 0.$$

In terms of the Green's function $G(r, t|r't')$ defined in the region r and $r' > R$ and satisfying the equations

$$\frac{\partial^2 G}{\partial r'^2} + \frac{1}{a^2} \frac{\partial G}{\partial t'} = -\delta(r - r')\delta(t - t'), \tag{2}$$

$$G(r, t|r', t') = 0 \qquad \text{when } t' > t \tag{3}$$

and

$$G(r, t|R, t') = 0, \tag{4}$$

the solution is

$$\Theta(r, t) = \int_0^\infty dt' \ \Theta(R, t') \frac{\partial G}{\partial R} (rt | Rt').$$

This may be proved by multiplying Eq. (1) by G, Eq. (2) by Θ and integrating over r' and t', making use of Eqs. (3) and (4). To calculate G, consider the function G_0 defined to satisfy Eqs. (2) and (3):

$$G_0(r - r', t - t') = \int_{-\infty}^\infty dk f(k, t - t')e^{ik(r-r')}.$$

Then $k^2 f + (1/a^2)(\partial f/\partial t) = \delta(t - t')/2\pi$ with solution

$$f(k, t - t') = \frac{a^2}{2\pi} \theta(t - t')e^{-k^2 a^2(t-t')},$$

where

$$\theta(t - t') = \begin{cases} 1 & \text{for} & t > t', \\ 0 & \text{for} & t < t'. \end{cases}$$

Thus we have

$$G_0 = \frac{a^2}{2\pi}\theta(t - t') \int dk\, e^{-k^2a^2(t-t')+ik(r-r')}$$

$$= \frac{a\theta(t - t')e^{-(r-r')^2/4a^2(t-t')}}{\sqrt{4\pi(t - t')}} ;$$

and if G is given by (method of images in electrostatics)

$$G = G_0(r - r', t - t') - G_0(2R - r - r', t - t'),$$

Eqs. (2) through (4) will be satisfied in the region $r, r' > R$. Finally then

$$\frac{\partial G}{\partial R}(r, t \,|\, R, t') = \frac{(r - R)\theta(t - t')}{2a\sqrt{\pi}\,(t - t')^{3/2}}\exp\left\{-\frac{(r - R)^2}{4a^2(t - t')}\right\},$$

and the temperature is given by

$$T(r, t) = T_0 + \frac{R(T_1 - T_0)(r - R)}{2ar\sqrt{\pi}}\int_0^t \frac{dt'\,e^{-(r-R)^2/4a^2(t-t')}}{(t - t')^{3/2}},$$

which can be simplified by a change of variables to read

$$T(r, t) = T_0 + \frac{2R(T_1 - T_0)}{r\sqrt{\pi}}\int_{(r-R)/2at^{1/2}}^{\infty} dx\, e^{-x^2}.$$

(b) In the limit as $t \to \infty$, the integral approaches

$$\int_0^\infty dx\, e^{-x^2} = \frac{\sqrt{\pi}}{2} ;$$

thus $T \to T_0 + (T_1 - T_0)(R/r)$, as would be expected from a static problem with the boundary conditions $T(R) = T_1$ and $T(\infty) = T_0$.

16. (a) We start from the energy equation with $U = EV$,

$$\left(\frac{\partial U}{\partial V}\right)_T = T\left(\frac{\partial P}{\partial T}\right)_V - P.$$

The pressure is given by $P = E/3$. Hence

$$\left(\frac{\partial P}{\partial T}\right)_V = \frac{\left(\frac{\partial E}{\partial T}\right)_V}{3}, \qquad \text{and} \qquad \left(\frac{\partial U}{\partial V}\right)_T = E.$$

Substituting these relations in the above equation yields $4E/T = \partial E/\partial T$. This has solution $E = kT^4$.

Alternate solution: Consider an infinitesimal Carnot cycle operating on the photon gas. In an infinitesimal displacement, $dP = dE/3$. The work done is the volume enclosed in the PV-plane, or $W = (dP)(dV) = dE\, dV/3$. The heat adsorbed is $Q = E\, dV + P\, dV = (4E/3)\, dV$. The efficiency of the engine is

$$\eta = \frac{W}{Q} = \frac{dE}{4E} = \frac{dT}{T}.$$

This has solution $E = kT^4$. The equating of η to dT/T follows, since all reversible heat engines have efficiency dT/T. The constant cannot be obtained from thermodynamics. It can be calculated from quantum-statistical mechanics; alternatively, it can be related to the Stefan-Boltzmann constant σ. Imagine a small hole in the cavity, from which energy is emitted. The energy flux is

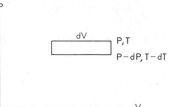

$$\frac{(cE)}{4} = \frac{(ckT^4)}{4} = W = \sigma T^4,$$

from which $k = 4\sigma/c$.

17. The power absorbed by the satellite is $(4\pi R^2 \sigma T_0^4)(\pi r^2/4\pi D^2)$. The power it radiates is $\sigma T^4 \cdot 4\pi r^2$. This gives an equilibrium temperature determined by $T^4 = T_0^4 R^2/4D^2 = \alpha^2 T_0^4/4$. The data in the problem give $T = 288°\text{K}$.

18. The probability of observing a magnetic-moment density between M and $(M + dM)$ is proportional to $dM\, e^{-A/kT}$. Thus the most probable magnetization M_0 is the one for which $A(M, T)$ has a minimum as a function of M. In addition, when one neglects the effects of fluctuations, the average magnetization,

$$\bar{M} \equiv \int_0^\infty dM\, M e^{-A/kT} \bigg/ \int_0^\infty dM\, e^{-A/kT}, \tag{1}$$

is equal to M_0. This is because the distribution function $e^{-A/kT}$ is large only in the vicinity of M_0. From the condition that $A(M, T)$ be a minimum, one finds

$$M_0 = \begin{cases} \left[\dfrac{\alpha}{2\beta}(T_c - T)\right]^{1/2} & \text{for} \quad T < T_c, \\ 0 & \text{for} \quad T > T_c. \end{cases}$$

To obtain an estimate of the effect fluctuations have on the calculation of $\bar{M}$, we consider the difference $\bar{M} - M_0$ for the two cases (a) $T < T_c$ and (b) $T_c < T$.

(a) Define a new variable x by

$$M = M_0 + x\left[\frac{kT}{2\alpha(T_c - T)}\right]^{1/2}, \qquad \text{where} \qquad M_0 = \left[\frac{\alpha}{2\beta}(T_c - T)\right]^{1/2}.$$

Then, expanding $A(M, T)$ about the value M_0, one finds

$$A = A_0 - \frac{\alpha^2}{4\beta}(T_c - T)^2 + kT\left(x^2 + \frac{x^3}{B} + \frac{x^4}{4B^2}\right),$$

with

$$B = [\alpha^2(T_c - T)^2/\beta\, kT]^{1/2}.$$

With this new variable, one finds from Eq. (1),

$$\frac{\bar{M} - M_0}{M_0} = \frac{\int_{-B}^{\infty} x \, dx \, e^{-(x^2 + x^3/B + x^4/4B^2)}}{B \int_{-B}^{\infty} dx \, e^{-(x^2 + x^3/B + x^4/4B^2)}}.$$

Exact evaluation of these integrals as a function of B is not possible. However, in the limit $B \gg 1$, one has, to lowest order in $1/B$, that

$$\int_{-B}^{\infty} x \, dx \, e^{-(x^2 + x^3/B + x^4/4B^2)} \approx -\frac{1}{B} \int_{-B}^{\infty} x^4 e^{-x^2} dx \approx -\frac{1}{B} \int_{-\infty}^{\infty} x^4 e^{-x^2} dx.$$

Thus, in this limit $B \gg 1$, one finds

$$\frac{\bar{M} - M_0}{M_0} \approx \frac{-\int_{-\infty}^{+\infty} dx \, x^4 e^{-x^2}}{B^2 \int_{-\infty}^{+\infty} dx \, e^{-x^2}} = -\frac{3}{4B^2} = -\frac{3\beta kT}{4\alpha^2(T_c - T)^2}.$$

(b) Define a new variable y by

$$M = y \left[\frac{kT}{\alpha(T - T_c)} \right]^{1/2};$$

then

$$A = A_0 + kT \left(y^2 + \frac{y^4}{B^2} \right),$$

and

$$\bar{M} = \left[\frac{kT}{\alpha(T - T_c)} \right]^{1/2} \frac{\int_0^{\infty} y \, dy \, e^{-(y^2 + y^4/B^2)}}{\int_0^{\infty} dy \, e^{-(y^2 + y^4/B^2)}},$$

where B is as defined in (a). In the limit $B \gg 1$, one obtains

$$\bar{M} \approx \left[\frac{kT}{\pi \alpha(T - T_c)} \right]^{1/2}.$$

In the presence of a magnetic field, one adds a term $-\mathbf{M} \cdot H$ to A. Thus above the Curie temperature, the average magnetization is given by

$$\bar{M}(H) = \left[\frac{kT}{\alpha(T - T_c)} \right]^{1/2} \frac{\int_0^{\infty} y \, dy \, e^{-(y^2 + yH/[kT\alpha(T-T_c)]^{1/2})}}{\int_0^{\infty} dy \, e^{-y^2}},$$

where terms of order $1/B \ll 1$ have been neglected. Thus to first order in H,

$$\bar{M}(H) = \frac{H}{2\alpha(T - T_c)} \qquad \text{(Curie-Weiss law)}.$$

The susceptibility is given by

$$\chi = \left. \frac{\partial \bar{M}}{\partial H} \right|_{H=0} = \frac{1}{2\alpha(T - T_c)}.$$

Note that this result could also have been obtained from the thermodynamic relation $H = \partial A/\partial M = 2\alpha(T - T_c)M + 4\beta M^3$. Thus

$$1 = [2\alpha(T - T_c) + 12\beta M^2]\frac{\partial M}{\partial H},$$

which implies

$$\frac{\partial M}{\partial H}\bigg|_{H=0} = \frac{1}{2\alpha(T - T_c)}.$$

19. The sound velocity is calculated from the adiabatic bulk modulus at constant magnetic field H:

$$v^2(H) = -\frac{V}{\rho}\left(\frac{\partial P}{\partial V}\right)_{S,H}.$$

We begin by showing that the familiar relation

$$\left(\frac{\partial P}{\partial V}\right)_{S,H} = \frac{C_{P,H}}{C_{V,H}}\left(\frac{\partial P}{\partial T}\right)_V$$

is true even in the presence of a magnetic field, the specific heats being calculated at constant H. Of course these specific heats will be a function of H, and it is because of this fact that the sound velocity is changed.

From the identities:

(a) $$\left(\frac{\partial P}{\partial V}\right)_{S,H} = \left(\frac{\partial P}{\partial V}\right)_{T,H} + \left(\frac{\partial P}{\partial T}\right)_{V,H}\left(\frac{\partial T}{\partial V}\right)_{S,H}$$

$$= \left(\frac{\partial P}{\partial V}\right)_{T,H}\left\{1 + \left(\frac{\partial V}{\partial P}\right)_{T,H}\left(\frac{\partial P}{\partial T}\right)_{V,H}\left(\frac{\partial T}{\partial V}\right)_{S,H}\right\},$$

(b) $$\left(\frac{\partial V}{\partial P}\right)_{T,H}\left(\frac{\partial P}{\partial T}\right)_{V,H} = -\left(\frac{\partial V}{\partial T}\right)_{P,H},$$

(c) $$\left(\frac{\partial T}{\partial V}\right)_{S,H}\left(\frac{\partial S}{\partial T}\right)_{V,H} = -\left(\frac{\partial S}{\partial V}\right)_{T,H},$$

one finds by substituting (b) and (c) into (a) that

$$\left(\frac{\partial P}{\partial V}\right)_{S,H} = \left(\frac{\partial P}{\partial V}\right)_{T,H}\left\{1 + \frac{T}{C_{V,H}}\left(\frac{\partial V}{\partial T}\right)_{P,H}\left(\frac{\partial S}{\partial V}\right)_T\right\},$$

where we have in addition used the result that $C_{V,\mathbf{H}} = T(\partial S/\partial T)_{V,\mathbf{H}}$. In addition, the definitions

$$C_{P,H} = T\left(\frac{\partial S}{\partial T}\right)_{P,H} = T\left(\frac{\partial S}{\partial T}\right)_{V,H} + T\left(\frac{\partial S}{\partial V}\right)_{T,H}\left(\frac{\partial V}{\partial T}\right)_{P,H}$$

and

$$C_{V,H} = T\left(\frac{\partial S}{\partial T}\right)_{V,H}$$

yield the relation

$$T\left(\frac{\partial S}{\partial V}\right)_{T,H}\left(\frac{\partial V}{\partial T}\right)_{P,H} = C_{P,H} - C_{V,H_3}$$

Thus

$$\left(\frac{\partial P}{\partial V}\right)_{S,H} = \left(\frac{\partial P}{\partial V}\right)_{T,H}\left[1 + \frac{C_{P,H} - C_{V,H}}{C_{V,H}}\right] = \frac{C_{P,H}}{C_{V,H}}\left(\frac{\partial P}{\partial V}\right)_{T,H}.$$

We must now calculate $C_{P,H}/C_{V,H}$. This is accomplished by calculating the entropy S and using

$$C_{P,H} = T(\partial S/\partial T)_{P,H} \quad \text{and} \quad C_V = T(\partial S/\partial T)_{V,H}.$$

From the $T\,dS$ relation,

$$T\,dS = dU - H\,dM + P\,dV,$$

one finds that the thermodynamic potential $\phi = U - TS - HM$ satisfies

$$d\phi = -M\,dH - S\,dT - P\,dV. \tag{1}$$

In addition, we are given $M = \gamma H/T$; thus, integrating the above equation at constant temperature and volume, one finds

$$\phi(T, V, H) = \phi_0(T, V) - \frac{\gamma H^2}{2T},$$

where $\phi_0(T, V)$ is independent of H. The entropy is then found from

$$S = -\left(\frac{\partial \phi}{\partial T}\right)_{V,H} = S_0(T, V) - \frac{\gamma H^2}{2T}.$$

Finally, then,

$$\frac{C_{P,H}}{C_{V,H}} = \frac{(C_P^0 + \gamma H^2/T^2)}{(C_V^0 + \gamma H^2/T^2)},$$

where C_P^0, C_V^0 are the specific heats in the absence of a magnetic field. Assuming that the magnetic field does not affect the equation of state relating P, V, and T (e.g. $PV = RT$ for an ideal gas), then $(\partial P/\partial V)_T$ is independent of H, and one has

$$v^2(H) = \frac{v^2(0)(1 + \gamma H^2/C_P^0 T^2)}{(1 + \gamma H^2/C_V^0 T^2)},$$

which, to lowest order in $\gamma H^2/CT^2$, yields

$$\frac{v(H) - v(0)}{v(0)} \approx \frac{-\gamma H^2}{2C_V^0 T^2}\frac{(C_P^0 - C_V^0)}{C_P^0}.$$

20. Because $M \approx 0$ for the normal state, the Gibbs function in the normal state may be assumed to be independent of H. Thus

$$G_N(T, H) = G_N(T, H_c). \tag{1}$$

Furthermore, there is no discontinuity in G across the phase boundary:

$$G_s(T, H_c) = G_N(T, H_c). \tag{2}$$

Inside the superconductor, $B = 0 = H + 4\pi M$, from which $M = -H/4\pi$. Therefore, at constant temperature, $dG_T = H\, dH/4\pi$. This may be integrated from $H < H_c$ to H_c, holding the temperature constant at $T < T_c$, to give

$$G_s(T, H) - G_s(T, H_c) = \frac{H^2}{8\pi} - \frac{H_c^2}{8\pi}.$$

However, from (1) and (2), $G_s(T, H_c) = G_N(T, H)$. This gives $G_s(T, H) - G_N(T, H) = H^2/8\pi - H_c^2/8\pi$. This is negative below the critical field, and positive above it. This shows that the superconductor is energetically favored, when $H < H_c$, and the normal state is favored when $H > H_c$.

Only the state (either normal or superconducting) with the lower Gibbs free energy at a given temperature and magnetic field is stable.

In order to compute the latent heat of transition, the analog to Clapeyron's equation is derived. For a change of H and T along the phase boundary, the changes in the Gibbs function for the normal and superconducting states must be the same, i.e., $dG_N = dG_s$ with $dG = -S\,dT - M\,dH$. Thus $dH_c/dT = (S_N - S_s)/(M_s - M_N)$ along the phase boundary. But the latent heat is given by $L = (S_N - S_s)T$, and hence

$$L = T(M_s - M_N)\left(\frac{dH_c}{dT}\right) = \frac{H_0^2}{2\pi}\left(\frac{T}{T_c}\right)^2\left[1 - \left(\frac{T}{T_c}\right)^2\right].$$

The discontinuity in the specific heat is given by

$$C_N - C_s = T\frac{d}{dT}(S_N - S_s) = T\frac{d}{dT}\left(\frac{L}{T}\right)$$

$$= \frac{H_0^2}{2\pi}\frac{T}{T_c^2}\left[1 - 3\left(\frac{T}{T_c}\right)^2\right].$$

Note that for $H_c = 0$, that is $T = T_c$, the latent heat vanishes while the discontinuity in the specific heat is nonzero. This indicates that the phase transition is second order at $H = 0$.

21. Any difference between P_r and P_∞ must be due to surface tension γ. For a process in which P and T are given, the Gibbs function G is a minimum. For a droplet of radius r and mass M_1 in equilibrium with its vapor, mass M_2,

$$G = M_1 g_1 + M_2 g_2 + 4\pi\gamma r^2.$$

Here g_1, g_2 are the Gibbs functions per unit mass. The last term represents the surface effects; it arises because the energy of a droplet is

$$U_0 + 4\pi\gamma r^2,$$

the first term giving the energy when surface effects can be ignored.

Setting $\delta G = 0$ and insisting that mass be conserved, one has

$$\delta G = 0 = \left[\delta M_1(g_1 - g_2) + 8\pi\gamma r \frac{\partial r}{\partial M_1}\right] = \delta M\left[g_1 - g_2 + \frac{2\gamma}{\rho r}\right],$$

where ρ is the mass density of the droplet. Hence $g_2 - g_1 = 2\gamma/\rho r$ in equilibrium. Now for a given phase:

$$\left(\frac{\partial G}{\partial P}\right)_T = V \qquad \text{[see Solution (7–1)]},$$

so $(\partial g/\partial P)_T = 1/\rho$. Hence, differentiating with respect to P at constant temperature yields

$$\frac{1}{\rho_{\text{vapor}}} - \frac{1}{\rho_{\text{drop}}} = -\frac{2\gamma}{\rho r^2}\frac{\partial r}{\partial P}.$$

On the assumptions that (1) $\rho_{(\text{vapor})} \ll \rho_{(\text{drop})}$, and (2) the vapor is a perfect gas,

$$P\left(\frac{\partial r}{\partial P}\right)_T = -\left(\frac{kT}{M}\right)\frac{\rho r^2}{2\gamma}.$$

When integrated,

$$P = P_\infty \exp\left(\frac{2M\gamma}{\rho k T r}\right), \tag{1}$$

where M is the mass of a single molecule of the substance. At a given temperature and pressure, only drops of radius r given by (1), are in equilibrium. Droplets that are too small will evaporate and disappear, in an attempt to increase P and reach equilibrium. But this only further decreases r. Similarly, large drops tend to become larger.

22. The universe is regarded as a gas of stars; in the long run, the average energy for each species of "molecule" will be approximately the same. Hence the kinetic energy of the rocket will approach the average kinetic energy of the stars:

$$\left(\frac{Mv^2}{2}\right)_{\text{star}} \simeq m_{\text{rocket}}(\gamma - 1)C^2.$$

Thus

$$\gamma \simeq (\gamma - 1) \simeq 5 \times 10^{17},$$

and $v_{\text{rocket}} \approx c$.

23. Statistical arguments show that ortho-hydrogen may have $J = 1$, $3, 5, \ldots$; and para-hydrogen may have $J = 0, 2, 4, \ldots$. The relative population of $(J = 2)$- to $(J = 0)$-levels is

$$\left\{(2J + 1) \exp\left[-\frac{J(J + 1)h^2}{m R^2 k T}\right]\right\}_{J=2} : 1$$

or

$$5 \exp\left[-\frac{6h^2}{m R^2 k T}\right] : 1 = n_2 : n_0 \qquad \text{where} \qquad m = \frac{M}{2}.$$

If we consider 1 mole of gas, the total number of molecules is N_0. Then $n_{J=0} + n_{J=2} = N_0(1 - x)$, and

$$n_2 = \frac{N_0(1 - x)}{1 + \frac{1}{5}\exp\left[6\hbar^2/mR^2kT\right]}.$$

(a) Take $E_{J=0} = 0$. Then $E_1 = 2\,\hbar^2/mR^2$ and $E_2 = 6\,\hbar^2/mR^2$, and the total rotational energy is given by

$$E_J \approx \frac{N_0\hbar^2}{mR^2}[2x + 30(1 - x)e^{-6\hbar^2/mR^2kT}],$$

$$C_J = \frac{\partial E_J}{\partial T} = 180R(1 - x)\left(\frac{\hbar^2}{mR^2kT}\right)^2 e^{-6\hbar^2/mR^2kT},$$

where $R = N_0k$. To this rotational specific heat, one must add the specific heat due to kinetic energy, $c_{\text{kin}} = 3R/2$.

(b) A simple one-dimensional model is used to estimate the thermal conductivity. Let λ be the mean free path. The number of molecules crossing a given plane in one direction, per unit time and area, is $\frac{1}{4}\,\rho v$, where ρ is the particle density, and v the mean velocity. The net energy transport is

$$\frac{1}{4}\rho v[E(\lambda) - E(-\lambda)] = \frac{\rho v\lambda}{2}\frac{dE}{dx} = \frac{\rho v\lambda}{2}\frac{dE}{dT}\frac{dT}{dx}.$$

By definition, $K = (\rho v\lambda/2)(dE/dT)$ is the conductivity. The contribution due to kinetic energy is $\frac{3}{4}\rho v\lambda k$. Since $\lambda = 1/\rho\sigma$ and $v = \sqrt{3kT/M}$, we have

$$K_1 = \frac{3k}{4\sigma}\sqrt{\frac{3kT}{M}}.$$

The contribution from the rotational levels is $K_2 = (\lambda/2)\rho v\,dE_J/dT$, which is negligible because $C_J \ll C_{\text{kin}}$, as shown in part (a).

Note that we have held ρv constant, because in equilibrium there must be no net transport of matter.

24. Since ratios are to be calculated, one need not worry about numerical factors. Let the temperature change in a distance of one mean free path, λ, be dT. Then the rate of energy transfer through unit area in the plane $Z = \text{const}$ is proportional to

$$(nv)\left(-\lambda\frac{dE}{dZ}\right) = \left(nv\lambda\frac{dE}{dT}\right)\left(-\frac{dT}{dZ}\right).$$

The first factor, nv, is the flux of molecules in either direction across the plane. The second factor, $(-\lambda\,dE/dZ)$, is the difference in average molecular energy in a distance of one mean free path. The coefficient of $(-dT/dZ)$ is the heat conductivity, K. Thus $K \sim nv\,\lambda C$ where C is the molecular specific heat.

The mean free path is $\lambda = 1/n\sigma$, where $\sigma =$ collision cross section. In addition, $C = 3k/2$ for a monatomic gas, and $v = \sqrt{3kT/M}$. Finally, then, $K \sim \sqrt{T}$ independent of pressure, and the desired ratio is 1.

The viscosity may be calculated by considering the net transverse momentum transferred across a unit area of a plane $Z =$ const, in unit time. In the same notation, this is proportional to $-(nv)(m\lambda\, dU/dZ)$, where m is the molecular mass and u is the transverse velocity. By definition, the coefficient of du/dZ is the viscosity, η. Thus, as was the case for the heat conductivity,

$$\eta \sim \sqrt{T} \qquad \text{(independent of pressure)},$$

and the desired ratio is 1. It can also be seen from the above considerations that $\eta/K =$ const, independent of both temperature and pressure, for an ideal gas.

25. The particles obey a Maxwell-Boltzmann distribution. The number of particles with velocities between $\mathbf{v}$ and $\mathbf{v} + d\mathbf{v}$ is

$$d^3vn(\mathbf{v}) = N\left(\frac{m}{2\pi kT}\right)^{3/2} \exp\left[-\frac{mv^2}{2kT}\right]d^3\mathbf{v}.$$

Note that $\int d^3vn(\mathbf{v}) = N$. The mean vector velocity is

$$\langle v_z \rangle = \frac{1}{N}\int_{-\infty}^{\infty} dv_x dv_y\, N\left(\frac{m}{2\pi kT}\right)^{3/2}\exp\left[-\frac{m}{2kT}(v_x^2 + v_y^2)\right]$$
$$\times \int_0^{\infty} v_z dv_z \exp\left(-\frac{mv_z^2}{2kT}\right) = \sqrt{\frac{kT}{2\pi m}}.$$

26. The distribution of Hg atoms in velocity is given by kinetic theory:

$$dN(\mathbf{v}) = N_0\rho(\mathbf{v})d^3\mathbf{v}, \qquad \text{with} \qquad \rho(\mathbf{v}) = \left(\frac{m}{2\pi kT}\right)^{3/2}\exp\left(-\frac{mv^2}{2kT}\right).$$

N_0 is the density of mercury atoms, and is found from the ideal gas equation of state, $N_0 = P/kT$. At time t, the number of atoms with speed v which have escaped into the solid angle $d\Omega$ at an angle θ is

$$dn = (\pi a^2 \cos\theta)(vt)\, N_0\, \rho(\mathbf{v})\, v^2\, dv\, d\Omega.$$

Of these, however, the number $dn' = (\pi a^2 \cos\theta)\, rN_0\rho(\mathbf{v})\, v^2\, dv\, d\Omega$ are still in flight, and have not actually struck the collector. In addition, the speed v must satisfy $vt > r$ or none of these will have struck the collector. Hence, in time t, the total number striking the collector in the solid angle $d\Omega$ is

$$dn(t) = (\pi a^2 \cos\theta)N_0 d\Omega \int_{r/t}^{\infty} dv\, v^2\rho(\mathbf{v})(vt - r).$$

After a little manipulation, the integral is shown to be

$$f(r, t) \equiv \int_{r/t}^{\infty} dv\, v^2\rho(\mathbf{v})(vt - r) = \frac{\bar{v}t}{4\pi}\exp\left(-\frac{4r^2}{\pi v^2 t^2}\right) - \frac{r}{4\pi}E\left(\frac{2r}{\bar{v}t\pi^{1/2}}\right),$$

where $\bar{v} = 4\pi \int_0^\infty v^3 \rho(v) \, dv = (8 \, kT/\pi m)^{1/2}$ is the mean speed, and $E(x)$ is the error function defined as

$$E(x) = \frac{2}{\pi^{1/2}} \int_x^\infty dy \, e^{-y^2} \qquad \text{with} \qquad E(0) = 1.$$

Note that in the limit $\bar{v}t \gg r$, the expression for $f(r, t)$ reduces to $(\bar{v}t - r)/4\pi$, which is what one would expect. Finally, the solid angle $d\Omega$ is related to an element of surface area, $d\sigma$, on the collector, through $d\Omega = d\sigma \cos \theta / r^2$. Thus the mass collected per unit area in time t is given by

$$\frac{dM}{d\sigma} = \frac{\pi m P a^2 \cos^4 \theta}{h^2 kT} f\left(\frac{h}{\cos \theta}, t\right).$$

8

STATISTICAL PHYSICS

1. At each corner the man may go either up or to the right. Therefore a particular path is specified by a sequence (u, u, r, u, ... , r) where the total number of u's and r's are n and m respectively. For example, the path shown in the figure has the sequence (r, u, u, u, r, r, u). The number of distinguishable ways of writing such a sequence and hence the total number of paths is $(m+n)!/n!m!$.

2. Assume that the radiation consists of standing waves enclosed in an N-dimensional cube of side a. Then the field vectors have the spatial variation

$$\prod_{i=1}^{N} \sin(k_i x_i) = \prod_{i=1}^{N} \sin\left(\frac{n_i \pi x_i}{a}\right), \qquad \text{with} \qquad \sum_{i=1}^{N} k_i^2 = \frac{\omega^2}{c^2}.$$

Here, the n_i are integers. Imagine an N-dimensional occupation space, in which each mode is assigned a point with coordinates $n_1, \ldots, n_N$. The density of points in this space is unity. The number of states with frequencies between ω and $d\omega$, regardless of direction, is then the differential element of volume in occupation space, integrated over angle. This must be doubled, because of the two polarization modes. Also the mode obtained when $n_i \longrightarrow -n_i$ is not independent of that for n_i; hence only $1/2^N$ of a spherical shell is included. Then

$$dV = 2^{-N+1} A R^{N-1} dR,$$

where A is a constant, determined in Problem 1. 34, and

$$R^2 = \sum_{i=1}^{N} (n_i^2) = \frac{\omega^2 a^2}{\pi^2 c^2}.$$

Then $dV = 2A(a/2\pi c)^N \omega^{N-1} d\omega$. The energy is

$$E = \int_0^\infty E(\omega) dV = \frac{2A a^N \hbar}{(2\pi c)^N} \int_0^\infty \frac{\omega^N d\omega}{e^{\hbar\omega/kT} - 1}$$

$$= 2A \frac{a^N (kT)}{(2\pi c)^N} \left(\frac{kT}{\hbar}\right)^N \int_0^\infty \frac{x^N dx}{e^x - 1}.$$

Hence $K = N + 1$.

3. The number of states per unit volume, in bandwidth df, at frequency f, is

$$2 \times \frac{4\pi p^2}{(2\pi\hbar)^3} dp = \frac{8\pi f^2}{c^3} df, \qquad \text{where} \qquad p = \frac{E}{c} = \frac{hf}{c},$$

and the extra factor of 2 comes from the two possible polarizations of the photon. For those modes such that $hf \ll kT$ (well satisfied for 3-cm radiation at these temperatures), the average energy is given by $E = kT$. Therefore, the energy density of the radiation field in the bandwidth df with $hf \ll kT$ is

$$dU = \frac{8\pi f^2}{c^3} kT \, df.$$

The power radiated from unit area is $dU(f)(c/4)$, and so the total power radiated is

$$dP = \frac{8\pi f^2}{c^3} kT \, df \left(\frac{c}{4}\right)(4\pi R^2)$$

$$= 18.5 \times 10^8 \, \text{W} \qquad \text{for } df = 1 \, \text{Mc}.$$

4. In terms of the total spin operator $\mathbf{S} = \frac{1}{2}(\boldsymbol{\sigma}_1 + \boldsymbol{\sigma}_2 + \boldsymbol{\sigma}_3)$, the energy is given by $H = (\lambda/6)(4S^2 - 9)$:

For $S = \frac{3}{2}$, $E = \lambda$, degeneracy $= 4$;

for $S = \frac{1}{2}$, $E = -\lambda$, degeneracy $= 4$.

The reason the latter degeneracy is 4 rather than 2 is because there are 2 independent ways to form a spin-$\frac{1}{2}$ state from three spin-$\frac{1}{2}$ particles.

The partition function $Z = \sum e^{-E_n/kT} = 8 \cosh(\lambda/kT)$.

5. The equation of state and all other thermodynamic quantities may be obtained from the partition function. We assume the particles do not interact. Then

$$Z = \int e^{-E/kT} dV_1 \cdots dV_n \frac{d^3 p_1 \cdots d^3 p_N}{h^{3N}}$$

$$= \left[\left(\frac{4\pi V}{h^3}\right) \int_0^\infty e^{-pc/kT} p^2 \, dp\right]^N$$

$$= \left[\left(\frac{kT}{c}\right)^3 \left(\frac{8\pi V}{h^3}\right)\right]^N.$$

Connection with thermodynamics is through the expression $F = -kT \log Z$ and $dF = -P \, dV - S \, dT$, from which

$$P = -\left(\frac{\partial F}{\partial V}\right)_T = kT\left[\frac{\partial \ln Z}{\partial V}\right] = \frac{NkT}{V}.$$

The internal energy is expressed as

$$U = -\frac{\partial(\ln Z)}{\partial\left(\frac{1}{kT}\right)} = 3NkT.$$

The pressure is the same as that of an ordinary gas; however, for an ordinary gas the energy is $U_0 = 3NkT/2$.

6.
$$Z = \left[e^{+U/kT} + \frac{4\pi V}{h^3}\int_0^\infty e^{-p^2/2mkT}\, p^2 dp\right]^N,$$

where U is 1 eV. Thus

$$Z = \left[e^{+U/kT} + V\left(\frac{2\pi mkT}{h^2}\right)^{3/2}\right]^N$$

and

$$P = kT\frac{\partial}{\partial V}(\log Z) = \frac{NkT}{V}\left[1 + \frac{h^3 e^{+U/kT}}{V(2\pi mkT)^{3/2}}\right]^{-1},$$

which is less than that for a free gas. The reader may carry on from here.

7. If an electron leaves a sodium atom and enters the metal, the gain in energy is $(\phi - W)$ where W is the work-function of the metal. Thus

$$\frac{n(\mathrm{Na}^+)}{n(\mathrm{Na})} = e^{(W-\phi)kT} = 10^2. \tag{1}$$

Likewise for an electron leaving the metal and joining a neutral chlorine atom to form Cl^-, the gain in energy is $(W - V)$ where V is the electron affinity of Cl. Thus

$$\frac{n(\mathrm{Cl}^-)}{n(\mathrm{Cl})} = e^{(V-W)/kT} = 10^{-6}. \tag{2}$$

We eliminate W between these two expressions and obtain

$$e^{(V-\phi)/kT} = 10^{-4} \qquad \text{or} \qquad (V - \phi) = -9.2\, kT.$$

Substituting $T = 1073°\mathrm{K}$, we find $V = 4.25$ V.

8. For processes where the number of particles is not constant, it is useful to introduce the chemical potential μ, defined as follows: If, in a process at constant T, V, the number of particles of a given system is increased by dN, then the change in free energy is $dF = \mu\, dN$. For the surface-gas system of the problem,

$$dF = dF_g + dF_s = \mu_g\, dN_g + \mu_s\, dN_s.$$

Equilibrium is reached when this vanishes. However, $dN_g = -dN_s$, and so the condition for equilibrium is

$$\mu_g = \mu_s, \qquad \text{or} \qquad \frac{\partial F}{\partial N_g} = \frac{\partial F}{\partial N_s}.$$

F is obtained from the partition function, which we now calculate.

$$Z_g = \frac{1}{N_g!}\left[\int \frac{d^3\mathbf{p}\,d^3\mathbf{r}}{h^3}e^{-mv^2/2kT}\right]^{N_g} = \frac{[V(2\pi mkT/h^2)^{3/2}]^{N_g}}{N_g!};$$

$$Z_s = \frac{1}{N_s!}\left[e^{\phi/kT}\int \frac{d^2\mathbf{p}\,d^2\mathbf{r}}{h^2}e^{-mv^2/2kT}\right]^{N_s} = \frac{[(2\pi AmkT/h^2)e^{+\phi/kT}]^{N_s}}{N_s!}.$$

The total $Z = Z_g Z_s$.

Note that we have used "correct Boltzmann counting" to avoid paradoxical results. (That is, we have treated the particles as indistinguishable.)

The free energy is

$$F = -kT\ln Z$$

$$= -kT\left[N_g\ln\left\{V\left(\frac{2\pi mkT}{h^2}\right)^{3/2}\right\}\right.$$

$$\left. + N_s\ln\left(\frac{2\pi AmkT}{h^2}e^{\phi/kT}\right) - \ln(N_g!) - \ln(N_s!)\right].$$

Using Stirling's formula to approximate $\ln N!$, one finds, upon setting $\mu_g = \mu_s$,

$$\frac{N_s}{N_g} = \left(\frac{Ah}{V}\right)\left(\frac{1}{2\pi mkT}\right)^{1/2}\exp\left(\frac{\phi}{kT}\right).$$

The number of atoms adsorbed per unit area is therefore

$$n = \frac{hP}{kT}\left(\frac{1}{2\pi mkT}\right)^{1/2}e^{\phi/kT}.$$

Had we not counted states correctly, the condition for equilibrium would not have involved N_s or N_g; we would have obtained a condition involving only the parameters of the problem. This is obviously wrong; the paradox is, in fact, equivalent to Gibbs paradox. Both are resolved by counting correctly. The reader should think about why similar paradoxes fail to appear in many of the calculations performed with the partition function (calculation of pressures, etc.).

9. In terms of the charge Q on the condenser, the Hamiltonian governing oscillations is

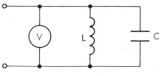

$$H = \frac{L}{2}\left(\frac{dQ}{dt}\right)^2 + \frac{1}{2C}Q^2.$$

It is easily seen that H represents a harmonic oscillator of frequency $\omega = (LC)^{-1/2}$. Thus the energy eigenvalues are $E_n = \hbar\omega(n+1/2)$, and the average energy in the circuit is

$$U = \langle E \rangle = \frac{\Sigma E_n e^{-E_n/kT}}{\Sigma e^{-E_n/kT}} = \frac{\hbar\omega}{2} + \frac{\hbar\omega}{\exp(\hbar\omega/kT) - 1}.$$

In addition, the energy U is given by

$$\frac{U}{2} = \left\langle \frac{CV^2}{2} \right\rangle = \left\langle \frac{LI^2}{2} \right\rangle.$$

Thus

$$\langle V^2 \rangle = \frac{\hbar\omega}{2C} \coth\left(\frac{\hbar\omega}{2kT}\right) \quad \text{and} \quad \langle I^2 \rangle = \frac{\hbar\omega}{2L} \coth\left(\frac{\hbar\omega}{2kT}\right).$$

In the classical limit, $kT \gg \hbar\omega$, these expressions reduce to

$$\langle V^2 \rangle = \frac{kT}{C} \quad \text{and} \quad \langle I^2 \rangle = \frac{kT}{L}.$$

In the limit $kT \ll \hbar\omega$, one obtains

$$\langle V^2 \rangle = \frac{\hbar\omega}{2C} \quad \text{and} \quad \langle I^2 \rangle = \frac{\hbar\omega}{2I}.$$

$\langle V^2 \rangle^{1/2}$ is the rms noise voltage.

10.
$$Z = \Sigma e^{-E/kT} = (1 + 2e^{-\epsilon/kT})^N,$$

$$F = -kT \ln Z = -NkT \ln(1 + 2e^{-\epsilon/kT}),$$

$$S = -\frac{\partial F}{\partial T} = Nk \ln(1 + 2e^{-\epsilon/kT}) + \frac{2N\epsilon}{T} \frac{e^{-\epsilon/kT}}{(1 + 2e^{-\epsilon/kT})},$$

and

$$U = \frac{2N\epsilon e^{-\epsilon/kT}}{(1 + 2e^{-\epsilon/kT})} \approx \frac{2N\epsilon}{3}\left(1 - \frac{\epsilon}{3kT}\right)$$

for $E \ll kT$. Then $C = \partial U/\partial T = 2N\epsilon^2/9kT^2 = (\frac{2}{9}) Nk (\epsilon/kT)^2$.

11. The partition function is given by

$$Z = \sum_{J=0} (2J + 1)e^{-A^2J(J+1)/kT} = 2e^{A^2/4kT}\sum_{J=0} (J + \tfrac{1}{2})e^{-(A^2/kT)(J + 1/2)^2},$$

where $A^2 \equiv \hbar^2/2I$, and I is the moment of inertia of the rotator.

Using Euler's formula we have

$$Z = 2e^{A^2/4kT}\left\{\left[\int_0^\infty u\, e^{-A^2u^2/kT}\, du\right] + \frac{1}{24}\right\} = 2e^{A^2/4kT}\left(\frac{kT}{2A^2} + \frac{1}{24}\right).$$

The desired thermodynamic quantities may be calculated from

(a) $F = -kT \ln Z$;

(c) $S = \dfrac{(U - F)}{T} = -\dfrac{\partial F}{\partial T}$;

(b) $U = kT^2 \dfrac{\partial}{\partial T} \ln Z$;

(d) $C = \dfrac{\partial U}{\partial T}$.

12. From $\langle (E - \langle E \rangle)^2 \rangle = \langle E^2 - 2E\langle E \rangle + \langle E \rangle^2 \rangle$, we have $\langle (E - \langle E \rangle)^2 \rangle = \langle E^2 \rangle - \langle E \rangle^2$. In addition

$$\langle E \rangle = \frac{\sum E_n e^{-E_n/kT}}{\sum e^{-E_n/kT}} = -\frac{1}{Z}\frac{\partial Z}{\partial \theta},$$

where E_n are the energy states of the system,

$$Z = \sum e^{-E_n/kT} \quad \text{and} \quad \left(\frac{1}{\theta}\right) \equiv kT.$$

Similarly

$$\langle E^2 \rangle = \frac{\sum E_n^2 e^{-E_n/kT}}{\sum e^{-E_n/kT}} = \frac{1}{Z}\frac{\partial^2 Z}{\partial \theta^2};$$

thus

$$\langle E^2 \rangle - \langle E \rangle^2 = \frac{\partial}{\partial \theta}\left(\frac{1}{Z}\frac{\partial Z}{\partial \theta}\right) = -\frac{\partial}{\partial \theta}\langle E \rangle,$$

but

$$\frac{\partial}{\partial \theta} = -kT^2\frac{\partial}{\partial T} \quad \text{and} \quad \frac{\partial \langle E \rangle}{\partial T} = C_v.$$

Finally one obtains $\langle (E - \langle E \rangle)^2 \rangle = kT^2 C_v$.

Now consider a macroscopic system with mean energy $\langle E \rangle$; then the fractional deviation in energy of the system is

$$\left[\frac{\langle E^2 \rangle - \langle E \rangle^2}{\langle E \rangle^2}\right]^{1/2} = \left[\frac{kT^2 C_v}{\langle E \rangle^2}\right]^{1/2}.$$

To estimate the size of this number, one expects the energy $\langle E \rangle$ to be of the magnitude NkT (especially at high temperatures); then $C_v = Nk$, and we have

$$\left[\frac{\langle E^2 \rangle - \langle E \rangle^2}{\langle E \rangle^2}\right]^{1/2} \approx N^{-1/2},$$

which is very small for systems of macroscopic size, i.e., $N \approx 10^{23}$.

13. This is the one-dimensional Ising model for ferromagnetism. As there are N spins, there are $(N - 1)$ interacting pairs. Of these, N_p is the number of parallel spins and N_a the number of antiparallel spins. Since

$$N_a + N_p = N - 1,$$

the energy of a given configuration is

$$E_{N_p, N_a} = J(N_p - N_a) = 2N_p + 1 - N.$$

The partition function is defined as

$$Z = \sum_{\substack{\text{All} \\ \text{states}}} e^{-E/kT}.$$

There are $(N - 1)!$ permutations of $N - 1$ pairs, but only $(N - 1)!/N_a!\, N_p!$

are distinguishable. Hence

$$Z = 2 \sum_{N_\mathrm{p}=0}^{N-1} \frac{(N-1)!}{N_\mathrm{a}! \, N_\mathrm{p}!} \exp\left[-\frac{J(2N_\mathrm{p} + 1 - N)}{kT} \right]$$

$$= 2 \exp\left[\frac{J(N-1)}{kT} \right] \sum_{N_\mathrm{p}=0}^{N-1} \frac{(N-1)!}{[(N-1)-N_\mathrm{p}]! \, N_\mathrm{p}!} \exp\left[-\frac{2J\,N_\mathrm{p}}{kT} \right].$$

The overall factor of 2 arises because reversing the direction of all spins does not change N_p or N_a but does give rise to a new configuration. In the above, the sum is the expansion of $[1 + \exp(-2J/kT)]^{N-1}$. The partition function then becomes

$$Z = 2 \exp\left[\frac{J(N-1)}{kT} \right]\left[1 + \exp\left(-\frac{2J}{kT} \right) \right]^{N-1},$$

$$Z = 2^N \left[\cosh\left(\frac{J}{kT} \right) \right]^{N-1}.$$

14. The entropy is related to the specific heat by

$$C = T \frac{dS}{dT};$$

thus the entropy change on going from $T = 0^\circ\mathrm{K}$ to a temperature $T > T_0$ is

$$\Delta S = \int_0^T \frac{C\,dT}{T} = C_\mathrm{max}(1 - \ln 2).$$

In order to calculate C_max, an independent calculation of ΔS is needed. This is furnished by the Boltzmann formula for entropy, $S = k \ln W$, where W is the number of distinguishable states of the system. Because of the ferromagnetic property, all the spins are lined up at $T = 0^\circ\mathrm{K}$. Hence $W(0^\circ\mathrm{K}) = 1$. However, for a temperature greater than T_0 we see that the system has maximum entropy, since it can no longer absorb heat. Thus all spins are uncorrelated and $W(T > T_0) = 2^N$, where N is the number of spins (Avogadro's number for one mole). Finally, then, one has $\Delta S = Nk \ln 2 = R \ln 2$, which, when combined with the previous expression for ΔS, yields

$$C_\mathrm{max} = \frac{R \ln 2}{(1 - \ln 2)}.$$

15. Consider a state described by a complete set of quantum numbers, with energy eigenvalue ϵ. If the state is occupied by p noninteracting particles, the energy is $p\epsilon$. Since Fermi statistics allows at most only one particle having a given set of quantum numbers, the partition function for this state is $Z_F = 1 + e^{-\epsilon/kT}$.

On the other hand, Bose statistics allows an unlimited number of particles in a given state. Hence, $Z_B = \sum_{n=0}^{\infty} e^{-n\epsilon/kT} = 1/(1 - e^{-\epsilon/kT})$. That form of para-statistics which allows two particles to a given state has

$$Z_P = 1 + e^{-\epsilon/kT} + e^{-2\epsilon/kT}.$$

The energy is found by computing

$$U = (kT)^2 \frac{\partial(\log Z)}{\partial(kT)}.$$

One finds

$$U_F = \epsilon \left[\frac{1}{e^{\epsilon/kT} + 1} \right],$$

$$U_B = \epsilon \left[\frac{1}{e^{\epsilon/kT} - 1} \right],$$

$$U_P = \epsilon \left[\frac{1 + 2e^{-\epsilon/kT}}{e^{\epsilon/kT} + 1 + e^{-\epsilon/kT}} \right].$$

The bracketed factor in each case is the statistical factor. (One should emphasize that no examples of para-statistics have as yet been discovered; nature so far seems to prefer the two extreme cases.)

16. Assume, to simplify the argument, that longitudinal and transverse modes of vibration propagate with the same speed c, and neglect dispersion (i.e., c does not depend on frequency). In the Debye theory with boson-phonons,

$$E = \frac{\hbar V}{\pi^2 c^3} \int_0^{\omega_{max}} \frac{\omega^3 d\omega}{e^{\hbar\omega/kT} - 1}. \tag{1}$$

The upper limit is found from $\hbar\omega_{max} = k\Theta$, where Θ is the Debye temperature. If the phonons were fermions, we would have

$$E = \frac{\hbar V}{\pi^2 c^3} \int_0^{\omega_{max}} \frac{\omega^3 d\omega}{[e^{\hbar\omega/kT} + 1]}. \tag{2}$$

At high temperatures (1) gives a constant specific heat, whereas (2) gives a specific heat going to zero. At low temperatures, the simple transformation $x = \hbar\omega/kT$ shows $c \sim T^3$ in both cases.

17. The number of particles per unit volume in the velocity interval $d^3\mathbf{v}$ is

$$n(v)d^3\mathbf{v} = \frac{2m^3}{h^3} d^3\mathbf{v} \frac{1}{\exp\{[(mv^2/2) - E_F]/kT\} + 1},$$

where E_F is the Fermi energy.

The current density leaving the cathode and entering the plate is

$$J = e\langle nv_x \rangle = 2e \left(\frac{m}{h}\right)^3 \int_{-\infty}^{\infty} dv_x\, dv_z \int_u^{\infty} \frac{v_x\, dv_x}{\exp\{[(mv^2/2) - E_F]/kT + 1\}},$$

where $mu^2/2 \equiv E_F + \phi + eV$ is the total kinetic energy an electron must have in order to reach the plate. Here ϕ is the work-function and V is the retarding voltage. Assuming $e(\phi + V) \gg kT$, the current density becomes:

$$J = 2e\left(\frac{m}{h}\right)^3 e^{E_F/kT} \int_u^\infty v_x e^{-mv_x^2/2kT} dv_x \int_{-\infty}^\infty dv_y\, dv_z\, e^{-m(v_y^2+v_z^2)/2kT}$$

$$= 2e\left(\frac{m}{h}\right)^3 e^{E_F/kT} 2\pi\left(\frac{kT}{m}\right)^2 e^{-mu^2/2kT},$$

$$= 4\pi\, em\frac{k^2 T^2}{h^3} \exp\left\{-\frac{(\phi + eV)}{kT}\right\}.$$

The reader should compare this result with that obtained by assuming Boltzmann statistics; note particularly the temperature dependence.

18. The Fermi energy, computed from the equation

$$V \int_0^{E_F/c} 2 \cdot 4\pi p^2\, dp = N,$$

is

$$E_F = \left(\frac{9Nc^3}{32\pi^2 R^3}\right)^{1/3},$$

where R is the radius of the star. If the star is sufficiently dense, the Fermi energy will be much larger than the average thermal energy, and few electrons will be excited above the Fermi level. Thus the star may be treated as a degenerate gas.

The ground-state energy is

$$U = \frac{8\pi V}{h^3} \int_0^{E_F/c} p^2(pc)dp = \frac{3}{4}NE_F,$$

and the pressure is $P = -(\partial E/\partial V) = NE_F/4V$.

But P can also be obtained by requiring hydrostatic equilibrium: if the star expands, the decrease in gravitational potential energy must be the work done by the pressure. Hence

$$P \cdot 4\pi R^2 \sim GM^2/R^2 \qquad \text{(order of magnitude)}.$$

Here M is the gravitational mass; it is different from zero because the neutrinos have a total kinetic energy $Mc^2 = U = 3NE_F/4$.

For equilibrium, the two expressions for pressure must be equal, and this leads to the condition $3GNE_F/4Rc^4 = 1$, or, in terms of the mass,

$$\frac{GM}{Rc^2} = 1.$$

The considerations developed here may be extended to the case in which the leptons have a rest mass (Chandrasekhar's theory of white dwarf stars), and it may be shown that there is an upper limit to the mass of the star.*

---

* S. Chandrasekhar, *Introduction to the Study of Stellar Structure.* U. of Chicago Press, 1939, Chap. XI.

19. For case (a), the net magnetic moment is

$$\langle M \cos\theta \rangle = \frac{M \int \cos\theta \; e^{MH\cos\theta/kT} \, d(\cos\theta)}{\int \exp\{MH\cos\theta/kT\}d(\cos\theta)}$$

$$= M\left[\coth\left(\frac{MH}{KT}\right) - \left(\frac{kT}{MH}\right)\right] \approx \frac{M^2 H}{3kT} \qquad \text{for} \qquad MH \ll kT.$$

Hence

$$\chi = \left(\frac{\partial M}{\partial H}\right)_{H=0} = \frac{M^2}{3kT}.$$

For case (b),

$$\langle \mathbf{M} \rangle = \frac{M(e^{MH/kT} - e^{-MH/kT})}{(e^{MH/kT} + e^{-MH/kT})} \approx \frac{M^2 H}{kT}$$

for small H. The magnetic susceptibility is $\chi = M^2/kT$.

20. The energy of a particle whose magnetic moment is parallel (anti-parallel) to H, is given by

$$U_\mp = \frac{p^2}{2m} \mp \mu H.$$

Since the energy levels of the system are populated according to the distribution function

$$f(U) = \frac{1}{\exp{(U - \xi)/kT} + 1},$$

and the density of levels is given by $(4\pi V/h^3)\, p^2 dp$, the total number of particles N is given by

$$N = \frac{4\pi V}{h^3} \int dp \, p^2[f(U_-) + f(U_+)] \tag{1}$$

and the magnetization/volume is

$$\frac{M}{V} = \frac{4\pi\mu}{h^3} \int dp \, p^2[f(U_-) - f(U_+)]. \tag{2}$$

Equation (1) may be solved for ξ in terms of N, T, and H, and ξ may then be substituted in Eq. (2) to determine M/V as a function of N, T, and H.

Upon defining a new variable of integration $E = p^2/2m$ and using the low-temperature expansion formula given, we find that Eq. (2) becomes

$$\frac{M}{V} = \frac{8\pi\mu(2m^3)^{1/2}}{3h^3}\left\{(\xi + \mu H)^{3/2}\left[1 + \frac{\pi^2}{8}\left(\frac{kT}{\xi + \mu H}\right)^2\right]\right.$$

$$\left. - (\xi - \mu H)^{3/2}\left[1 + \frac{\pi^2}{8}\left(\frac{kT}{\xi - \mu H}\right)^2\right]\right\},$$

which, after expanding in powers of H and keeping only the leading term, becomes

$$\left(\frac{M}{V}\right) = \frac{8\pi\mu^2(2m^3)^{1/2}}{h^3}\xi^{1/2}H\left\{1 - \frac{\pi^2}{24}\left(\frac{kT}{\xi}\right)^2 + \cdots\right\} + \text{terms of order } H^3.$$

Equation (1) (for $H = 0$) becomes

$$n = \frac{N}{V} = \frac{16\pi}{3h^3}(2m^3)^{1/2}\xi^{3/2}\left\{1 + \frac{\pi^2}{8}\left(\frac{kT}{\xi}\right)^2 + \cdots\right\}.$$

Solving for ξ, one obtains

$$\xi = \xi_0\left\{1 - \frac{\pi^2}{12}\left(\frac{kT}{\xi_0}\right)^2 + \cdots\right\},$$

where ξ_0 is the Fermi energy at $T = 0°K$, and $\xi_0^{3/2} = 3h^3n/16\pi(2m^3)^{1/2}$. The susceptibility then becomes

$$\chi \equiv \frac{M}{VH} = \left(\frac{3\mu^2n}{2\xi_0}\right)\left\{1 - \frac{\pi^2}{12}\left(\frac{kT}{\xi_0}\right)^2 + \cdots\right\}.$$

21. Consider the force equation for the particle in one dimension,

$$M\frac{d^2x}{dt^2} = -\beta\frac{dx}{dt} + F(t),\tag{1}$$

where $\beta = 6\pi R\eta$ (Stokes' law) and $F(t)$ is the force due to fluctuations of the molecular collisions.

Equation (1) may be rewritten as

$$\frac{M}{2}\frac{d^2}{dt^2}(x^2) - M\dot{x}^2 = -\left(\frac{\beta}{2}\right)\frac{d}{dt}(x^2) + xF(t).$$

When an average is taken over a large number of drops (ensemble average), this becomes

$$\frac{1}{2}\frac{d}{dt}\left\langle\frac{d}{dt}x^2\right\rangle - \frac{kT}{M} = -\left(\frac{\beta}{2M}\right)\left\langle\frac{d}{dt}x^2\right\rangle,\tag{2}$$

where $\langle xF(t)\rangle = 0$ because x and $F(t)$ are uncorrelated and the average value of $F(t)$ is zero.

Integrating Eq. (2), one finds

$$\left\langle\frac{d}{dt}x^2\right\rangle = \frac{2kT}{\beta} + Ce^{-\beta t/M}.\tag{3}$$

Neglecting the transient term, we find that Eq. (3) has the solution $\langle x^2\rangle = 2kTt/6\pi R\eta$. Since $\langle x^2\rangle = \langle y^2\rangle = \langle z^2\rangle$ because of symmetry, the total mean square displacement $\langle r^2\rangle = \langle x^2\rangle + \langle y^2\rangle + \langle z^2\rangle$ is given by $\langle r^2\rangle = kTt/\pi R\eta$.

Substituting $T = 300°K$; $t = 10$ sec; $\eta = 189 \times 10^{-6}$ P; and $R = 10^{-4}$ cm,

$$\langle r^2\rangle^{1/2} = 2.7 \times 10^{-3} \text{ cm}.$$

22. The plasma is electrically neutral, as a whole. Nevertheless local deviations in density appear. Consider the electrical potential $\phi(r)$ in the vicinity of a particular ion. The energy of another ion, of charge e, in that potential, is $e\phi(r)$. Therefore the density near the ion has the dependence

$$n(r) = ne^{-e\phi(r)/kT}. \tag{1}$$

The constant n must be the average density of the plasma, because the influence of the potential energy is expected to disappear as the thermal energy kT increases indefinitely. Each species of ion obeys an equation of the form (1) with density $n_\alpha(r)$ and charge e_α.

Another relation between $\phi(r)$ and n_α is provided by Poisson's equation relating the potential to the charge density:

$$\nabla^2\phi(r) = -4\pi \sum_\alpha e_\alpha n_\alpha. \tag{2}$$

On the assumption that the plasma is very hot, we may write, for Eq. (1),

$$n_\alpha(r) = n_\alpha\left[1 - \frac{e_\alpha\phi}{kT}\right].$$

We substitute in Eq. (2) and obtain the Helmholtz equation

$$\nabla^2\phi(r) = 4\pi\left\{\sum_\alpha \frac{n_\alpha e_\alpha^2}{kT}\right\}\phi \equiv k^2\phi,$$

which has the solution:

$$\phi(r) = e_{\text{ion}}\frac{e^{-kr}}{r}.$$

The effects of the electromagnetic interaction are therefore limited to a sphere of radius k^{-1} (Debye-Hückel radius).

9

ATOMIC PHYSICS

1. The energy levels of the hydrogen atom are given by

$$E_n = -\frac{(13.6 \text{ eV})}{n^2}, \qquad (n = 1, 2, \ldots).$$

In the absence of interactions among the electrons, the energy levels of helium for single-electron excitation would be those of the hydrogen atom, with nuclear charge $Z = 2$. However, interactions modify this. The strongest of these is the Coulomb repulsion e^2/r_{12} between the electrons. The qualitative effect of this is seen by taking the expectation value of e^2/r_{12} between unperturbed wave functions. The effects of spin are taken into account only so far as statistics are concerned.

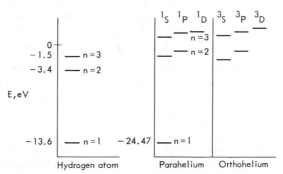

The $(n = 2, l = 0)$-level [that is, one electron in the $(n = 1, l = 0)$-level, the other in the $(n = 2, l = 0)$-level] has, in the absence of Coulomb repulsion, two degenerate spin states: $S = 1$ and $S = 0$. The Coulomb repulsion lifts the degeneracy, giving the energy shifts $(J + K)$, $(J - K)$, where

$$J = \int d^3r_1 d^3r_2 u_{10}(r_1)u_{20}(r_2)\left(\frac{e^2}{r_{12}}\right)u_{10}(r_1)u_{20}(r_2)$$

and

$$K = \int d^3r_1 d^3r_2 u_{10}(r_1)u_{20}(r_2)\left(\frac{e^2}{r_{12}}\right)u_{10}(r_2)u_{20}(r_1);$$

the former corresponds to the spatially symmetric, $(S = 0)$-state and the latter to the spatially antisymmetric triplet, $(S = 1)$-state. Since $K > 0$, the triplet lies lower. The physical origin of this is obvious: the Coulomb repulsion implies higher energy for states in which the electrons have a high probability for being found close to one another.

The P levels split similarly, as do the $(n = 3)$-levels, etc. In addition, the radiation interaction commutes with the total spin of the electrons; hence, radiative transitions between singlet and triplet levels are absolutely forbidden, in the absence of spin-orbit coupling. Thus there are two independent spectra for helium.

2. Carbon has six electrons. The Pauli Principle requires a ground-state configuration $1s^{(2)}2s^{(2)}2p^{(2)}$. The two electrons in the p shell may combine to form $S = 0, 1$ and $L = 0, 1, 2$. The spin-zero state is antisymmetric under interchange, the spin-one state is symmetric; $L = 0, 2$ are symmetric, $L = 1$ antisymmetric; the allowed states are $^1D_2$ with $L = 2$, $S = 0$, $^3P_{0,1,2}$ with $L = 1$, $S = 1$, and $^1S_0$ with $L = S = 0$.

The lowest terms are given by Hund's Rules:
  (1) maximize S,
  (2) maximize L,
  (3) minimize J.
The first two rules serve to minimize the Coulomb repulsion of the electrons, while the last minimizes the energy due to spin-orbit coupling. We ignore the latter; this amounts to neglecting the energy differences among the ^3P states. In order of increasing energy, the lowest terms are ^3P, $^1D_2$, $^1S_0$.

It remains to find the wave functions. First the orbital part of the wave function will be constructed. Consider the D state. We have to combine two $(L = 1)$-states to get $L = 2$. There will be five wave functions, one for each of the possible values of $m_L = 0, \pm 1, \pm 2$. In general

$$|2, m_L\rangle = \sum_{m_1 + m_2 = m_L} C(2, m_L \,|\, m_1, m_2)\phi(1, m_1)\psi(1, m_2),$$

where the coefficients C must be determined, and where ϕ and ψ are single-particle wave functions. Some of the coefficients can be written by inspection. The $(L = 2)$ wave function must be symmetric, so

$$|2, 2\rangle = \phi(1, 1)\psi(1, 1),$$

$$|2, -2\rangle = \phi(1, -1)\psi(1, -1),$$

$$|2, 1\rangle = \frac{1}{\sqrt{2}}\{\phi(1, 0)\psi(1, 1) + \phi(1, 1)\psi(1, 0)\},$$

$$|2, -1\rangle = \frac{1}{\sqrt{2}}\{\phi(1, 0)\psi(1, -1) + \phi(1, 1)\psi(1, 0)\}.$$

To obtain the remaining state, operate on $|2, 1\rangle$ with the lowering operator

$$L_-|J, M + 1\rangle = \sqrt{J(J + 1) - M(M + 1)}\,|J, M\rangle.$$

Then

$$|2, 0\rangle = \sqrt{\tfrac{1}{6}}\, \phi(1, 1)\psi(1, -1) + \sqrt{\tfrac{2}{3}}\, \phi(1, 0)\psi(1, 0) + \sqrt{\tfrac{1}{6}}\phi(1, -1)\psi(1, 1).$$

The singlet spin wave function, which in the absence of LS-coupling simply multiplies the orbital part of the wave function, is

$$(1/\sqrt{2})\{\alpha(1)\beta(2) - \alpha(2)\beta(1)\}.$$

We repeat for the P states. Here we must add 1 and 1 to get 1. This is simple because the states must be antisymmetric. Hence

$$|1, 1\rangle = \frac{1}{\sqrt{2}}\{\phi(1, 0)\psi(1, 1) - \phi(1, 1)\psi(1, 0)\},$$

$$|1, 0\rangle = \frac{1}{\sqrt{2}}\{\phi(1, 1)\psi(1, -1) - \phi(1, -1)\psi(1, 1)\},$$

$$|1, -1\rangle = \frac{1}{\sqrt{2}}\{\phi(1, 0)\psi(1, -1) - \phi(1, -1)\psi(1, 0)\}.$$

Note that the $\phi(1, 0)$ states are not used in constructing $|1, 0\rangle$, because there is no way of using $\phi(1, 0)\psi(1, 0)$ in the construction and, at the same time, maintaining the antisymmetric character of the state.

The $|0, 0\rangle$ state must be symmetric: we may write

$$|0, 0\rangle = a\phi(1, 1)\psi(1, -1) + b\phi(1, 0)\psi(1, 0) + a\phi(1, -1)\psi(1, 1).$$

Also $|0, 0\rangle$ must be orthogonal to $|2, 0\rangle$; this shows $b = -a$. Therefore

$$|0, 0\rangle = \frac{1}{\sqrt{3}}\{\phi(1, 1)\psi(1, -1) - \phi(1, 0)\psi(1, 0) + \phi(1, -1)\psi(1, 1)\}.$$

The $|0, 0\rangle =$ state combines with the $(S = 0)$-spin state to make a $^1S_0$ state. The three P states must be combined properly with the three triplet-amplitudes

$$|1, 1\rangle = \alpha(1)\alpha(2),$$

$$|1, 0\rangle = \frac{1}{\sqrt{2}}\{\alpha(1)\beta(2) + \alpha(2)\beta(1)\},$$

and

$$|1, -1\rangle = \beta(1)\beta(2)$$

to make the correct J states. This is again the addition of 1 plus 1 to get 0, 1, 2. For example, the $^3P_2$ state with $M_z = 2$ is just $|1, 1\rangle|1, 1\rangle$ and so on.

3. Nitrogen has the ground-state configuration $1s^2 2s^2 2p^3$. The various terms arise from the three p electrons, each having $l = 1$, ($m_l = 0, \pm1$) and $s = \tfrac{1}{2}$ ($m_s = \pm\tfrac{1}{2}$).

Each electron may be in one of six states, labeled by (m_l, m_s). These are

$$a = (1, \tfrac{1}{2}), \qquad b = (1, -\tfrac{1}{2}), \qquad c = (0, \tfrac{1}{2}),$$
$$d = (0, -\tfrac{1}{2}), \qquad e = (-1, \tfrac{1}{2}), \qquad f = (-1, -\tfrac{1}{2}).$$

A particular state is obtained by combining three of these to form a state with quantum numbers $(M_L, M_S) = (\sum M_l, \sum M_s)$. The exclusion principle is satisfied by choosing no two states having identical quantum numbers. One obtains

$$a + b + c = (2, \tfrac{1}{2}), \qquad a + c + d = (1, \tfrac{1}{2}) \qquad (a + c + e) = (0, \tfrac{3}{2})$$
$$a + b + e = (1, \tfrac{1}{2}), \qquad (a + d + e) = (0, \tfrac{1}{2})$$
$$(b + c + e) = (0, \tfrac{1}{2})$$
$$(a + c + f) = (0, \tfrac{1}{2}).$$

States with negative values of M_L or M_S have been omitted as uninformative —they give nothing new.

The presence of a $(2, \tfrac{1}{2})$ state, and the absence of states with higher values of M_L or M_S indicates a ^2D term. Associated with it is one state having quantum numbers $(1, \tfrac{1}{2})$ and one with $(0, \tfrac{1}{2})$. The other state $(1, \tfrac{1}{2})$ implies the existence of a ^2P term, which accounts for another $(0, \tfrac{1}{2})$ state. Remaining are the levels $(0, \tfrac{3}{2})$, $(0, \tfrac{1}{2})$, explained by a ^4S state. Hund's Rules imply $E(^2P) > E(^2D) > E(^4S)$.

4. Each level labeled by principal quantum number n has sublevels labeled by l, which takes on values from 0 to $n - 1$. Each sublevel is allowed $2(2l + 1)$ electrons by the exclusion principle. This information is sufficient to build up the atoms specified. The configurations are

$$\text{Zr:} \quad 4s^2\, 4p^6\, 4d^4$$

and

$$\text{Hf:} \quad 5s^2\, 5p^6\, 5d^4.$$

Each has four electrons in an unfilled d shell; chemically the two elements are very similar because these levels are so similar.

5. (a) The energy of a 4123-Å photon (expressed in cm^{-1}) is 24,300 cm^{-1}. Since a photon must be absorbed in its entirety, there will be no excitation of the atom.

(b) The energy of a 3.3-eV electron (expressed in cm^{-1}) is

$$E/hc = 28,000 \text{ cm}^{-1}.$$

However, the electron may give up only part of its energy, and the transitions $3s \longrightarrow 3p$, and $3s \longrightarrow 4s$ will take place. There is not enough energy to excite the atom to higher levels.

6. The electrostatic potential energy of a point electron and a uniform charge distribution of radius R, is

$$U = -eV(r) = -e \begin{bmatrix} \dfrac{e}{r} & \text{when } r > R, \\[2ex] \dfrac{3}{2}\dfrac{e}{R}\left(1 - \dfrac{r^2}{3R^2}\right) & \text{when } r < R. \end{bmatrix}$$

To calculate the approximate energy shift, we use first-order perturbation theory:

$$\Delta E = \langle \psi | \delta U | \psi \rangle \qquad \text{and} \qquad \delta U = U - \text{(interaction of electron with point nuclear charge)},$$

$$\delta U = \begin{cases} \dfrac{e^2}{r} - \dfrac{3}{2}\dfrac{e^2}{R}\left(1 - \dfrac{1}{3}\dfrac{r^2}{R^2}\right) & \text{when } r \leq R, \\[2ex] 0 & \text{when } r > R. \end{cases}$$

Hence

$$\Delta E = \frac{e^2}{\pi a^3} \int_0^R e^{-2r/a}\left\{\frac{1}{r} + \frac{1}{2}\frac{r^2}{R^3} - \frac{3}{2}\frac{1}{R}\right\} 4\pi r^2\, dr$$

$$= \frac{2e^2}{a^3} \int_0^R e^{-2r/a}\left\{2r + \frac{r^4}{R^3} - \frac{3r^2}{R}\right\} dr.$$

Also, since $R = 10^{-13}$ cm $\ll \frac{1}{2} \times 10^{-8}$ cm $= a$, we may take $e^{-2r/a} \approx 1$. Therefore

$$\Delta E \approx \frac{2e^2}{a^3}\left(R^2 + \frac{1}{5}R^2 - R^2\right) = \frac{2}{5}\frac{e^2}{a}\left(\frac{R}{a}\right)^2 \approx \frac{4}{5} \times (13.5 \text{ eV}) \times (2 \times 10^{-5})^2$$

$$\approx 4.4 \times 10^{-9} \text{ eV}.$$

7. In spherical coordinates,

$$V(r) = -\frac{e^2}{r} + r^2[\alpha + (\beta - \alpha)\cos^2\theta],$$

where θ is the polar angle. The inequalities $0 < \alpha < -\beta \ll e^2/a_0^3$ permit treatment of the harmonic term as a small perturbation, with the unperturbed Hamiltonian that of the hydrogen atom. The unperturbed wave functions u_{nlm} are

$$u_{100} = \frac{1}{(\pi a^3)^{1/2}} e^{-r/a},$$

$$u_{200} = \left(\frac{1}{32\pi a^3}\right)^{1/2}\left(2 - \frac{r}{a}\right)e^{-r/2a},$$

$$u_{210} = \left(\frac{1}{32\pi a^3}\right)^{1/2} e^{-r/2a}\left(\frac{r}{a}\right)\cos\theta,$$

$$u_{2,1,\pm1} = \left(\frac{1}{64\pi a^3}\right)^{1/2}\left(\frac{r}{a}\right)e^{-r/2a}\sin\theta\, e^{\pm i\phi}.$$

The perturbation will not connect levels of different m since V commutes with

$$L_z = \frac{\hbar}{i} \frac{\partial}{\partial \phi}.$$

In first order,

$$\Delta E_{100} = (2\alpha + \beta)a^2, \qquad \Delta E_{200} = 14(2\alpha + \beta)a^2,$$

$$\Delta E_{210} = 6(2\alpha + 3\beta)a^2, \qquad \Delta E_{2,1,\pm1} = 6(4\alpha + \beta)a^2.$$

The levels are therefore split as shown:

$$\left.\begin{array}{l} \text{——} \; l = 1, \, m_l = \pm 1 \\ \text{——} \; l = 0 \\ \text{——} \; l = 1, \, m_l = 0 \end{array}\right\} n = 2,$$

$$\text{——} \; n = 1, \, l = 0.$$

When the magnetic field is along the z-axis, it splits the ($m_l = 1$)- from the ($m_l = -1$)-level. This splitting is determined by adding the term $E_{\text{mag}} = (eB\hbar/2mc)m_l$ to the above energy levels. When B is parallel to x, the magnetic interaction is given by $E_{\text{mag}} = (eB\hbar/2mc)L_x$. Since matrix elements of L_x between states of definite m_z are nonzero only if $\Delta m = \pm 1$, this interaction has no expectation value in the previous states, nor does it mix the ($m_l = \pm 1$) degenerate levels. Thus in this case there is no linear Zeeman effect.

8. If we neglect all states other than the $2S$ and $2P$, the complete Hamiltonian in the applied field is

$$H = \begin{bmatrix} 0 & -aE \\ -aE & \Delta \end{bmatrix},$$

where the energy of the $2P$ state has been taken as 0 for convenience, and $a = \langle 2S \, | \, ez \, | \, 2P, \, m_z = 0 \rangle$. This expression for a follows from the perturbation interaction $V = eEz$. Other matrix elements of V vanish, either because of parity or because of conservation of J_z. As a consequence, the levels $| 2P, \, m_z = \pm 1 \rangle$ are unshifted.

Upon diagonalizing this matrix, one finds that the $2P$ level is shifted by the amount

$$\frac{\Delta - (\Delta^2 + 4a^2E^2)^{1/2}}{2},$$

while the shift of the $2S$ level is $[(\Delta^2 + 4a^2E^2)^{1/2} - \Delta]/2$.

Note that for strong fields, $aE \gg \Delta$, one obtains a linear Stark effect, i.e. shifts proportional to E. However, for weak fields, $aE \ll \Delta$, the shifts are quadratic in E.

9. The energy due to the presence of an orbital angular momentum in a magnetic field is found from the principle of minimal electromagnetic coupling, which provides the rule $\mathbf{p} \rightarrow (\mathbf{p} - e\mathbf{A}/c)$ in the Hamiltonian. To first order in the field, the additional energy is $(e/mc)\mathbf{A}\cdot\mathbf{P}$. But for a uniform field, one may take

$$\mathbf{A} = \frac{\mathbf{B} \times \mathbf{r}}{2},$$

so

$$\Delta E = \frac{e}{2mc}(\mathbf{B} \times \mathbf{r})\cdot\mathbf{p} = \frac{e}{2mc}\mathbf{B}\cdot(\mathbf{r} \times \mathbf{p}) = \frac{e\hbar}{2mc}Bm_l,$$

where B is the axis of quantization for L_z.

Electric dipole transitions obey the selection rule $\Delta m_l = 0, \pm 1$, according to which the components of the normal Zeeman effect are separated by $\Delta\omega = eB/2mc = 8.8 \times 10^9 \text{ sec}^{-1}$. This is, in fact, the separation observed here. The argument given above must be augmented for transitions in which spin plays a role. In that case the Zeeman effect is called anomalous.

10. The interaction with the magnetic field responsible for the linear Zeeman effect is

$$H = -\boldsymbol{\mu} \cdot \mathbf{B}, \quad \text{where } \boldsymbol{\mu} = 2\mu_0 (\mathbf{S}^+ - \mathbf{S}^-);$$

$\mathbf{S}^+$ and $\mathbf{S}^-$ are spin operators for the positron and electron respectively. The spin wave functions of the unperturbed state are:

$$\text{Triplet:} \begin{bmatrix} \alpha(+)\alpha(-), \\ \dfrac{\alpha(+)\beta(-) + \alpha(-)\beta(+)}{\sqrt{2}}, \\ \beta(+)\beta(-); \end{bmatrix}$$

$$\text{Singlet:} \left[\frac{\alpha(+)\beta(-) - \alpha(-)\beta(+)}{\sqrt{2}}\right].$$

The shift in energy due to a linear Zeeman effect is $\langle\psi| H |\psi\rangle$, but this vanishes for all spin states above, because the operator $\mathbf{S}^+ - \mathbf{S}^-$ is odd under interchange. Alternatively, one may compute the expectation value of H directly, using

$$S_z\alpha = \tfrac{1}{2}\alpha, \qquad S_z\beta = -\tfrac{1}{2}\beta,$$

where the z axis has been taken along $\mathbf{B}$. Needless to say, this explicit computation also gives zero for all $\langle\psi| H |\psi\rangle$.

11. In terms of the raising and lowering operators $S_\pm = S_x \pm iS_y$, and $L_\pm = L_x \pm iL_y$, the Hamiltonian may be written

$$H = \left\{\frac{2\epsilon}{3}S_z L_z + \mu_0(L_z + 2S_z)B\right\} + \frac{\epsilon}{3}(S_+L_- + S_-L_+).$$

If B is taken along the z-axis, J_z commutes with H and may be used to label states. We will denote by m_J the eigenvalues of J_z. For the cases $m_J = \pm\frac{3}{2}$, matrix elements of the second term vanish, and the first term gives only diagonal matrix elements. The energy levels for the two cases are $E_{\pm 3/2} = \epsilon/3 \pm 2\mu_0 B$. For the cases $|m_J| = \frac{1}{2}$, there are two states,

$$\begin{pmatrix} \psi_1 \\ \psi_2 \end{pmatrix} = \begin{pmatrix} \alpha y_1^{m_J - 1/2} \\ \beta y_1^{m_J + 1/2} \end{pmatrix} \qquad \text{for each } m_J.$$

Here L_z and S_z are chosen diagonal and y_1^m is the orbital wave function for a P state.

In the basis

$$\begin{pmatrix} \psi_1 \\ \psi_2 \end{pmatrix},$$

matrix elements of the first term are purely diagonal while the last term has no diagonal matrix elements. The off-diagonal elements are given by

$$\langle \psi_2 | H | \psi_1 \rangle = \langle \psi_1 | H | \psi_2 \rangle = \frac{\epsilon}{3} \langle \alpha | S_+ | \beta \rangle \langle m_J - \tfrac{1}{2} | L_- | m_J + \tfrac{1}{2} \rangle,$$

which, upon using the matrix elements of the raising and lowering operators,

$$\langle J, m_J \pm 1 | J_\pm | J, m_J \rangle = \sqrt{J(J+1) - m_J(m_J \pm 1)},$$

yields

$$\langle \psi_1 | H | \psi_2 \rangle = \frac{\epsilon}{3} \sqrt{2 - (m_J^2 - \tfrac{1}{4})}.$$

We indicate briefly how this result is obtained. In a basis with J, J_z diagonal, the commutation relation

$$[J_\pm, J_z] = \pm J_\pm$$

shows that $J_+ (J_-)$ connects the state $|J, M\rangle$ only to the state $|J, M + 1\rangle$ ($|J, M - 1\rangle$). All other matrix elements of $J_\pm$ vanish. The value of the nonvanishing matrix element is determined by taking the matrix element $\langle JM | J^2 | JM \rangle$ and using the relations

$$[J^+, J^-] = 2J_z, \qquad J^2 = \frac{J_+ J_- + J_- J_+}{2} + J_z^2.$$

Finally then, the Hamiltonian to be diagonalized is

$$H = \begin{pmatrix} \dfrac{\epsilon}{3}(m_J - \tfrac{1}{2}) + \mu_0 B(m_J + \tfrac{1}{2}) & \dfrac{\epsilon}{3}\sqrt{2} \\[2ex] \dfrac{\epsilon}{3}\sqrt{2} & -\dfrac{\epsilon}{3}(m_J + \tfrac{1}{2}) + \mu_0 B(m_J - \tfrac{1}{2}) \end{pmatrix}.$$

Carrying through the diagonalization, one finds the eigenvalues

$$E^{\pm}_{m_J} = \frac{-\frac{\epsilon}{3} + 2\mu_0 B m_J \pm \sqrt{\left(\frac{2\epsilon}{3}m_J + \mu_0 B\right)^2 + \frac{8}{9}\epsilon^2}}{2}. \tag{1}$$

In the case of a weak field $\mu_0 B \ll \epsilon$, these energies reduce to

$$E^{\pm}_{m_J} \text{ (weak)} \approx -\frac{\epsilon}{6} + \mu_0 B m_J \pm \frac{1}{2}\left(\epsilon + \frac{2\mu_0 B m_J}{3}\right).$$

While for the strong field case, E_{m_J} (strong) $\approx \mu_0 B(m_J \pm \frac{1}{2})$. One might further check that these limits are also obtained by a perturbative calculation. The appropriate basis in the weak field case is one in which $\mathbf{J}^2$ and J_z are diagonal while in the strong field case the basis with L_z and S_z diagonal is the more appropriate.

12. Consider the proton in the magnetic field produced by the electron. The interaction energy is

$$H = -\boldsymbol{\mu}_p \cdot \int d^3 r \rho(\mathbf{r}) \mathbf{B}(\mathbf{r}),$$

where $\boldsymbol{\mu}_p$ is the total magnetic moment of the proton, and $\rho(\mathbf{r})$ is the distribution of the magnetic-moment density throughout the proton, with $\int d^3 r \rho(\mathbf{r}) = 1$. The distribution function ρ will be taken to be spherically symmetric. The magnetic field produced by the electron has as sources the orbital motion of the electron and its intrinsic magnetic moment. The former is the current $\mathbf{J} \propto \psi^* \nabla \psi - \psi \nabla \psi^*$, which vanishes for an S state since ψ may be chosen to be purely real.

However, a point magnetic dipole produces a vector potential

$$\mathbf{A}(\mathbf{r}) = \frac{\boldsymbol{\mu} \times \mathbf{r}}{r^3} = -\boldsymbol{\mu} \times \nabla\left(\frac{1}{r}\right) \quad \text{with } \mathbf{B}(\mathbf{r}) = \nabla \times \mathbf{A}(\mathbf{r}).$$

Thus a magnetic dipole density $\mathbf{M}(\mathbf{r}')$ produces a vector potential

$$\mathbf{A}(\mathbf{r}) = -\int d^3 r' \mathbf{M}(\mathbf{r}') \times \nabla\left(\frac{1}{|\mathbf{r} - \mathbf{r}'|}\right),$$

which for an electron in a state $\psi(\mathbf{r}')$ becomes

$$\mathbf{A}(\mathbf{r}) = -\boldsymbol{\mu}_e \times \nabla \int d^3 r' \frac{|\psi(\mathbf{r}')|^2}{|\mathbf{r} - \mathbf{r}'|} \equiv -\boldsymbol{\mu}_e \times \nabla\phi(\mathbf{r})$$

with

$$\phi(\mathbf{r}) \equiv \frac{\int d^3 r' |\psi(\mathbf{r}')|^2}{|\mathbf{r} - \mathbf{r}'|}.$$

The expression for the energy then becomes

$$H = \boldsymbol{\mu}_p \cdot \int d^3 r \rho(\mathbf{r}) \nabla \times (\boldsymbol{\mu}_e \times \nabla\phi) = \boldsymbol{\mu}_p \cdot \int d^3 r \rho(\mathbf{r})\{\boldsymbol{\mu}_e \nabla^2 \phi - (\boldsymbol{\mu}_e \cdot \nabla)\nabla\phi\}.$$

Note at this point that, since $|\psi(r')|^2$ is spherically symmetric, one finds that $\phi(\mathbf{r})$ is spherically symmetric. Then taking $\rho(\mathbf{r})$ to be spherically symmetric, we integrate as follows:

$$\int d^3r\,\rho(\mathbf{r})\frac{\partial^2\phi}{\partial x_i\,\partial x_j} = \frac{\delta_{ij}}{3}\int d^3r\,\rho(\mathbf{r})\nabla^2\phi.$$

Thus

$$H = \tfrac{2}{3}\,\boldsymbol{\mu}_{\mathrm{p}}\!\cdot\!\boldsymbol{\mu}_{\mathrm{e}}\int d^3r\,\rho(\mathbf{r})\nabla^2\phi,$$

and when one uses $\nabla^2\phi = -4\pi\,|\psi(\mathbf{r})|^2$, this reduces to

$$H = -\frac{8\pi}{3}\boldsymbol{\mu}_{\mathrm{p}}\!\cdot\!\boldsymbol{\mu}_{\mathrm{e}}\int d^3r\,\rho(\mathbf{r})\,|\psi(\mathbf{r})|^2.$$

In addition, $|\psi(\mathbf{r})|^2$ is slowly varying over the dimensions of the proton and

$$H = -\frac{8\pi}{3}\boldsymbol{\mu}_{\mathrm{p}}\!\cdot\!\boldsymbol{\mu}_{\mathrm{e}}\,|\psi(0)|^2 \qquad \text{(Fermi's formula)}$$

$$= -\frac{8\pi}{3}\frac{\mu_{\mathrm{p}}\mu_{\mathrm{e}}}{\pi a^3 n^3}\,\boldsymbol{\sigma}_{\mathrm{p}}\!\cdot\!\boldsymbol{\sigma}_{\mathrm{e}},$$

where n is the principal quantum number, and a is the Bohr radius ($= \hbar^2/me^2$), with $\mu_{\mathrm{p}} = 2.79\,\mu_{\mathrm{N}}$, and

$$(\boldsymbol{\sigma}_{\mathrm{p}}\!\cdot\!\boldsymbol{\sigma}_{\mathrm{e}}) = \begin{cases} +1 & \text{triplet} \\ -3 & \text{singlet} \end{cases}.$$

The energy difference between the singlet and triplet is

$$E_{\mathrm{t}} - E_{\mathrm{s}} = \frac{8\times(2.79)}{3}\left(\frac{m_{\mathrm{e}}}{m_{\mathrm{p}}}\right)\alpha^4 m_{\mathrm{e}}c^2, \qquad \text{where } \alpha = \frac{e^2}{\hbar c} \approx \frac{1}{137}.$$

13. Due to the interaction of the spin angular momentum $S = \tfrac{1}{2}$ ($L = 0$) with the nuclear spin $I = \tfrac{1}{2}$, the total spin state of the hydrogen atom ground state is shifted by $4A\mu_{\mathrm{p}}\mu_{\mathrm{e}}\mathbf{S}_{\mathrm{p}}\!\cdot\!\mathbf{S}_{\mathrm{e}}$, while $\Delta E(D) = 2A\mu_{\mathrm{D}}\mu_{\mathrm{e}}\mathbf{S}_{\mathrm{D}}\!\cdot\!\mathbf{S}_{\mathrm{e}}$. The splitting in hydrogen is thus $\Delta E(H) = 4A\mu_{\mathrm{p}}\mu_{\mathrm{e}}$ and in deuterium, $\Delta E = 3A\mu_{\mathrm{D}}\mu_{\mathrm{e}}$. Taking the ratio gives, for the hyperfine splitting in deuterium,

$$\Delta E = 0.33 \times 10^9 \text{ cps.}$$

14. The electronic angular momentum is $J = \tfrac{5}{2}$. If I is the nuclear spin, the multiplicity is $(2J + 1)$ or $(2I + 1)$, whichever is smaller. But $2J + 1 = 6$, and since only four terms are found, $2I + 1 = 4$, and $I = \tfrac{3}{2}$.

The energy shift in hyperfine structure arises because the nucleus interacts, through its magnetic moment, with the magnetic field created by the electron. This suggests

$$\Delta E \sim 2\mathbf{I}\cdot\mathbf{J} = F(F+1) - I(I+1) - J(J+1),$$

where $\mathbf{F} = \mathbf{I} + \mathbf{J}$ takes on integral values from 1 to 4. One finds

$$\Delta E(F = 1) = 2 + C,$$
$$\Delta E(F = 2) = 6 + C,$$
$$\Delta E(F = 3) = 12 + C,$$
$$\Delta E(F = 4) = 20 + C.$$

The differences are 4, 6, 8; this suggests the interval ratios $2:3:4$ in the hyperfine quadruplet.

15. The Hamiltonian for the electron in the presence of both the external magnetic field and the Coulomb field of the nucleus is

$$H = \frac{(\mathbf{p} - e\mathbf{A}/c)^2}{2m} - \frac{e^2}{r}.$$

Choosing a gauge such that $\mathbf{A} = \frac{1}{2}\mathbf{B} \times \mathbf{r}$, where $\mathbf{B}$ is the constant magnetic field, and defining $H_0 = \mathbf{p}^2/2m - e^2/r$, the Hamiltonian becomes $H = H_0 + H_1$, where

$$H_1 = -\frac{e(\mathbf{L} \cdot \mathbf{B})}{2mc} + \frac{e^2(\mathbf{B} \times \mathbf{r})^2}{8mc^2},$$

and $\mathbf{L} = \mathbf{r} \times \mathbf{p}$. Treating H_1 as a perturbation, the shift in the ground-state energy is given by

$$\Delta E = \langle 1S | H_1 | 1S \rangle = \frac{e^2 \langle 1S | (\mathbf{B} \times \mathbf{r})^2 | 1S \rangle}{8mc^2},$$

when use is made of the relation $\mathbf{L} | 1S \rangle \equiv 0$. Upon using $| 1S \rangle = (\pi a^3)^{-1/2} e^{-r/a}$ (a = Bohr radius),

$$\Delta E = \frac{e^2 B^2}{12\pi mc^2 a^3} \int 4\pi r^4 e^{-2r/a} \, dr = \frac{e^2 B^2 a^2}{4mc^2}.$$

The diamagnetic moment is given by

$$\mu \equiv -\frac{\partial(\Delta E)}{\partial \mathbf{B}} = \frac{-e^2 a^2}{2mc^2} \mathbf{B}.$$

16. (a) The Doppler shift in the limit of small velocities is $\nu' = \nu(1 \pm \beta)$; hence $\Delta\lambda/\lambda = 2\beta$. The average value of β is determined from kinetic theory:

$$\beta^2 = \frac{3kT}{mc^2}.$$

At room temperatures $\beta = 1.4 \times 10^{-6}$; thus $\Delta\lambda = 0.7 \times 10^{-2}\,\text{Å}$.
(b) From the uncertainty principle, $(\Delta\omega)(\Delta\tau) \approx 1$. Due to collisions in the gas, $\Delta\tau = L/v$, where L is the mean free path $L = 1/4\pi\rho R^2$; so $\Delta\lambda/\lambda^2 = 2PR^2\beta/kT$. This equals the Doppler broadening when $P = kT/R^2\lambda$.

17. Each molecule is homonuclear; those which have nuclei with integral (half-integral) spin obey Bose (Fermi) statistics, and must have total nuclear wave functions which are symmetric (antisymmetric) under exchange of nuclei. Hence the product of rotational and spin wave functions must be symmetric (antisymmetric). The vibrational part of the wave function is symmetric in all cases, since it is a function of the magnitude of the nuclear separation only.

We treat the integral spin case first. Nuclear states with total nuclear spin $N = 0, 2, 4 \ldots$ are symmetric and thus require a rotational angular momentum $J = 0, 2, 4 \ldots$ in order to have a totally symmetric wave function. Likewise states with $N = 1, 3, 5 \ldots$ are antisymmetric and require $J = 1, 3, 5 \ldots$. Of the $(2I + 1)^2$ nuclear spin states, $(2I + 1)(I + 1)$ are symmetric while $I(2I + 1)$ are antisymmetric. Remembering that a rotational state of angular momentum J has degeneracy $(2J + 1)$, and that the intensity is proportional to the degeneracy (i.e. $e^{-E_J/kT} \approx 1$), we have

$$\frac{\text{Intensity } (2J \longrightarrow 2J - 2)}{\text{Intensity } (2J - 1 \longrightarrow 2J - 3)} = \frac{(I + 1)(4J + 1)}{I(4J - 1)}.$$

For the half-integer case exactly similar arguments yield

$$\frac{\text{Intensity } (2J - 1 \longrightarrow 2J - 3)}{\text{Intensity } (2J \longrightarrow 2J - 2)} = \frac{I\,(4J + 1)}{(I + 1)(4J - 1)}.$$

For large J values, which are abundant at high temperatures, these ratios reduce to $(I + 1)/I$ and $I/(I + 1)$ respectively.

(a) $(H^1)_1$ *Protons* have $I = \frac{1}{2}$. The ratio is $1 : 3$.

(b) $(H^2)_2$ *Deuterons* have $I = 1$. The ratio is $2 : 1$.

(c) $(He^3)_2$ $I = \frac{1}{2}$. The ratio is $1 : 3$.

(d) $(He^4)_2$ *Each nucleus* is an α particle; $I = 0$.

Here there are no antisymmetric nuclear wave functions and only $J = 0, 2, 4,$ are allowed. Thus every other line in the spectrum is absent.

18. With the introduction of the auxiliary function $u(r) = r\psi(r)$, the wave equation is

$$-\frac{\hbar^2}{2\mu} \frac{d^2u}{dr^2} + \left[-2V_0\left(\frac{1}{\rho} - \frac{1}{2\rho^2}\right) + \frac{L(L + 1)\hbar^2}{2\mu a^2 \rho^2} \right] u = Eu,$$

where μ is the reduced mass $(\mu = M_1 M_2/(M_1 + M_2))$.

The effective potential is the term in brackets; it has a minimum when

$$\rho = \rho_0 \equiv 1 + \frac{L(L + 1)\hbar^2}{2\mu a^2 V_0} \equiv 1 + B.$$

Expanding about this point,

$$V(\rho) = -V_0(1 + B)^{-1} + V_0(1 + B)^{-3}(\rho - \rho_0)^2.$$

The wave equation then takes the form

$$\frac{d^2u}{dr^2} + \frac{2\mu}{\hbar^2}[E + V_0(1 + B)^{-1} - V_0(1 + B)^{-3}(\rho - \rho_0)^2]u = 0.$$

This has the form of the one-dimensional harmonic oscillator equation; hence,

$$E + V_0(1 + B)^{-1} = \hbar\sqrt{\frac{2V_0}{\mu a^2}(1 + B)^{-3}}\left(n + \frac{1}{2}\right);$$

and for small B,

$$E = -V_0 + \frac{L(L + 1)\hbar^2}{2\mu a^2} + \hbar\omega_0\left(n + \frac{1}{2}\right) - \frac{3}{2}\frac{\hbar^3 L(L + 1)(n + \frac{1}{2})}{\mu^2 a^4 \omega_0}$$

where $\omega_0 = \sqrt{2V_0/\mu a^2}$.

19. The far-infrared spectrum is rotational. The rotational energies are

$$E_J = \frac{\hbar^2 J(J + 1)}{2\mu r^2} \qquad \text{where } \mu \text{ is the proton mass.}$$

The lines are then separated by an amount $\hbar^2/\mu r^2 = 2\pi\hbar c\,\Delta(1/\lambda)$ which gives $r = 1.4 \times 10^{-8}$ cm.

20. The energy of a diatomic molecule may be approximated by

$$E = -A + \hbar\omega(n + \tfrac{1}{2}) + BJ(J + 1) + \text{higher order terms.}$$

The dissociation energy is the difference between the energy of the ground-state molecule ($n = 0$, $J = 0$) and the energy of the two noninteracting atomic systems. Thus the dissociation energy is given by $E' = +A - \tfrac{1}{2}\hbar\omega$. The coefficient A depends only on the internuclear separation and the respective charges of the two nuclei. Indeed, in the adiabatic approximation, one may neglect the motion of the nuclei. In that approximation, the masses of the nuclei are irrelevant. As that is the only difference in the properties of the two nuclei, it follows that $A(\mathrm{H}) = A(\mathrm{D})$. Therefore

$$E'(\mathrm{D}) - E'(\mathrm{H}) = -\frac{1}{2}\hbar[\omega(\mathrm{D}) - \omega(\mathrm{H})].$$

Now the frequencies are given by $\omega = \sqrt{k/\mu}$ where k is the force constant, depending on the charge distribution, and μ is the reduced mass of the two nuclei. Hence

$$\frac{\hbar\omega(\mathrm{H})[1 - \omega(\mathrm{D})/\omega(\mathrm{H})]}{2} = \frac{\hbar\omega(\mathrm{H})[1 - 1/\sqrt{2}\,]}{2} = 0.08 \text{ eV,}$$

and the zero-point energy of H_2 is given by $\hbar\omega(\mathrm{H})/2 = 0.27$ eV.

21. An accurate solution requires the use of computers*; we shall therefore be content with dimensional arguments. The nuclei in a diatomic molecule are bound by electrostatic forces. This guarantees that the internuclear distance be of the same magnitude as the atomic radius of the valence elec-

---

* W.Kolos, C.C.J.Roothaan and R.A.Sack, *Rev. Mod. Phys.* **32**, 178 (1960).

trons. Using the Bohr model to estimate this, we have

$$\frac{r(\mu)}{r(e)} = \frac{m(e)}{m(\mu)} \approx \frac{1}{200}.$$

Thus the dimensions of the mulecule are extremely small compared to the dimensions of an ordinary molecule. Similarly, the ratio of the electronic energies (for identical quantum numbers) is

$$\frac{E_{el}(\mu)}{E_{el}(e)} = \frac{m(\mu)}{m(e)} \approx 200.$$

The vibrational energies of the molecule may be described by a "force constant," whose magnitude is given by simple arguments as approximately $k = e^2/r^3$. Therefore $k(\mu)/k(e) = (r_e/r_\mu)^3 = 8 \times 10^6$, and $E_0(\mu)/E_0(e) = \sqrt{k(\mu)/k(e)} = 2.84 \times 10^3$ is the ratio of zero-point energies.

Thus $r(\mu) = 5 \times 10^{-11}$ cm and $E_0(\mu) = 400$ eV. The data given for H_2^+ allow the energy spectrum for H_2^+ to be written

$$U = -A(e) + E_0(e)(2n + 1) + \text{smaller terms},$$

with $A(e) = 2.84$ eV and $E_0(e) = 0.14$ eV. Since A arises from electronic energies and includes a contribution from the energy of the electron in the field of the two nuclei, as well as a contribution from the energy due to the Coulomb repulsion of the two protons, one has $A(\mu)/A(e) \approx 200$. The energy spectrum for the mulecule is then

$$U = -A(\mu) + E_0(\mu)(2n + 1),$$

with $A(\mu) = 568$ eV. The binding energy is therefore $D(\mu) = A(\mu) - E_0(\mu) = 168$ eV.

10

SOLID STATE

1. Let A_{fc} and A_{bc} represent the lattice distances for the two cases. From simple geometry,

$$D_{fc} = \frac{\sqrt{2}}{2} A_{fc} \quad \text{and} \quad D_{bc} = \frac{\sqrt{3}}{2} A_{bc}.$$

Now, the face-centered crystal has four molecules per unit cell, while the body-centered crystal has only two. Since there is no volume change, we have, for the volume per molecule, $\frac{1}{4}A_{fc}^3 = \frac{1}{2}A_{bc}^3$. Thus

$$D_{fc}/D_{bc} = \sqrt{2/3}\,(2)^{1/3} = 1.029.$$

2. Suppose a sample of the metal is subject to an acceleration $-\mathbf{a}$. From the viewpoint of the metal, the electrons experience a backward acceleration $\mathbf{a}$, and hence an equivalent electric field $\mathbf{E} = m\mathbf{a}/e$. This produces a current density $\mathbf{J} = m\sigma\mathbf{a}/e$ which can be measured. Since σ is known from electrical measurements, one obtains a value for e/m.

3. Consider one mole of solid lithium reacting with one-half mole of chlorine gas, giving rise to lithium chloride. The reaction may be broken down into several steps (Born-Haber cycle):

$$Li_{solid} + D \longrightarrow Li_{vapor};$$

$$Li_{vapor} + NC \longrightarrow Li_+ + N \text{ electrons};$$

$$\frac{1}{2} Cl_2 + \left(\frac{1}{2}\right)E \longrightarrow Cl;$$

$$Cl + N \text{ electrons} \longrightarrow Cl_- + NF;$$

$$Li_+ + Cl_- \longrightarrow LiCl_{solid} + A.$$

Here N is Avogadro's number. We add:

$$Li_{solid} + \left(\frac{1}{2}\right)Cl_2 + D + NC + \left(\frac{1}{2}\right)E \longrightarrow LiCl_{solid} + NF + A.$$

Also,

$$Li_{solid} + \left(\frac{1}{2}\right)Cl_2 \longrightarrow LiCl_{solid} + B.$$

Subtraction yields $NF = B - A + D + NC + E/2$. The addition can be performed only when all the quantities are expressed in the same units. Converting calories to electron-volts through

$$1 \text{ cal} = 4.2 \times 10^7 \text{ ergs and } 1 \text{ eV} = 1.6 \times 10^{-12} \text{ ergs,}$$

with $N = 6 \times 10^{23}$, one finds $F = 4.07$ V.

4. Consider current flowing in the y-direction with a magnetic field along the z-axis. The fact that germanium shows no Hall effect means that the current in the x-direction is zero. Consider an electric field E along the y-axis. The velocity of electrons and holes in the y-direction is given by $v_e = -\mu_e E$ and $v_h = \mu_h E$. Because of the magnetic field these charges experience a force, on electrons and holes respectively, of

$$F_e = -ev_e H \quad \text{and} \quad F_h = ev_h H$$

along the x-direction, thus inducing transverse velocities

$$v'_e = -\mu_e v_e H = +\mu_e^2 HE \quad \text{and} \quad v'_h = \mu_h^2 HE,$$

and producing a current, I' along the x-axis, equal to:

$$I' = en_h v'_h - en_e v'_e = eHE(n_h \mu_h^2 - n_e \mu_e^2).$$

This vanishes when

$$n_h \mu_h^2 = n_e \mu_e^2. \tag{1}$$

The total current in the y-direction is $I = e(\mu_h n_h + \mu_e n_e)E$, and the fraction of this current which is due to electrons is

$$f = \frac{\mu_e n_e}{(\mu_e n_e + \mu_h n_h)} = \left(1 + \frac{\mu_h n_h}{\mu_e e_e}\right)^{-1}$$

Using Eq. (1) this becomes

$$f = \left(1 + \frac{\mu_e}{\mu_h}\right)^{-1} = \frac{2}{7}.$$

5. Imagine the two phases in equilibrium at temperature T. For equilibrium. the chemical potentials of the two phases must be equal, i.e.

$$\mu_{\text{solution}} = \mu_{\text{bcc}}.$$

We assume that carbon is perfectly soluble in the fcc phase, but not soluble at all in the bcc phase. We assume also that the carbon and iron atoms are chemically inert with respect to one another. Then μ_{bcc} is obtained from the Gibbs function for pure bcc iron, but μ_{solution} has an additional term due to the presence of carbon.

In order to compute this extra term, it is first necessary to find the entropy increase due to the carbon. If the total number of iron atoms is N, and n is the number of carbon atoms, then the number of ways in which the carbon atoms in the fcc phase may be arranged is

$$P = \frac{(N + n)!}{n! \, N!}.$$

The entropy is then given by the Boltzmann formula,

$$S = k \, lnP \approx k[(N + n) \log (N + n) - N \log N] \quad \text{for } N \gg n.$$

Now, in the absence of impurities, $\mu_{\text{fcc}} = \mu_{\text{bcc}}$, at equilibrium. However, introducing the impurity shifts the equilibrium temperature from T to $(T + \Delta T)$, changing the chemical potentials from μ to $[\mu + (\partial \mu / \partial T) \, \Delta T]$. Therefore

$$\Delta T \left(\frac{\partial u}{\partial T} \right)_{\text{bcc}} = \Delta T \left(\frac{\partial u}{\partial T} \right)_{\text{fcc}} + \mu',$$

where μ' is the chemical potential due to the mixing:

$$\mu' = \left(\frac{\partial G}{\partial N} \right)_{\text{mixing}} \approx -\frac{\partial (TS)}{\partial N} \approx -\frac{nkT}{N}.$$

Thus

$$\Delta T \left[\left(\frac{\partial \mu}{\partial T} \right)_{\text{bcc}} - \left(\frac{\partial \mu}{\partial T} \right)_{\text{fcc}} \right] = -\frac{nkT}{N}. \tag{1}$$

In addition,

$$-\frac{\partial \mu}{\partial T} = \frac{-\partial}{\partial N} \left(\frac{\partial G}{\partial T} \right) = \frac{\partial S}{\partial N},$$

and this is the entropy per molecule. If we multiply Eq. (1) by Avogadro's number, we obtain $\Delta T[S_{\text{fcc}} - S_{\text{bcc}}] = -cRT$, the S's being molar entropies. But the latent heat is defined by $L = T[S_{\text{fcc}} - S_{\text{bcc}}]$. Thus

$$\Delta T = -\frac{cRT^2}{L}, \quad \text{where } c = \frac{n}{N} = 10^{-3}.$$

The transition temperature is therefore lowered by $11°C$.

6. A harmonic oscillator may have energies $E_n = n\hbar\omega$. Note that we neglect the zero-point energy as a nonessential complication. The probability for a given level n to be occupied is

$$\langle n \rangle = \frac{\sum\limits_{n} ne^{-E_n/kT}}{\sum\limits_{n} e^{-E_n/kT}} = \frac{\sum\limits_{n} ne^{-n\hbar\omega/kT}}{\sum\limits_{n} e^{-n\hbar\omega/kT}}$$

$$= \frac{-\dfrac{d}{d(\hbar\omega/kT)} \sum\limits_{n=0}^{\infty} e^{-n\hbar\omega/kT}}{\sum\limits_{n=0}^{\infty} e^{-n\hbar\omega/kT}} = \frac{e^{-\hbar\omega/kT} (1 - e^{-\hbar\omega/kT})^{-2}}{(1 - e^{-\hbar\omega/kT})^{-1}}$$

$$= \frac{1}{e^{\hbar\omega/kT} - 1}.$$

Each molecule has three modes of vibration; hence

$$\langle E \rangle = \frac{3N\hbar\omega}{e^{\hbar\omega/kT} - 1}.$$

At high temperatures, $E = 3NkT = 3nRT$, where n is the number of moles and R the gas constant; while at low temperatures

$$E = 3N\hbar\omega e^{-\hbar\omega/kT}.$$

Hence, at high temperatures,

$$C = \frac{1}{n}\frac{\partial E}{\partial T} = 3R \approx 6 \text{ cal/mole·deg.}$$

At low temperatures,

$$C = 3R\left(\frac{\hbar\omega}{kT}\right)^2 e^{-\hbar\omega/kT}.$$

At high temperatures, the specific heat is in agreement with the empirical law of Dulong and Petit; at low temperatures $C \to 0$, in agreement with observation. However, the temperature dependence of C for small T predicted here, is not in agreement with experiment. A more sophisticated calculation (Debye model) gives better agreement.

7. The energy due to lattice vibrations is given by

$$U = \int \frac{\hbar\omega g(\omega)\,d\omega}{\exp\left(\hbar\omega/kT\right) - 1},$$

where $g(\omega)$ is the density of states for phonons. At low temperatures the low frequency dependence of $g(\omega)$ is all that is needed to calculate the energy as a function of temperature. For a three-dimensional lattice,

$$g(\omega)\,d\omega = \frac{d^3k}{(2\pi)^3} = \frac{4\pi k^2 dk}{(2\pi)^3} = \frac{4\pi\omega^2\,d\omega}{(2\pi)^3 c^3},$$

where c is the speed of sound. Thus $g(\omega) \sim \omega^2$, yielding an energy proportional to T^4 and a specific heat proportional to T^3. However, if the solid is composed of two-dimensional crystals (as graphite is), then $g(\omega) \sim \omega$, yielding a T^2 dependence for specific heat. Thus the T^2 dependence indicates that this phase of carbon is composed of two-dimensional crystals.

8. The conductivity is given by $\sigma = e(n_e\mu_e + n_h\mu_h)$, where n_e, μ_e (n_h, μ_h) are the density and mobility of the electrons (holes). Moreover, in a pure semiconductor, $n_e = n_h$, with

$$n_e = \frac{1}{h^3}\int \frac{p^2\,dp}{\exp\left\{(\Delta + p^2/2m_e - \xi)/kT\right\} + 1}.$$

The probability $\rho_h(E)$ for a hole to occupy a state of energy E is equal to the probability that *an electron does not* occupy that state; i.e.,

$$\rho_{\text{h}}(E) = 1 - \rho_{\text{e}}(E),$$

where one must take $E = -p^2/2m_{\text{h}}$. Therefore,

$$n_{\text{h}} = \frac{1}{h^3} \int \frac{p^2\,dp}{\exp\{(p^2/2m_{\text{h}} + \xi)/kT\} + 1}.$$

In the above expressions Δ is the energy gap (0.1 eV) and ξ is the Fermi energy. If $(\Delta - \xi)/kT \gg 1$ and $\xi \gg kT$, as is the case in this problem, the above expressions simplify to

$$n_{\text{e}} = \frac{1}{h^3} \int dp\,p^2 \exp\left[-\frac{(\Delta + p^2/2m_{\text{e}} - \xi)}{kT}\right]$$

and

$$n_{\text{h}} = \frac{1}{h^3} \int dp\,p^2 \exp\left[-\frac{(p^2/2m_{\text{h}} + \xi)}{kT}\right].$$

The equation $n_{\text{e}} = n_{\text{h}}$ implies

$$\xi = \frac{\Delta}{2} + kT \log \left(\frac{m_{\text{h}}}{m_{\text{e}}}\right)^{3/4}.$$

The density of conduction electrons is then

$$n_{\text{e}} = A T^{3/2} e^{-\Delta/2kT},$$

where A is a constant independent of temperature. Finally, the conductivity as a function of temperature is $\sigma = \sigma_0 T^{3/2} \exp\left(-\Delta/2kT\right)$, and

$$\frac{\sigma(T_2)}{\sigma(T_1)} = \left(\frac{T_2}{T_1}\right)^{3/2} \exp\left[\frac{\Delta(T_2 - T_1)}{2kT_2T_1}\right].$$

Converting Δ to a temperature, one finds $\Delta/k = 1160°\text{K}$. Hence

$$\sigma(T_2)/\sigma(T_1) = 6.5.$$

9. In vacuum, the electron velocity is determined by $\frac{1}{2}mv_0^2 = 25$ eV, while in the crystal, the velocity is increased, according to $\frac{1}{2}mv_c^2 = 25 + \phi$. Thus the quantity $\sqrt{(25 + \phi)/25}$ acts as an index of refraction for the electrons. Upon emerging from the crystal, the electrons are diffracted according to Snell's Law, which gives

$$\sin \theta_1 = \sqrt{\frac{25}{25 + \phi}} \sin (60°),$$

and $\theta_1 = 2\theta$ from the geometry.

Diffraction from a set of crystalline planes with Miller indices (h, k, l) is governed by the von Laue equation,

$$\sin^2 \theta = \frac{\lambda_c^2}{4a^2}(h^2 + k^2 + l^2) \equiv \frac{\lambda_c^2 A}{4a^2}.$$

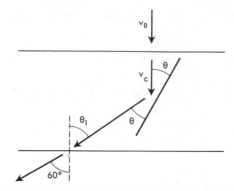

However, the appropriate wavelength is not that of the electrons in vacuum. Rather,

$$\frac{\lambda_c}{\lambda_0} = \frac{p_0}{p_e} = \sqrt{\frac{25}{25 + \phi}}.$$

Inserting the numerical values of the problem, von Laue's equation becomes $\sin^2 \theta = 1.5A/(25 + \phi)$. Now

$$\cos \theta_1 = 1 - 2 \sin^2 \theta = 1 - \frac{3A}{25 + \phi}.$$

Then the restriction $\cos^2 \theta_1 + \sin^2 \theta_1 = 1$ gives the condition

$$25 + \phi = \frac{12A^2}{8A - 25}.$$

Since ϕ is positive, we must have $8A > 25$, or $A > 3$. Further, the condition $12A^2 - 25(8A - 25) > 0$, or (equivalently), $(6A - 25)(2A - 25) > 0$, gives rise to the conditions $A < \frac{25}{6}$ or $A > 12.5$. Thus the smallest ring occurs for an A which represents reflection from the (200) plane. The corresponding depth of the potential well is $\phi = 17/7$ eV. The fact that (100) reflection did not occur (if it did, it would be at a smaller angle) is explained if the crystal is *face-centered* cubic; then the maximum separation between adjacent planes is *half* the lattice constant.

10. Conservation of energy implies $d\epsilon = e\mathbf{E} \cdot d\mathbf{x}$. Thus

$$\epsilon(t) - \epsilon(0) = e\mathbf{E} \cdot (\mathbf{x} - \mathbf{x}_0).$$

Now $\epsilon(0) = \epsilon(\mathbf{k}_0)$, and $\epsilon(t) = \epsilon(k(t))$. Also $\hbar \dot{\mathbf{k}} = e\mathbf{E}$, from which $\mathbf{k} = (e\mathbf{E}t/\hbar) + \mathbf{k}_0$. Thus

$$\epsilon\left(\mathbf{k}_0 + \frac{e\mathbf{E}t}{\hbar}\right) - \epsilon(\mathbf{k}_0) = e\mathbf{E} \cdot (\mathbf{x} - \mathbf{x}_0).$$

Note that the functional dependence of ϵ on $\mathbf{k}$ need not be that of a free particle; it depends on the band structure. A typical dispersion curve is shown in the accompanying figure. Consider an electron, which, at time $t = 0$, is at the center of the Brillouin zone, i.e. $\mathbf{k}_0 = 0$. As t increases, so does $\mathbf{k}$ and ϵ; hence $\Delta\mathbf{x} = \mathbf{x} - \mathbf{x}_0$ increases. $\Delta\mathbf{x}$ attains its maximum value when the electron reaches the end of the Brillouin zone, i.e. $\mathbf{k} = \mathbf{k}_B$. The electron then suffers Bragg reflection; its momentum is $\mathbf{k} = -\mathbf{k}_B$. As $\mathbf{k}$ increases from this value, ϵ decreases and so does $\Delta\mathbf{x}$. The amplitude of oscillation is $\Delta x = \epsilon(k_B)/eE$, while the period τ satisfies $eE\tau/\hbar = 2k_B$; i.e. $\tau = 2k_B\hbar/eE$. For an order-of-magnitude estimate, one may take $\epsilon(k_B) \sim \hbar^2/ma^2$, with $k_B = \pi/a$. Here $a = $ lattice constant $\approx 10^{-8}$ cm   and $m = $ electron mass. Then $\epsilon(k_B) \sim 10^{-11}$ erg ~ 6 eV. If E is measured in V/cm, one then finds

$$\Delta x \approx \left(\frac{6}{E}\right) \text{ cm,}$$

and

$$\tau \approx \left(\frac{4}{E}\right) \times 10^{-7} \text{ sec.}$$

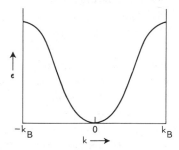

The above discussion of the oscillations is legitimate only when the length of the crystal is very much larger than Δx. For a crystal of length 6 cm, this means $E \gg 1$ V/cm.

11. (a) Let one of the photons [call it photon (1)] make an angle θ with respect to the velocity of the electron being annihilated, assuming the positron to be at rest (see figure). If the velocity v of the electron were zero, conservation of momentum would require the two photons to be antiparallel, each photon having a momentum $p = mc$. For $v \neq 0$, there is a momentum $\Delta p = mv \sin\theta$, perpendicular to photon (1), which must be carried by the other photon. Thus to first order in v/c, the photons deviate from collinearity by the angle

$$\delta = \frac{\Delta p}{p} = \frac{v}{c} \sin\theta.$$

One must now estimate the magnitude of v/c to be expected. An electron bound in an atom has a mean kinetic energy of $\langle \frac{1}{2}mv^2 \rangle$ equal to the binding energy, $Z^2\alpha^2mc^2/2n^2$ where $\alpha = e^2/\hbar c \approx 1/137$ and n is the principal quantum number. Taking $Z = 1$, since we are assuming that the outer shell electrons dominate the annihilation, we obtain $\delta_{max} = v/c \sim \alpha/n \sim 1/137$.

An alternative method is to calculate the Fermi velocity, v_F, in terms of the density of conduction electrons, with $\delta_{max} \sim v_F/c$. Since this calculation of v_F is needed later, it will be done here. For a free-electron gas at $T = 0°$K, the density of electrons is given by

$$n = \frac{1}{(2\pi\hbar)^3} \int_0^{p_F} d^3p = \left(\frac{m}{2\pi\hbar}\right)^3 \int_0^{v_F} 4\pi v^2\, dv = \frac{4\pi}{3}\left(\frac{mv_F}{2\pi\hbar}\right)^3,$$

or, alternatively,

$$v_F = \left(\frac{2\pi\hbar}{m}\right)\left(\frac{3n}{4\pi}\right)^{1/3}. \tag{1}$$

As an estimate for v_F we choose $n = 1/a^3$ where $a = \hbar^2/me^2$; that is, a is the Bohr radius. Then $v_F/c = (6\pi^2)^{1/3}\alpha \approx \alpha = 1/137$.

(b) The transition rate in a metal, Γ, is given by the product $\Gamma = \sigma \times \text{flux} \times \text{phase-space density}$. The product ($\sigma \times \text{flux}$) does not depend on velocity, so Γ depends only on the phase space. Since the probability of a given angular separation depends only on the probability of an electron having the appropriate velocity in the direction bisecting the angle between the photon momenta, the direction of $\Delta\mathbf{p}$ makes a convenient choice for the z-axis. Then $W(\delta)$ depends only on the probability distribution $\rho(v_z)$. Hence we write $v^2 = v_\perp^2 + v_z^2$, where $v_\perp$ is in the plane normal to v_z. Note that we have assumed implicitly that all measurements are made in a plane. This is in fact necessary for accurate measurements, as δ is small.

The phase-space factor is

$$dv_\perp dv_z \rho(\mathbf{v}_\perp, v_z) = v_\perp dv_\perp d\phi\, dv_z \rho(v_\perp, \phi, v_z),$$

where ρ vanishes outside the Fermi surface. Since the angular correlation depends only on v_z, we compute

$$dv_z \rho(v_z) = dv_z \int v_\perp dv_\perp d\phi \rho(v_\perp, \phi, v_z).$$

C_1, C_2 are counters

At $T = 0°\mathrm{K}$, ρ is constant inside the Fermi surface. Therefore $\rho(v_z)$ is proportional to the area of a section of the Fermi surface normal to the direction of v_z. Thus the angular correlation, which measures $\rho(v_z)$, gives information about the Fermi surface. For an isotropic metal,

$$\rho(v_z) = A \int v_\perp dv_\perp = A \int_{v_z}^{v_F} v\, dv = A(v_F^2 - v_z^2).$$

Thus

$$W(\delta) = \text{const} \times (v_F^2 - c^2\delta^2), \qquad \text{with } \delta \le \frac{v_F}{c}.$$

There is a small temperature effect because at nonzero temperature the Fermi distribution is

$$\rho(\mathbf{v}) = \frac{1}{\exp\left[m(v^2 - v_F^2)/2kT\right] + 1}.$$

At ordinary temperatures this differs from the previous distribution only at velocities $v \approx v_F$. One thus expects the following qualitative behavior for $W(\delta)$ as a function of temperature:

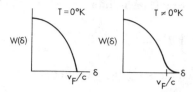

In reality, however, thermal motion of the positron is the dominant factor.

12. Consider the long-wavelength plasma oscillations of the lattice. The motion of an individual ion satisfies $M\dot{\mathbf{v}} = Ze\mathbf{E}$, which, in terms of the current density of the ions $\mathbf{j} = Ze N\mathbf{v}$, reads

$$\frac{d}{dt}\mathbf{j} = \frac{Z^2e^2N}{M}\mathbf{E}. \tag{1}$$

In addition, the motion must satisfy the charge conservation equation

$$\nabla\cdot\mathbf{j} + \dot{\rho} = 0. \tag{2}$$

Combining Eqs. (1) and (2), we obtain

$$\frac{Z^2e^2N}{M}(\nabla\cdot\mathbf{E}) + \ddot{\rho} = 0. \tag{3}$$

To proceed further, we must represent the screening effect of the electrons in the dielectric constant. Thus, for a plane wave $\mathbf{E} = \mathbf{E}(k, \omega)e^{i(\mathbf{k}\cdot\mathbf{x}-\omega t)}$ and lattice charge density $\rho = \rho(k, \omega)e^{i(\mathbf{k}\cdot\mathbf{x}-\omega t)}$, we have $\nabla\cdot\mathbf{E} = (4\pi/\epsilon(k, \omega))\rho$, which when substituted in Eq. (3) yields

$$\omega^2 = \frac{\Omega^2}{\epsilon(k, \omega)} \quad \text{with } \Omega^2 \equiv \frac{4\pi Z^2e^2N}{M}.$$

For a heavy ion, we expect the electrons to follow the ion motion adiabatically, and thus $\omega^2 \approx \Omega^2/\epsilon(k, 0)$.

The sound velocity is defined as the limit

$$v_s = \lim_{k\to 0}\left(\frac{\omega}{k}\right).$$

Thus

$$v_s^2 = \lim_{k\to 0}\frac{\Omega^2}{k^2\,\epsilon(k, 0)}.$$

One must now calculate the dielectric constant $\epsilon(k, 0)$. Consider a static external charge density $\rho_0(x) = \rho_0e^{i\mathbf{k}\cdot\mathbf{x}}$ embedded in the electron gas. Then the total charge density is ρ_0/ϵ, and the induced electron charge density is expressed as $\rho_e = \rho_0(1 - \epsilon)/\epsilon$. The electrostatic potential satisfying $\nabla^2 V = -4\pi\rho_0/\epsilon$ is then given by

$$V = \frac{4\pi\rho_e}{k^2(1 - \epsilon)}. \tag{4}$$

In order to solve for ϵ, we need another relation between V and ρ_e. This can be obtained by the Fermi-Thomas method where the density of electrons is taken as

$$n(x) = \frac{1}{(2\pi\hbar)^3} \int d^3p \, \frac{1}{\exp\left[(p^2/2m - eV(x) - E_F)/kT\right] + 1}.$$

In the degenerate gas limit $E_F \gg kT$, we obtain

(a) $n(x) \propto [E_F + eV(x)]^{3/2}$,

while in the classical limit, appropriate for a *hot* plasma,

(b) $n(x) \propto \exp(eV(x)/kT)$.

Expression (b) is not appropriate for electrons in a metal, which behave like a degenerate gas; it was discussed in Problem (8–22) in connection with the Debye-Hückel radius.

From expression (a) one obtains, to first order in V,

$$\rho_e(\mathbf{x}) = -e\delta n(x) = -\frac{3e^2 V(x) n}{2E_F}.$$

When this equation is combined with Eq. (1), we obtain

$$\epsilon(\mathbf{k}, 0) = 1 + \frac{6\pi e^2 n}{E_F k^2}.$$

Finally, then, the sound velocity is given by

$$v_s = \left(\frac{2ZE_F}{3M}\right)^{1/2} = v_F \left(\frac{Z}{3}\frac{m}{M}\right)^{1/2}.$$

(Note $n = ZN$.) Since $m/M \ll 1$, the sound velocity will in general be much less than the Fermi velocity.

11

NUCLEAR PHYSICS

1. Cerenkov radiation appears when the velocity of the particle exceeds the phase velocity of light in the material, c/n, when n is the index of refraction. Therefore the μ's alone will give Cerenkov radiation for those μ such that $c/v_\pi > n > c/v_\mu$. The velocity $pc(p^2 + m^2c^2)^{-1/2}$ of the π and μ are $c/\sqrt{2}$ and $c/[1 + (106/140)^2]^{1/2}$ respectively. Therefore the desired range of n is

$$1.414 = \sqrt{2} > n > [1 + (\tfrac{106}{140})^2]^{1/2} = 1.26.$$

2. According to the uncertainty principle, the momentum and position of a particle are not completely well defined: $\Delta x \cdot \Delta p \approx \hbar$. When a meson is emitted by one nucleon and absorbed by another, its position during transit is uncertain by an amount equal to the range of the force. Thus for an order-of-magnitude estimate,

$$\Delta x \approx r \qquad \text{and} \qquad \Delta p \approx mc,$$

where r is the range. Hence $r \approx \hbar/mc$.

3. Let N = number of protons/cm^3; then in terms of $\lambda = 1/N\sigma$ and $\lambda_e = 1/N\sigma_e$, the probability that the antiproton is elastically scattered in the intervals dx_1 and dx_2 is

$$dp = e^{-\lambda x_1}(\lambda_e dx_1)e^{-\lambda(x_2 - x_1)}(\lambda_e dx_2)e^{-\lambda(l - x_2)},$$

where $0 \leq x_1 \leq l$ and $x_1 \leq x_2 \leq l$. Integrating over x_1, x_2, one obtains

$$P_2(l) = \left(\frac{\lambda_e^2}{2!}\right)e^{-\lambda l}.$$

In general $P_n(l) = (\lambda_e^n/n!)\,e^{-\lambda l}$, and $\sum_{n=0}^{\infty} P_n(l) = e^{-(\lambda - \lambda_e)l}$ represents the probability that the antiproton will not be absorbed.

4. $C = Ae^{-\lambda t}$, so $\lambda = \ln(C_1/C_2)/(t_2 - t_1)$. If time T_1 (T_2) is spent counting C_1 (C_2), then the total number of counts is $N_1 = C_1 T_1$ $(N_2 = C_2 T_2)$, and the mean square deviation of N_1 (N_2) is $\sigma_1^2 = N_1$ $(\sigma_2^2 = N_2)$.

Now the error in a function $f(x, y)$ due to the errors in x and y is determined by

$$(\delta f)^2 = \left(\frac{\partial f}{\partial x}\right)^2 (\delta x)^2 + \left(\frac{\partial f}{\partial y}\right)^2 (\delta y)^2.$$

Hence

$$\sigma^2\left(\frac{C_1}{C_2}\right) = \sigma^2\left(\frac{T_2}{T_1}\frac{N_1}{N_2}\right)$$

and

$$\frac{\sigma^2(C_1/C_2)}{(C_1/C_2)^2} = \frac{1}{N_1} + \frac{1}{N_2} = \frac{1}{C_1 T_1} + \frac{1}{C_2 T_2}.$$

This is minimized subject to the constraint $T_1 + T_2 = \text{const}$. Hence $T_1/T_2 = (C_2/C_1)^{1/2}$. This shows that more time should be spent measuring the lower counting rate. The error, if T is the total time available, is then

$$\frac{\sigma^2(C_1/C_2)}{(C_1/C_2)^2} = \frac{1}{T}\left(\frac{1}{\sqrt{C_1}} + \frac{1}{\sqrt{C_2}}\right)^2$$

and

$$\sigma^2(\lambda) = \frac{1}{T}\left[\frac{(1/\sqrt{C_1}) + (1/\sqrt{C_2})}{(t_2 - t_1)}\right]^2.$$

But $C_1 \sim e^{-\lambda t_1}$ and $C_2 \sim e^{-\lambda t_2}$, thus $\sigma^2(\lambda)$ is smallest when $t_1 = 0$. Then $\sigma^2(\lambda)$ has a minimum when $[(\lambda t_2/2) - 1]e^{\lambda t_2/2} = 1$. This has solution

$$\lambda t_2/2 = 1.28, \qquad \text{or} \qquad t_2 = 2.56/\lambda.$$

5. The ground state of O^{16} has no spin, whereas the neutron has spin $\frac{1}{2}$. A neutron with $l = 2$ must, therefore, carry angular momentum $J = \frac{3}{2}$ or $\frac{5}{2}$. The spin of O^{17} must be one of these. The first excited state is made from s-wave neutrons, from which $J = \frac{1}{2}$. This is the spin of O^{17*}.

The nucleons have even parity, as does the ground state of oxygen. The parity of the initial state (free neutron $+ O^{16}$) is therefore the parity of the angular part of the neutron wave function. Since $Y_{lm}(-\cos\theta) = (-1)^l Y_{lm}(\cos\theta)$, the parity in both cases is even. Because the strong and electromagnetic interactions conserve parity, this is also the parity of the O^{17} levels in question.

6. Pions have isospin $I = 1$; two pions may exist in an $I = 0, 1$ or 2 state. Therefore none of the decays is forbidden by isospin conservation alone.

Because of Bose statistics, the total wave function must be symmetric under interchange of the two pions. The $(I = 0)$-state is symmetric under interchange; therefore the spatial part of the wave function for the pion system must be symmetric. (There is no spin wave function because pions are spinless.) This allows states with $L = 0, 2, 4 \ldots$

The parity of these states is $(-1)^L$, because all pions have the same intrinsic parity. Therefore two-pion decay may result from particles with quantum numbers $J = (0^+)$, (2^+), etc. Thus the only decay allowed is $f^0 \longrightarrow \pi^+ + \pi^-$.

If we drop the assumption of isospin conservation, the decay $\omega^0 \longrightarrow \pi + \pi$ is allowed; it may proceed via the electromagnetic interactions which don't conserve isospin, while the decay $\eta \longrightarrow \pi + \pi$ is forbidden merely by conservation of parity and angular momentum.

7. Describing the initial spin-$\frac{1}{2}$ state by a spinor, $\psi = \binom{\alpha}{\beta}$, the most general expression for the probability (summed over the final spin projections of B and C) that a given particle is emitted in the direction $\hat{n}$ is

$$P(\hat{n}) = \psi^+ M(\hat{n})\psi.$$

M is a (2×2) Hermitian matrix because the probability must be real. Since any (2×2) matrix may be expanded in terms of the unit matrix and the $\boldsymbol{\sigma}$ matrices, M must be of the form $M = A + B \, \boldsymbol{\sigma} \cdot \hat{n}$, where A and B are real constants, independent of $\hat{n}$ so as to guarantee rotational invariance. If the initial state were polarized along the z-axis, then $P(\hat{n}) = A + B \cos \theta$. However, since the decay proceeds via electromagnetic or strong interactions, parity is conserved and $P(\hat{n})$ must be even under parity. Thus $B = 0$ since $\psi^+ \boldsymbol{\sigma} \cdot \hat{n}\psi$ is odd under parity ($\hat{n} \longrightarrow -\hat{n}$ and $\boldsymbol{\sigma} \longrightarrow \boldsymbol{\sigma}$; hence $\boldsymbol{\sigma} \cdot \hat{n} \longrightarrow -\boldsymbol{\sigma} \cdot \hat{n}$), and the decay is isotropic. In weak decays, where parity is not conserved, the coefficient B will not vanish, and asymmetry in the decay will be observed. A typical example is the weak decay $\Lambda \longrightarrow p\pi^-$.

8. The relative parity of the $(\pi^- d)$ system is $P(\pi^- d) = (-1)^{P_\pi} (-1)^{P_d} (-1)^l$. However, $l = 0$, since the pion is captured at rest. In addition, the deuteron has positive parity; hence $P(\pi^- d) = P(\pi^-)$. The total angular momentum of the initial state is $J(\pi^- d) = \mathbf{S}_\pi + \mathbf{S}_d + \mathbf{l}$, from which $J(\pi^- d) = 1$.

Since the reaction conserves angular momentum, $J(nn) = 1$. The resultant spin of the two neutrons may be $S = 0$, or 1, and when combined with the relative orbital angular momentum, L, must result in a $J = 1$ state. Thus the only possibilities are

$$^3S_1, \, ^3P_1, \, ^1P_1, \text{ and } ^3D_1.$$

In addition the two neutrons, being fermions, must be in a state antisymmetric under interchange of the neutrons. The spin states ($S = 0$) and ($S = 1$) are antisymmetric and symmetric respectively under exchange. Similarly S and D states are symmetric while P states are antisymmetric. Thus, of the above possibilities only $^3P_1$ is allowed on the basis of Fermi statistics. The parity of this final state is (-1), since the product of the intrinsic parities of two identical particles is even. The parity of the allowed final state $^3P_1$ is odd, and since parity is conserved in this reaction (as in all strong interaction processes), one concludes that the parity of the pion is negative, i.e. $P(\pi^-) = -1$.

9. The isospin of Λ is 0, while the $p\pi^-$ and $n\pi^0$ states are linear combinations of states with definite isospin; that is, they are not themselves eigenstates of the operator $\mathbf{I}^2$. The nucleons have isospin $\frac{1}{2}$, while the pions have isospin 1. The pion-nucleon system may have those values of isospin resulting from the combination of $I = 1$ with $I = \frac{1}{2}$, according to the rules of combining isospin (or angular momentum): $1 \otimes \frac{1}{2} = \frac{1}{2} \oplus \frac{3}{2}$. Because of the $\Delta I = \frac{1}{2}$ rule, the Λ decays *only to the* $(I = \frac{1}{2})$-*state*.

We may expand
$$|\tfrac{3}{2}, -\tfrac{1}{2}\rangle = a\,|p\pi^-\rangle + b\,|n\pi^0\rangle,$$

$$|\tfrac{1}{2}, -\tfrac{1}{2}\rangle = b\,|p\pi^-\rangle - a\,|n\pi^0\rangle,$$

where a and b are the amplitudes for the $p\pi^-$ and $n\pi^0$ states in the $(I = \frac{3}{2})$-combination. The coefficients in the expansion of the $(I = \frac{1}{2})$-state have been chosen to make the $(I = \frac{1}{2})$- and $(I = \frac{3}{2})$-states mutually orthogonal. The normalization condition is $a^2 + b^2 = 1$. The coefficients, once determined, then give the ratio

$$A = \frac{\text{Rate}(\Lambda \longrightarrow p\pi^-)}{\text{Rate}(\Lambda \longrightarrow n\pi^0)} = \frac{|b|^2}{|a|^2}.$$

The coefficients a and b depend not on the detailed properties of the elementary particles, but only on the isospin properties. We emphasize this by introducing the notation

$$|p\pi^-\rangle = |\tfrac{1}{2}, \tfrac{1}{2}\rangle\,|1, -1\rangle$$

and

$$|n\pi^0\rangle = |\tfrac{1}{2}, -\tfrac{1}{2}\rangle\,|1, 0\rangle.$$

In this notation,

$$|\tfrac{3}{2}, -\tfrac{1}{2}\rangle = a\,|\tfrac{1}{2}, \tfrac{1}{2}\rangle\,|1, -1\rangle + b\,|\tfrac{1}{2}, -\tfrac{1}{2}\rangle \times |1, 0\rangle.$$

Now the coefficients, depending only on the transformation properties of isospin, cannot depend on whether the basis states are "elementary" or composite. For example, the coefficients cannot depend on whether we describe a state by writing $|1, -1\rangle$ or $|\tfrac{1}{2}, -\tfrac{1}{2}\rangle|\tfrac{1}{2}, -\tfrac{1}{2}\rangle$. We therefore write each state with $I = 1$ as the appropriate linear combination of $(I = \frac{1}{2})$-states:

$$|1, -1\rangle = |\tfrac{1}{2}, -\tfrac{1}{2}\rangle|\tfrac{1}{2}, -\tfrac{1}{2}\rangle \equiv \beta(2)\beta(3),$$

$$|1, 0\rangle = \frac{1}{\sqrt{2}}[|\tfrac{1}{2}, \tfrac{1}{2}\rangle|\tfrac{1}{2}, -\tfrac{1}{2}\rangle + |\tfrac{1}{2}, -\tfrac{1}{2}\rangle|\tfrac{1}{2}, \tfrac{1}{2}\rangle]$$

$$\equiv \frac{1}{\sqrt{2}}[\alpha(2)\beta(3) + \beta(2)\alpha(3)].$$

In this decomposition,

$$|\tfrac{3}{2}, -\tfrac{1}{2}\rangle = a\,\alpha(1)\beta(2)\beta(3) + \frac{b}{\sqrt{2}}\beta(1)[\alpha(2)\beta(3) + \beta(2)\alpha(3)],$$

where the index (1), (2), or (3) labels the states. We recall, at this point, the result of Problem (6–5), that when n particles of spin $\tfrac{1}{2}$ combine to make a total spin $n/2$, then all $(2n + 1)$ states are symmetric under interchange of any of the spin-$\tfrac{1}{2}$ particles. The same result applies to addition of states with isospin $\tfrac{1}{2}$.

Now the $|\tfrac{3}{2}, -\tfrac{1}{2}\rangle$-state written above is already symmetric under the interchange of particles (2, 3). The requirement that it be symmetric under the interchanges (1, 2) or (1, 3), imposes the condition $a = b/\sqrt{2}$. Therefore $A = b^2/a^2 = 2$.

The transitions $\pi^-p \longrightarrow K^0\Lambda$, $\pi^0 n \longrightarrow K^0\Lambda$ go via strong interactions; therefore they conserve isospin. Since the π^-p and $\pi^0 n$ systems may have $I = \tfrac{1}{2}$, or $I = \tfrac{3}{2}$, the K^0 must have $I = \tfrac{1}{2}$ or $I = \tfrac{3}{2}$. In the former case, the ratio is expressed as

$$B = \frac{(\pi^-p \longrightarrow K^0\Lambda^0)}{(\pi^0 n \longrightarrow K^0\Lambda^0)} = 2$$

while in the latter case, $B = \tfrac{1}{2}$.

10. For low excitations, one would expect, from the shape of $V(r)$, that the levels of the three-dimensional harmonic oscillator approximate those of $V(r)$. The energy levels of a three-dimensional harmonic oscillator are given by $E = \hbar\omega(m_1 + m_2 + m_3 + \tfrac{3}{2})$. The ground state corresponding to $m_1 = m_2 = m_3 = 0$ is spherically symmetric and, therefore, has orbital angular momentum $L = 0$. The first excited state corresponding to $m_1 + m_2 + m_3 = 1$ is threefold-degenerate corresponding to the $(2L + 1)$ degeneracy of an $L = 1$ state. One also sees that the spin-orbit coupling gives a lower energy to states with S and L parallel; therefore the $P_{3/2}$ is lower than the $P_{1/2}$. The level scheme of the first few states is thus

Energy	Degeneracy	Parity
$P_{1/2}$	2	-1
$P_{3/2}$	4	-1
$S_{1/2}$	2	$+1$

According to the model given, the neutrons and protons independently fill these levels.

(a) The two neutrons completely fill the $S_{1/2}$ level, combining to form a $J_n = 0$. The proton in the $S_{1/2}$ has a $J_p = \tfrac{1}{2}$, and the total spin and parity of $_1H^3$ is $J^P = \tfrac{1}{2}^+$.

(b) The $S_{1/2}$ is completely filled with protons and neutrons. In the $P_{3/2}$ there is one proton and two neutrons. Since the neutron wave function must be antisymmetric, the only allowed J_n values are $J_n = 2$ or 0. The angular momentum of the protons is $J_p = \frac{3}{2}$ and the total spin and parity is $J^P = \frac{7}{2}^-, \frac{5}{2}^-, \frac{3}{2}^-, \frac{1}{2}^-$. If one also uses the fact that the neutrons in the same level pair off to form $J_n = 0$ in the ground state, then one has $J^P = \frac{3}{2}^-$ for the ground state.

(c) The six neutrons completely fill the $S_{1/2}$ and $P_{3/2}$ levels while the five protons fill the $S_{1/2}$ level and have a "hole" in the $P_{3/2}$ level. Thus the neutron angular momentum is zero, while that of the protons is $\frac{3}{2}$. The angular momentum and parity is $J^P = \frac{3}{2}^-$.

(d) The eight neutrons completely fill the $S_{1/2}$, $P_{3/2}$ and $P_{1/2}$ levels, while there is only one proton in the $P_{1/2}$. Thus $J^P = \frac{1}{2}^-$.

11. The maximum energy of the total β-spectrum may be measured with a Geiger counter and an energy spectrometer consisting of varying thickness of aluminum shielding (the range-energy relation being known). This determines the presence of β_2^- with maximum energy 1.54 MeV. The presence of two γ's may be detected with a scintillation counter and the energies measured. The maximum energy of β_1 may be determined as was that of β_2, but by counting only those β's which are in coincidence with the 1.3-MeV γ-ray, thus eliminating the β_2 background. To complete the scheme it should be shown by the above procedure that the γ of 2.1 MeV is in coincidence with β_2.

12. From the fact that there are two β groups, the level scheme follows. The three possible gamma transitions are seen, and no others.

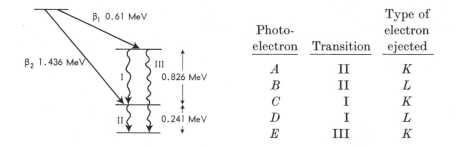

Photo-electron	Transition	Type of electron ejected
A	II	K
B	II	L
C	I	K
D	I	L
E	III	K

13. Consider the transition in the rest frame of the e^+e^- pair. By definition, there is no net momentum in this frame; hence the photon has no momentum in this frame. But, for a photon, energy and momentum are proportional. Thus the photon has neither energy nor momentum, and hence does not exist. Thus all transitions to states having massive particles are forbidden.

14. The π^+ has the same charge as the charge (Ze) of the nucleus. Thus upon disintegration the π^+ is given additional kinetic energy (of magnitude $\sim Ze^2/R$) due to Coulomb repulsion. One would therefore expect few low-energy π^+ mesons.

15. From charge independence of nuclear forces, the only difference in mass between $_{14}Si^{27}$ and $_{13}A^{27}$ is due to Coulomb energies. Taking a nucleus to be a homogeneously charged sphere of radius R and charge Ze the Coulomb energy becomes

$$E_c = 3Z^2e^2/5R.$$

Therefore the expected mass difference between these two mirror nuclei is

$$\frac{3e^2}{5R}[(14)^2 - (13)^2] = \frac{81e^2}{5R}.$$

Assuming that $R = r_0A^{1/3}$, this energy becomes $27e^2/5r_0$. Interpreting the positron energy to include the rest mass, we find $r_0 \sim (\frac{27}{35})r_c$, where r_c is the classical electron radius, $r_c = e^2/mc^2 = 2.8 \times 10^{-13}$ cm.

16. (a) From the data given,

$$N^{17} = O^{16*} + n + T_n + 3.72 \text{ MeV},$$

and

$$F^{17} = O^{17} + 1.72 \text{ MeV}.$$

(Recoil corrections to the energies are neglected.) O^{16*} may be an excited state of O^{16}, and the problem is to find the kinetic energy T_n of the neutron. These equations, together with Eqs. (1) through (3), yield

$$T_n = 1.99 \text{ MeV} - (O^{16*} - O^{16}).$$

Since $(O^{16*} - O^{16}) \geq 6.05$ MeV, the emission of neutrons only occurs in the transition to the O^{16} ground state, and neutrons are emitted with only one energy, 1.99 MeV.

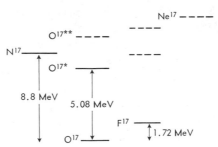

(b) and (c) The data given allow the construction of the above level diagram. (The dashed levels are the levels expected on the basis of charge independence.) The fact that the energies of O^{17} and F^{17} are so close is due to charge independence, O^{17} and F^{17} forming an isospin doublet. The energy difference is due to Coulomb repulsion, which associates a higher energy with higher Z. As a partner to $O^{17}*$, one expects an excited state of F^{17} with an energy $F^{17}* - O^{17}* = 1.72$ MeV. The N^{17} has a $T_Z = -\frac{3}{2}$; thus at least three more states are expected to form a $T = \frac{3}{2}$ isospin multiplet. These three states are another excited state of O^{17}, F^{17}, and the ground state of Ne^{17}.

17. To calculate the energy of the alpha from element Y, it will be assumed that the lifetime of an alpha emitter and the energy of the decay are related by the Gamow theory in the limit of large Coulomb barrier; i.e. $1/\tau = \alpha e^{-\beta E^{-1/2}}$, where α and β are taken to be constants. Actually, β is proportional to Z, but we are given no information in this problem about Z, so it will be assumed that the fractional change in Z from one emitter to another is small and can be neglected. A more convenient form for the purposes of this problem is to write $\log_{10}(\tau) = A + BE^{-1/2}$. Then the data given for X and Z allow A and B to be determined from

$$10 = A + B(5)^{-1/2} \quad \text{and} \quad 3 = A + B(10)^{-1/2}, \quad (1)$$

while the energy of the alpha from Y is determined from

$$A + BE^{-1/2} = 0. \quad (2)$$

From Eqs. (1) and (2), one finds $E = 14.8$ MeV.

The binding energy of a neutron is equal to the center-of-mass kinetic energy of the n, N^{14} system at threshold, i.e.

$$B = (\tfrac{14}{15})\ 10.6\ \text{MeV} \approx 10\ \text{MeV}.$$

The N^{14} nucleus is assumed to be at rest in the laboratory. The center-of-mass energy of the α, N^{14} system is

$$T = (\tfrac{14}{18})\ 14.8\ \text{MeV} \approx 11.5\ \text{MeV}.$$

Thus the reaction $N^{14}(\alpha, \alpha n)N^{13}$ is energetically possible. It should be remarked that there is a Coulomb barrier of the order $(4 \times 7)e^2/R \approx 14$ MeV, and the rate is inhibited by this barrier, but nevertheless the reaction is possible, since in quantum mechanics particles may tunnel through such barriers.

18. The reaction cross section for the neutron is simply $\sigma_n = \pi a^2$, which is the cross-sectional area of the target. If a neutron in the beam is within this area, it will strike the target nucleus and thus cause a reaction. However, a proton which is initially far removed from the target and within this cross-sectional area may not strike the nucleus because of the Coulomb repulsion. Thus the reaction cross section will be smaller than πa^2. To calculate this reaction cross section, consult the figure. Here b is the largest impact para-

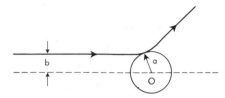

meter possible which will allow the proton to strike the stationary target. For this b the proton will strike the nucleus tangential to the surface, and thus the flight path is perpendicular to the radius drawn to this point. If E is the total kinetic energy of the protons at infinity, then conservation of energy requires that

$$Ka = \frac{1}{2} m v_a^2 = E - \frac{Ze^2}{a}$$

at the point a.

In addition, conservation of angular momentum about the point O requires that the initial angular momentum $L = pb = b\sqrt{2mE}$ be the same as the angular momentum when the particle is at a, that is, $L = a\sqrt{2mK_a}$. Thus

$$a\sqrt{2mK_a} = b\sqrt{2mE}.$$

This equation, together with the energy conservation equation, allows one to solve for b. Thus

$$b = a\sqrt{1 - \frac{Ze^2}{aE}} \qquad \text{and} \qquad \sigma_\text{p} = \pi b^2 = \pi a^2 \left(1 - \frac{Ze^2}{aE}\right).$$

Finally then

$$\frac{\sigma_\text{p}}{\sigma_\text{n}} = \begin{cases} 1 - \dfrac{Ze^2}{aE} & \text{for} \quad E > \dfrac{Ze^2}{a}, \\[2ex] 0 & \text{for} \quad E < \dfrac{Ze^2}{a}. \end{cases}$$

19. The grain density increases as the electron slows down; this is not so for the nucleus of charge $Z = 11$, because the nucleus begins to pick up electrons as it slows; these screen the nuclear charge, diminishing the ionization rate.

20. In the rest frame of the nucleus, there is an electric field $\mathbf{E} = Ze\,\mathbf{r}/r^2$. The force on a magnetic monopole of strength g is

$$\mathbf{F} = g\left(\mathbf{B} - \frac{\mathbf{v}}{c} \times \mathbf{E}\right).$$

Therefore the force equation in the nuclear rest frame is

$$\frac{d\mathbf{p}}{dt} = -Z\frac{eg}{c}\frac{\mathbf{v} \times \mathbf{r}}{r^3}.$$

For small-angle scattering, the right side of Eq. (1) is computed by taking

$$\mathbf{v} = (v, 0, 0) \qquad \text{and} \qquad \mathbf{r} = (vt, a, 0).$$

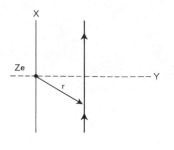

Then (1) reduces to

$$\frac{dv_z}{dt} = -\frac{Zeg}{mc}\frac{av}{(a^2 + v^2t^2)^{3/2}},$$

$$\frac{dv_x}{dt} \approx \frac{dv_y}{dt} \approx 0.$$

Integrating, subject to $v_z(-\infty) = 0$, we obtain

$$v_z(t) = -\frac{Zeg}{mca}\left[1 + \frac{vt}{(a^2 + v^2t^2)^{1/2}}\right].$$

Then

$$v_z(\infty) = -\frac{2Zeg}{mca} \qquad \text{and} \qquad \theta \approx \frac{v_z(\infty)}{v} = -\frac{2Zeg}{mcav}.$$

This provides a unique relationship between the impact parameter a and the scattering angle θ.

Now if f is the incident flux of particles, then conservation of particles requires that

$$f \cdot 2\pi a \, da = -f\frac{d\sigma}{d\Omega}2\pi d(\cos \theta).$$

The left side represents the number of particles incident per unit time on an annulus of radius a and width da, while the second is the scattering rate into solid angle $d\Omega = -2\pi d (\cos \theta)$. We solve for the scattering cross section, to obtain

$$\frac{d\sigma}{d\Omega} = \frac{a}{\sin \theta}\left|\frac{da}{d\theta}\right| \approx \frac{a}{\theta}\left|\frac{da}{d\theta}\right| = \left(\frac{2Zeg}{mcv}\right)^2\frac{1}{\theta^4}$$

$$= \left(\frac{g}{e}\right)^2\left(\frac{v}{c}\right)^2\frac{d\sigma}{d\Omega} \qquad \text{(Rutherford)}.$$

Charged particles lose energy primarily to the electrons, due to the small electronic mass, and the same is true for monopoles. The momentum transfer to an electron accompanying a small-angle scattering is $2eg/ca$, and the corresponding energy transfer is $(\Delta p)^2/2m = 2(eg/mca)^2$. The number of electrons in a cylindrical annulus of radius a, width da, and length dx, is $2\pi NZa \, da \, dX$. The energy loss, obtained by integrating over a, is

$$\frac{dE}{dx} = -\frac{4\pi NZ(eg)^2}{mc^2}\ln\left(\frac{a_{\max}}{a_{\min}}\right).$$

For an adequate discussion of the determination of $a_{\max}$ and $a_{\min}$ the reader should consult the textbooks in which energy loss by charged particles is treated.

21. Consider the s-wave Schrödinger equation $u'' + (k^2 - 2mV)u = 0$, where the wave function $\psi = u/r$. If V is a short-range potential, then in the region outside the potential, the solution is $\sin(kr + \delta)$ where $\delta = 0$ if $V = 0$. The scattering amplitude is given by $f = (e^{2i\delta} - 1)/2ik$; and the differential cross section, $d\sigma/d\Omega = \sin^2\delta/k^2$. To study the effect of a nearby bound state on the scattering, write the solution u in the form

$$u = \frac{\phi(k)e^{ikr} - \phi(-k)e^{-ikr}}{2i},$$

with u taken real, that is $\phi(k)^* = \phi(-k^*)$. When $k = i\sqrt{2mB} = i\gamma$, corresponding to a bound state of energy, $-B$, we know that $\phi(-i\gamma)$ must be zero, in order that a solution $u \to e^{-\gamma r}$, as $r \to \infty$, exist. Taylor-expanding $\phi(-k)$ about this point, we have $\phi(-k) \sim ia(-k + i\gamma)$, where a is real. [This follows from $\phi(k) = \phi^*(-k^*)$.] Thus the phase shift is given by

$$\tan\delta = \frac{\text{Im } \phi(k)}{\text{Re } \phi(k)} = -\frac{k}{\gamma},$$

which, for small k, reduces to $\delta = -k/\gamma$. Defining a scattering length, a, by $\delta = -ka$, we find $a = 1/\gamma$. The total cross section is

$$\sigma = \frac{4\pi}{k^2}\sin^2\delta = \frac{4\pi}{k^2 + 1/a^2}.$$

22. From the preceding problem,

$$\sigma = \frac{4\pi}{k^2 + \gamma^2} \longrightarrow \frac{4\pi}{\gamma^2} \qquad \text{as} \qquad k^2 \longrightarrow 0.$$

Here

$$\gamma^2 = -\frac{2mE_{\text{bound}}}{\hbar^2} = 5 \times 10^{21}\,\text{cm}^{-2};$$

hence

$$\sigma = 2.5 \times 10^{-21}\,\text{cm}^2 = 2500\,\text{barns}.$$

In this solution the correction for the range has been omitted, since $r_0\gamma \ll 1$.

23. One expects only s-wave scattering to be important at this energy; therefore

$$I = I_0\,e^{-Nt\sigma} \qquad \text{with} \qquad \sigma = \frac{4\pi\sin^2\delta}{k^2}.$$

From the data given, $Nt = 5 \times 10^{23}$ atoms/cm^2; thus $\sigma = \ln 2/Nt = 1.4$ barns. Also $k^2 = 2mE = 5 \times 10^{23}$ cm^{-2}. Finally, then, $\sin\delta = \pm 0.23$.

24. Lead is a doubly-magic nucleus; therefore one expects the cross section for inelastic scattering to be small. Assuming only elastic scattering and s-wave dominance, the cross section is $\sigma = 4\pi R^2$, estimating $R \approx (1.4 \times 10^{-13}\,\text{cm})A^{1/3}$; with $A = 208$, we obtain $\sigma = 8.6 \times 10^{-24}\,\text{cm}^2$. The number of lead atoms/cm³ is given by

$$n = \frac{\rho N_0}{A} \quad \text{where} \quad N_0 = 6.02 \times 10^{23} \quad \text{and} \quad \rho = 11 \text{ gm/cm}^3.$$

The intensity as a function of thickness is thus

$$I = I_0 \exp(-n\sigma t) = I_0 \exp(-0.55) = 0.58 \, I_0.$$

Hence 42% of the incident neutrons are scattered out of the beam. In this treatment the effects of multiple scatterings have been neglected.

25. Define the following quantities:

$$A = \text{area being irradiated};$$
$$\phi = \text{flux of particles} = i/Ae,$$

where e is the charge of the deuteron;

$$N_{56}(t) = \text{total number of Mn}^{56} \text{ present at time } t;$$
$$\lambda = (\log 2)/T_{1/2}.$$

The initial number of Mn⁵⁵ nuclei with which the deuteron beam may interact within a range R is $N_{55} = RA/M$, where M is the mass of Mn⁵⁵. Note that the range is given in terms of gm/cm². The equations governing the production of Mn⁵⁶ are

$$\frac{dN_{56}}{dt} = \phi \, \sigma \, N_{55} - \lambda \, N_{56},$$

and

$$\frac{dN_{55}}{dt} = -\frac{dN_{56}}{dt}.$$

The solution to these equations satisfying $N_{55}(0) = RA/M$, and $N_{56}(0)$ is

$$N_{56}(t) = \frac{RA \, \phi \, \sigma}{M(\lambda - \phi\sigma)}(e^{-\phi\sigma t} - e^{-\lambda t}).$$

Numerically,

$$A\phi = i/e = 4.8 \times 10^{-6}/1.6 \times 10^{-19} = 3 \times 10^{13}/\text{sec},$$
$$R/M = (6.02 \times 10^{23}) \times (0.11)/55 = 1.2 \times 10^{22}/\text{cm}^2,$$
$$\lambda = (\log 2)/(2.6 \text{ hr}) = 7.4 \times 10^{-5} \text{ sec}.$$

Note that for reasonable areas, i.e. $A \geq 1 \text{ cm}^2$, one has $\lambda \gg \phi\sigma$ and that for $t \approx 5$ hr, $\phi\sigma t \approx 0$. This means physically that for this relatively short-time duration and low flux, one may neglect the depletion of Mn⁵⁶. Thus

$$N_{56}(5.2 \text{ hr}) = \frac{RA \, \phi \, \sigma}{M\lambda}\left(1 - \frac{1}{4}\right) = 3.7 \times 10^{14}.$$

26. Consider first the scattering of neutrons on *atomic* hydrogen. The scattering amplitude may be written (considered in the limit $k \to 0$)

$$A = -\frac{a_t(3 + \boldsymbol{\sigma}_n \cdot \boldsymbol{\sigma}_h)}{4} - \frac{a_s(1 - \boldsymbol{\sigma}_n \cdot \boldsymbol{\sigma}_h)}{4}.$$

Here a_t and a_s are the scattering lengths for the triplet and singlet states of the n-p system. Note that in the triplet (singlet) state $\boldsymbol{\sigma}_n \cdot \boldsymbol{\sigma}_h = 1\ (-3)$. The differential cross section for unpolarized neutrons on uncorrelated hydrogen nuclei is thus

$$\sigma = \frac{4\pi}{2} \sum_{i,f} \langle f|A|i\rangle\langle f|A|i\rangle^* = 2\pi\, \mathrm{Tr}\{AA^+\},$$

where the sum is over the initial and final spin states of the neutron only. Using the properties $\mathrm{Tr}\,\{\sigma_i(n) = 0$ and $\mathrm{Tr}\,\{\sigma_i(n)\,\sigma_j(n)\} = 2\,\delta ij$, one finds

$$\frac{\sigma}{4\pi} = \frac{(3a_t + a_s)^2}{16} + \frac{(a_s - a_t)^2\,\boldsymbol{\sigma}_h \cdot \boldsymbol{\sigma}_h}{16}$$

$$= \frac{(3a_t + a_s)^2}{16} + \frac{3(a_s - a_t)^2}{16} = \frac{(3a_t^2 + a_s^2)}{4}.$$

Thus a measurement of the cross section on a single hydrogen nucleus does not determine the relative sign of a_s to a_t. However, the scattering from *molecular* hydrogen is coherent, since the neutron wavelength as $k \to 0$ is very much greater than the nuclear separation in H_2. Then the total coherent-scattering amplitude from the two protons is (neglecting recoil effects)

$$A_2 = -\frac{(3a_t + a_s)}{2} - \frac{(a_t - a_s)\boldsymbol{\sigma}_h \cdot (\boldsymbol{\sigma}_1 + \boldsymbol{\sigma}_2)}{4}. \tag{1}$$

The total cross section is given by

$$\frac{\sigma}{4\pi} = \frac{1}{2}\,\mathrm{Tr}(S_2^+ S_2) = \frac{(3a_t + a_s)^2}{4} + \frac{(a_t - a_s)^2 \mathbf{S}^2}{4},$$

where $\mathbf{S}$ is the total nuclear spin of the two protons,

$$\mathbf{S} = \frac{(\boldsymbol{\sigma}_1 + \boldsymbol{\sigma}_2)}{2}.$$

Thus, by measuring the cross section for ortho- $(S = 1)$ and para- $(S = 0)$ hydrogen, the relative sign of a_s to a_t may be determined, since these cross sections depend on the terms $(3a_t + a_s)^2$ and $(a_s - a_t)^2$ respectively; these are sensitive to the relative sign. Actually Eq. (1) for A_2 is not quite correct, since it neglects an overall factor due to a reduced-mass effect arising from the fact that the mass of the H_2 molecule is twice that of a single proton. However, this overall factor cancels when we consider the ratio

$$\frac{\sigma_{\text{ortho}}}{\sigma_{\text{para}}} = 1 + \frac{2(a_t - a_s)^2}{(3a_t + a_s)^2},$$

which, together with a knowledge of a_s^2 and a_t^2 obtainable by fitting the low-energy total cross section on single protons,

$$\sigma_t = \pi\left(\frac{3}{k^2 + (1/a_t^2)} + \frac{1}{k^2 + (1/a_s^2)}\right),$$

is sufficient to determine the relative sign of a_s to a_t.

27. Let the original reaction be given as

$$A + B \longrightarrow C + D, \tag{I}$$

and the inverse reaction as

$$C + D \longrightarrow A + B. \tag{II}$$

Omitting constant factors, we have

$$\frac{d\sigma}{d\Omega} = \sum_{\text{spins}} \frac{|\text{Matrix element}|^2 \, \delta^{(4)}(p_A + p_B - p_C - p_D) \, d^3p_C \, d^3p_D}{\text{Incident flux}}.$$

We shall assume that the matrix elements for reactions I and II are invariant under the time reversal operation, i.e. the operation which flips spins and reverses the directions of momenta. Hence the amplitude for the process

$$(\mathbf{p}_A, \mathbf{p}_B, \mathbf{s}_A, \mathbf{s}_B) \longrightarrow (\mathbf{p}_C, \mathbf{p}_D, \mathbf{s}_C, \mathbf{s}_D)$$

is equal to the amplitude for the process

$$(-\mathbf{p}_C, -\mathbf{p}_D, -\mathbf{s}_C, -\mathbf{s}_D) \longrightarrow (-\mathbf{p}_A, -\mathbf{p}_B, -\mathbf{s}_C, -\mathbf{s}_D).$$

In the cross-section measurement, the spins are not detected. On the assumption that initial beams are unpolarized, the spin sum includes a sum over the final spin states, and an average taken over the initial spins. But for every $\mathbf{s}$ taken in the sum, $-\mathbf{s}$ is taken also; hence

$$\sum_{\text{spins}} |\text{M.E.}|_{\text{I}}^2 = \sum_{\text{spins}} |\text{M.E.}|_{\text{II}}^2.$$

Because of this, the ratio of the cross sections does not depend on the details of the interaction. For the original reaction the incident flux is given by $(p_A/E_A + p_B/E_B)_{\text{I}}$, while for the inverse reaction it is $(p_C/E_C + p_D/E_D)_{\text{II}}$.

The phase space, for the original reaction, is proportional to

$$p_f^2 \frac{dp_f}{dE} = \frac{(p_f^{\text{I}})^2}{(p_C/E_C) + (p_D/E_D)},$$

where

$$p_f^{\text{I}} = |\mathbf{p}_C| = |\mathbf{p}_D|.$$

Similarly,

$$p_f^{\text{II}} = |\mathbf{p}_A| = |\mathbf{p}_B|.$$

For the inverse reaction the phase space is

$$\frac{(p_f^{\text{II}})^2}{(p_A/E_A) + (p_B/E_B)}.$$

Because of the average-over-initial spins, the spin degeneracy of each final state must be included. Then

$$\frac{d\sigma_{\text{I}}/d\Omega}{d\sigma_{\text{II}}/d\Omega} = \frac{(p_f^{\text{I}})^2(2s_C + 1)(2s_D + 1)}{(p_f^{\text{II}})^2(2s_A + 1)(2s_B + 1)} = \frac{3}{4} \frac{p_C^2}{p_A^2}$$

$$= \frac{3}{4} \left\{ \frac{(E - m_\pi^2 - m_D^2)^2 - 4m_\pi^2 m_D^2}{E^2(E^2 - 4m_p^2)} \right\}.$$

28. We shall use units $\hbar = c = 1$, and treat the scatterer as stationary. The Hamiltonian of the system (to lowest order in v) is

$$H = -\boldsymbol{\mu}\cdot\mathbf{B} \quad \text{where} \quad \mathbf{B} = -\mathbf{v} \times \mathbf{E}.$$

The magnetic moment of the neutron is

$$\boldsymbol{\mu} = \mu_0\boldsymbol{\sigma},$$

where σ_x, σ_y, σ_z are the Pauli matrices. The Hamiltonian may be conveniently written as

$$H = -\mu_0\,\boldsymbol{\sigma}\cdot(\mathbf{E} \times \mathbf{p})/m.$$

The scattering amplitude is

$$A = \langle f|H|i\rangle = \frac{Ze\,\mu_0}{m}\boldsymbol{\sigma}_{fi}\cdot \int d^3\mathbf{x}\, \psi_f^*\boldsymbol{\nabla}\!\left(\frac{1}{r}\right) \times (-i\boldsymbol{\nabla})\psi_i,$$

where

$$\boldsymbol{\sigma}_{fi} = \langle\chi_f|\boldsymbol{\sigma}|\chi_i\rangle \quad \text{and} \quad \psi_i = e^{i\mathbf{k}_i\cdot\mathbf{x}}, \psi_f = e^{i\mathbf{k}_f\cdot\mathbf{x}}.$$

$|\chi_i\rangle$ and $|\chi_f\rangle$ are the initial and final spin states. Finally A reduces to

$$\frac{i\,4\pi\,Z\,e\,\mu_0}{m\,|\mathbf{k}_f - \mathbf{k}_i|^2}[\boldsymbol{\sigma}_{fi}\cdot(\mathbf{k}_f \times \mathbf{k}_i)].$$

The definition of differential cross section is $d\sigma/d\Omega = \omega/\phi$ where ϕ = flux = k_i/m and ω is the transition rate into $d\Omega$: $\omega = 2\pi\,|A|^2\,\rho(E)$. The density of states is

$$\rho(E) = \frac{1}{(2\pi)^3}k_f^2\frac{dk_f}{dE} = \frac{mk_f}{(2\pi)^3}.$$

Thus

$$\frac{d\sigma}{d\Omega} = \left\{\frac{2Ze\,\mu_0}{|\mathbf{k}_f - \mathbf{k}_i|^2}\right\}^2|\boldsymbol{\sigma}_{fi}\cdot(\mathbf{k}_f \times \mathbf{k}_i)|^2.$$

Summing over final spin states and using

$$|\mathbf{k}_i - \mathbf{k}_f|^2 = 2k^2\,(1 - \cos\theta),$$

we finally obtain

$$\frac{d\sigma}{d\Omega} = (\mu_0\,Ze)^2\frac{\cos^2\,(\theta/2)}{\sin^2\,(\theta/2)}.$$

29. The interaction Hamiltonian consists of a nuclear and a magnetic term. The nuclear scattering can be described by a constant scattering length, a. For very slow neutrons, where s-wave scattering dominates, this is a good approximation. However, the magnetic scattering will be calculated in the Born approximation. Thus the total scattering amplitude, with $\hbar = 1$, is given by

$$f(\mathbf{q}) = a - \left(\frac{m}{2\pi}\right) \int d^3\mathbf{r}H_m(\mathbf{r})e^{i\mathbf{q}\cdot\mathbf{r}}, \quad \text{where} \quad \mathbf{q} = \mathbf{k}_0 - \mathbf{k}_f,$$

and $H_m(\mathbf{r})$ is the magnetic interaction (magnetic moment of neutron $= \boldsymbol{\mu}$):

$$H_m(\mathbf{r}) = -\int d^3\mathbf{y}\, g(\mathbf{r} + \mathbf{y})\boldsymbol{\mu}_I \cdot \boldsymbol{\nabla} \times \left[\frac{\boldsymbol{\mu} \times \mathbf{y}}{y^3}\right].$$

Upon completing the integrations, one finds

$$\langle f|f(q)|i\rangle = a\langle f|i\rangle + \frac{2m\tilde{g}(\mathbf{q})}{q^2}\boldsymbol{\mu}_I \cdot [\mathbf{q} \times (\boldsymbol{\mu}_{fi} \times \mathbf{q})],$$

where

$$\tilde{g}(\mathbf{q}) = \int d^3\mathbf{r}\, g(\mathbf{r})e^{i\mathbf{q}\cdot\mathbf{r}} \qquad \text{and} \qquad \boldsymbol{\mu}_{fi} = \langle f|\boldsymbol{\mu}|i\rangle.$$

The initial and final spin states of the neutron are $|i\rangle$ and $|f\rangle$ respectively. Since $\tilde{g}(0) = 1$ and the range of q being considered is such that $|\mathbf{q}| \times$ (Bohr radius) $\ll 1$, the approximation $\tilde{g}(\mathbf{q}) \approx 1$ will be used.

The spin-flip amplitude then becomes

$$\frac{2m}{q^2}\boldsymbol{\mu}_I \cdot [\mathbf{q} \times (\boldsymbol{\mu}_{fi} \times \mathbf{q})],$$

which equals $(-2m/q^2)(\boldsymbol{\mu}_I\cdot\mathbf{q})(\boldsymbol{\mu}_{fi}\cdot\mathbf{q})$ since $\boldsymbol{\mu}_I\cdot\boldsymbol{\mu}_{fi} = 0$ in the spin-flip case (one may take a Cartesian system of coordinates for which k_0 is along k, $\boldsymbol{\mu}_I$ is along $\hat{\mathbf{i}}$, and then take $\boldsymbol{\mu}_{fi}/\mu = (\hat{\mathbf{j}} + i\hat{\mathbf{k}})$). The total spin-flip cross section is

$$\sigma_s = \int d\Omega\, |f|^2 = \frac{32}{15}\pi(m\mu_I\mu)^2.$$

The no-spin-flip amplitude is

$$f_{\mu s} = a + \frac{2m\mu_I\mu(\mathbf{i} \times \mathbf{q})^2}{q^2},$$

and the total no-spin-flip cross section is

$$\sigma_{\mu s} = 4\pi\left[a^2 + \frac{4am\mu_I\mu}{3} + \frac{32(m\mu_I\mu)^2}{15}\right].$$

30. The matrix element for the transition is

$$M = \langle\gamma\Lambda|H|\Sigma^0\rangle = \frac{ge\hbar}{(M_\Sigma + M_\Lambda)c}\left[ic\sqrt{\frac{\hbar}{2\omega V}}(\mathbf{k} \times \boldsymbol{\epsilon})\cdot u_\Lambda^+\boldsymbol{\sigma} u_\Sigma\right].$$

The modulus-squared of this matrix element, summed over the spin population of the Λ in the final state, is

$$\sum_{\sigma_{\Lambda z}=\pm 1}|M|^2 = \frac{g^2 e^2\hbar^3}{2\omega V(M_\Sigma + M_\Lambda)^2}\sum_{\sigma_{\Lambda z}=\pm 1}u_\Sigma^+(\boldsymbol{\sigma}\cdot\mathbf{k} \times \boldsymbol{\epsilon})u_\Lambda u_\Lambda^+(\boldsymbol{\sigma}\cdot\mathbf{k} \times \boldsymbol{\epsilon})u_\Sigma.$$

When using the completeness relation $\sum_{\sigma_{\Lambda z}=\pm 1} u_\Lambda u_\Lambda^+ = 1$, which may be verified directly:

$$\begin{bmatrix}1\\0\end{bmatrix}[1 \quad 0] + \begin{bmatrix}0\\1\end{bmatrix}[0 \quad 1] = \begin{bmatrix}1 & 0\\0 & 1\end{bmatrix},$$

and the property $(\boldsymbol{\sigma} \cdot \mathbf{A})^2 = \mathbf{A}^2$, valid for any vector $\mathbf{A}$ for which $\mathbf{A} \times \mathbf{A} = 0$, the sum reduces to

$$\sum_{\sigma_{\Lambda_z}=\pm 1} |M|^2 = \frac{g^2 e^2 \hbar^3}{2\omega V} \frac{(\mathbf{k} \times \boldsymbol{\epsilon})^2}{(M_\Sigma + M_\Lambda)^2},$$

independent of the initial Σ^0 polarization. Because of the transversality condition $\mathbf{k} \cdot \boldsymbol{\epsilon} = 0$, one has $(\mathbf{k} \times \boldsymbol{\epsilon})^2 = k^2 = \omega^2/c^2$. The phase-space factor is $\rho(E) = (V/(2\pi\hbar)^3)(d^3 p/dE)$. Because the decay is isotropic, one may perform the trivial angular integration, and write

$$\rho(E) = \frac{V\omega^2}{2\pi^2 c^3} \frac{d\omega}{dE}.$$

Conservation of energy and momentum require

$$\hbar\omega = \frac{(M_\Sigma^2 - M_\Lambda^2)c^2}{2M_\Sigma};$$

thus

$$\frac{d\omega}{dE} = \frac{d\omega}{d(M_\Sigma c^2)} = \frac{(M_\Sigma^2 + M_\Lambda^2)}{2M_\Sigma^2 \hbar}.$$

The transition rate is given by the Golden Rule,

$$\Gamma = \frac{2\pi}{\hbar} \rho(E) \sum_{\epsilon,\sigma_{\Lambda_z}} |M|^2,$$

where the sum is over the Λ polarization and the two polarization states of the photon. Combining the factors above, one finds

$$\frac{1}{\tau} \equiv \Gamma = \frac{g^2}{4}\left(\frac{e^2}{4\pi\hbar c}\right)\left[\frac{(M_\Sigma^2 + M_\Lambda^2)(M_\Sigma^2 - M_\Lambda^2)^3 c^2}{\hbar\, M_\Sigma^5 (M_\Sigma + M_\Lambda)^2}\right].$$

This gives $\tau \approx 3 \times 10^{-19}$ sec, upon using $\hbar = 6.6 \times 10^{-22}$ MeV/sec. Note that we have treated the Λ nonrelativistically, by using Pauli spinors; this, however, is a very good approximation because the kinetic energy of the Λ is only 2.5 MeV.

31. In nonrelativistic Fermi theory of β-decay, the weak interaction matrix element for $n \longrightarrow p + e^- + \bar{\nu}$ is

$$H_{fi} = g \int dV\, \psi_p^*(x)\psi_n(x)\psi_e^*(x)\psi_\nu(x).$$

If the nucleons are in nuclei, then the appropriate generalization is

$$H_{fi} = \frac{g}{2} \int d^3\mathbf{x}_1 \cdots d^3\mathbf{x}_n\, \psi_f^*(x_1 \cdots x_n)(\tau_1 + i\tau_2)\psi_i(x_1 \cdots x_n)\psi_e^*(x_1)\psi_2(x_1)$$

$$\equiv g \int d^3\mathbf{x}\, M(x)\psi_e^*(x)\psi_\nu(x),$$

where $(\tau_1 + i\tau_2)/2$ converts a neutron into a proton (i.e. isotopic spin raising operator) and ψ_i, and ψ_f are the initial and final nuclear wave functions.

An allowed transition is one for which

$$M \equiv \int d^3\mathbf{x} M(x) \neq 0.$$

For the process $Z \longrightarrow (Z - 1) + e^+ + \nu$, the rate is given by perturbation theory:

$$d\Gamma_1 = \frac{2\pi}{\hbar} |H_{fi}|^2 \frac{dN}{dW},$$

where dN/dW is the density of final states

$$dN = \frac{d^3\mathbf{p}_e\, d^3\mathbf{p}_\nu V^2}{(2\pi\hbar)^6} \quad \text{and} \quad W = \left[\frac{p_e^2}{2m} + p_\nu c\right]. \qquad \text{(N.R.)}$$

Therefore, neglecting nuclear recoil, we have

$$\frac{dN}{dW} = \frac{(4\pi)^2 (W - p_e^2/2m) V^2 p_e^2 dp_e}{(2\pi\hbar)^6 c^3}.$$

Since the nuclear volume is small

$$\int d^3\mathbf{x} M(x)\, \psi_e^*(x)\, \psi_\nu(x) \approx \frac{M}{V},$$

when the wave functions of the electron and neutrinos are plane waves, that is $\psi = e^{i\mathbf{k}\cdot\mathbf{r}}/\sqrt{V}$. The total rate is then given by

$$\Gamma_1 = \frac{4g^2 |M|^2}{(2\pi)^3 \hbar^7 c^3} \int_0^{p_0} p_e^2 dp_e \left(W - \frac{p_e^2}{2m}\right)^2 = \frac{8g^2 |M|^2 p_0^7}{(2\pi)^3 (105) c^3 \hbar^7 m^2},$$

where p_0 is the maximum momentum of the emitted electrons.

For the process $Z + e^- \longrightarrow (Z - 1) + \nu$ and the electron is captured from a $1S$ hydrogen-like state; the matrix element is given by

$$H_{fi} = \frac{gM\psi(0)}{\sqrt{V}},$$

where $\psi(0) = (Z^3/\pi a^3)^{1/2}$ is the value of the electron wave function at the nucleus. (Here $a = \hbar^2/me^2$ is the Bohr radius.) The density of final states is

$$\frac{4\pi p_\nu^2 dp_\nu V}{(2\pi\hbar)^3} dW = \frac{4\pi p_\nu^2 V}{(2\pi\hbar)^3 c},$$

where p_ν is the momentum of the emitted neutrino; p_ν is found from the two conservation-of-energy equations,

(a) positron emission, $E_i - E_f = p_0^2/2m + p_0 c + m_e c^2$; and

(b) K-capture, $E_i + m_e c^2 - B = E_f + p_\nu c,$

where E_i and E_f are the initial and final nuclear energies and B is the binding energy of the electron in a $1S$ level. Thus

$$p_\nu c = 2m_e c^2 - B + p_0 c + \frac{p_0^2}{2m} \approx 2m_e c^2.$$

The transition rate is given by

$$\Gamma_2 = \frac{2\pi}{\hbar} |H_{fi}|^2 \frac{dN}{dW} = \left(\frac{2\pi}{\hbar}\right) |M|^2 g^2 |\psi(0)|^2 \, 4\pi/(2\pi\hbar)^3 c^3.$$

Hence

$$\frac{\Gamma_2}{\Gamma_1} = 420\pi Z^3 \left(\frac{e^2}{\hbar c}\right)^3 \left(\frac{m_e c}{p_0}\right)^7,$$

which, in terms of $W = p_0^2/2m$, may be written

$$\frac{\Gamma_2}{\Gamma_1} = 420\pi Z^3 \left(\frac{e^2}{\hbar c}\right)^3 \left(\frac{mc^2}{2W}\right)^{7/2}.$$

32. The transition rate is given by the Golden Rule as

$$\Gamma = \frac{2\pi}{\hbar} |M|^2 \frac{dn}{dE}.$$

The dynamics is contained in M; for the purposes of this problem it is a constant. The final factor, dn/dE, is the density of states in phase space. Momentum need not be conserved among the emitted particles; the infinite mass of the recoil particle can absorb any momentum without increasing its energy. Hence

$$E = (p_1^2 + m_1^2)^{1/2} + (p_2^2 + m_2^2)^{1/2}.$$

The differential volume in phase space is

$$d^2 n = 16\pi^2 \, p_1^2 \, dp_1 \, p_2^2 \, dp_2 / h^6.$$

However, to compute dn/dE, it is more convenient to choose, as variables, not p_1 and p_2, but E and p_1. Holding p_1 fixed, we have

$$\frac{dE}{p_1} = \frac{p_2 dp_2}{(p_2^2 + m_2^2)^{1/2}},$$

from which

$$\frac{d^2 n}{dE} = 16\pi^2 p_1^2 p_2 \sqrt{p_2^2 + m_2^2} \frac{dp_1}{h^6} = 16\pi^2 p_1^2 p_2 [E - \sqrt{p_1^2 + m_1^2}] \frac{dp_1}{h^6}$$

or

$$\frac{d^2 n}{dE} = \frac{16\pi^2 p_1^2}{h^6} \left(E - \sqrt{p_1^2 + m_1^2}\right) \left[\left(E - \sqrt{p_1^2 + m_1^2}\right)^2 - m_2^2\right]^{1/2} dp_1.$$

This exhibits the energy dependence of the decay rate. In the case of β-decay (e.g. $n \to p + e^- + \bar{\nu}$), the decaying nuclei may be regarded as infinitely heavy compared to the leptons. Also,

$$m_2 = m(\bar{\nu}) = 0.$$

Then

$$\frac{1}{p_1^2} \frac{d_n^2}{dE \, dp_1} \propto (E - E_e)^2. \tag{1}$$

This suggests the introduction (due to Kurie) of the variable

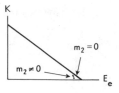

$$K = \left(\frac{1}{p_1^2} \frac{d^2 n}{dE\, dp_1} \right)^{1/2} \propto E - E_e.$$

This gives a straight-line plot as shown in the accompanying diagram.

One can see the effect of a small neutrino mass m_2 from (1): when $E - \sqrt{p_1^2 + m_1^2}$ is small, dn/dE becomes smaller for $m_2 \neq 0$. This would cause a dip near the endpoint of the spectrum, indicated with dotted lines in the figure. Before any inferences could be made from such a plot concerning possible nonzero rest mass of the neutrino, corrections for Coulomb distortion must be made. These corrections apply mainly to the low-energy end of the spectrum, however.

33. (a) The number of states for fixed $\mathbf{p}_e$ is given by $dn = V/(2\pi\hbar)^3\, d^3 p_{v_1}$. In order to calculate dn/dw, express $d^3 p_{v_1}$ in terms of $|\,\mathbf{p}_{v_1}| = p_{v_1}$, $\cos\theta$ and w, where $w = (|\,\mathbf{p}_e + |\,\mathbf{p}_{v_1}| + |\,\mathbf{p}_{v_2}|)c$, $\cos\theta = \mathbf{p}_e \cdot \mathbf{p}_{v_1}/p_e\, p_{v_1}$, and $\mathbf{p}_{v_2}$ is determined by the momentum conservation equation

$$\mathbf{p}_{v_1} + \mathbf{p}_{v_2} + \mathbf{p}_e = 0.$$

In this way we obtain

$$\frac{dn}{dW} = 2\pi V \sin\theta\, d\theta\, p_{v_1}^2 \frac{dp_1}{dW}$$

$$= 2\pi V \sin\theta\, d\theta \frac{(Mc)^2}{8c} (Mc - 2p_e)^2 \left\{ \frac{2(Mc \rightarrow p_e)(p_e \cos\theta - p_e) + Mc^2}{(Mc - p_e + p_e \cos\theta)^4} \right\},$$

where the total energy release $W = Mc^2$ (M is the muon mass).

Integrating finally over θ, we obtain

$$\omega = \frac{2\pi}{\hbar} g^2 \frac{p^2 dp}{2\pi^2 \hbar^3} \frac{1}{(2\pi\hbar)^3} \frac{\pi}{6c} (3M^2 c^2 - 6Mcp + 2p^2).$$

The number of electrons emitted in the momentum interval between p and $p + dp$ is thus $dn_e = Ap^2(3W^2 - 6Wp + 2p^2)\, dp$, with a maximum attainable $p = Mc/2$.

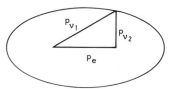

An alternative solution, due to Fermi, and involving less algebra, is to note that $n(W)$ is proportional to $\int d^3 p_{v_1}$ with $\mathbf{p}_{v_1} + \mathbf{p}_{v_2} + \mathbf{p}_e = 0$ and $|\,\mathbf{p}_{v_1}| + |\,\mathbf{p}_{v_2}| + |\,\mathbf{p}_e| = W/c$. Hence $n(W)$ is proportional to the volume of the ellipsoid in the figure above. This ellipsoid has principal axes

$a = (W - p)/2$, $b = c = \frac{1}{2}\sqrt{W^2 - 2Wp_e}$. The volume is

$$\frac{4\pi}{3}abc = \frac{\pi}{6}(W - p_e)(W^2 - 2Wp_e).$$

Thus dn/dW is proportional to

$$\frac{d}{dW}[(W - p_e)(W^2 - 2Wp_e)] = (3W^2 - 6Wp_e + 2p_e^2).$$

(b) Integrating the above expression for Γ over the allowed range of p_e $(0 \le p_e \le W/2)$ we obtain

$$\frac{1}{\tau} = \frac{7M^5c^4g^2}{7680\pi^3\hbar^7}.$$

This leads to $g \approx 3 \times 10^{-49}$ erg $=$ cm^3.

34. Denote the heavy particle of mass M by μ and the lighter one of mass m by e. To obtain the threshold frequency, ω_0, for the reaction e $+ \gamma \rightarrow \mu$, the center-of-mass energy of the photon and electron must be M. Therefore, from the Lorentz invariance of $(p_e + p_\gamma)^2 = -M^2$, one obtains $(\omega_0 + m)^2 - \omega_0^2 = M^2$; i.e. $\omega_0 = (M^2 - m^2)/2m$.

The key to finding the rate e $+ \gamma \rightarrow \mu$ is time-reversal invariance. Denote by $S_{s_\mu, s_e, \epsilon}$ ($\mu \rightarrow$ e $+ \gamma$) the matrix element for the decay of a μ at rest (polarization s_μ) into an electron and photon of polarization s_e and ϵ respectively (Fig. 1). Also denote by $S_{s_\mu, s_e, \epsilon}$ (e $+ \gamma \rightarrow \mu$) the matrix element for the inverse process (Fig. 2).

Time-reversal invariance requires

$$\sum_{s_\mu, s_e, \epsilon} |S_{s_\mu, s_e, \epsilon}(\mu \rightarrow \text{e} + \gamma)|^2 = \sum_{s_\mu, s_e, \epsilon} |S_{s_\mu, s_e, \epsilon}(\text{e} + \gamma \rightarrow \mu)|^2.$$

The lifetime of a μ is given by

$$\frac{1}{\tau} = \frac{2\pi}{(2s_\mu + 1)} \sum_{s_\mu, s_e, \epsilon} \int \frac{d\Omega}{(2\pi)^3} \frac{p^2 dp}{dE} |S_{s_\mu, s_e, \epsilon}(\mu \rightarrow \text{e} + \gamma)|^2,$$

where p is the momentum of one of the final particles, and $E = M$ is the total energy, $E = p + \sqrt{p^2 + m^2}$. Thus

$$\frac{1}{\tau} = \frac{1}{\pi(2s_\mu + 1)}\left(\frac{M^2 - m^2}{2M}\right)^2\left(\frac{M^2 + m^2}{2M^2}\right)\sum_{s_\mu, s_e, \epsilon} |S_{s_\mu, s_e, \epsilon}(\mu \rightarrow \text{e} + \gamma)|^2.$$

For the inverse process *in the rest frame of the* μ, the unpolarized rate is given by

$$\Gamma_0 = \frac{2\pi}{2(2s_e + 1)} \sum_{s_\mu, s_e, \epsilon} |S_{s_\mu, s_e, \epsilon}(\text{e} + \gamma \rightarrow \mu)|^2\frac{dN}{dE},$$

where dN/dE is the density of occupied *initial* states in the center-of-mass frame and E is the center-of-mass energy. Thus from

$$dN = \sqrt{(1 - v)/(1 + v)}\,[U(\omega)\,d\omega/\omega]$$

and $E = \omega' + \sqrt{m^2 + \omega'^2}$, and $d\omega'/d\omega = \sqrt{(1 - v)/(1 + v)}$, where ω' is the frequency in the center-of-mass frame, one finds

$$\frac{dN}{dE} = \frac{(M^2 + m^2)}{2M^2} \frac{U(\omega_0)}{\omega_0}.$$

The appearance of the reduction factor $\sqrt{(1 - v)/(1 + v)}$ in the expression for dN occurs because the density of photons transforms like the fourth component of a four-vector on going to the center-of-mass frame (just like the Doppler shift in frequency). This is because the total number of photons is invariant under Lorentz transformations. Now the differential element of volume $d^3\mathbf{x}$ in Euclidean three-space is a scalar under orthogonal rotations in three-space, and the same is true in four-space of the appropriate unit of volume $d^4x = d^3\mathbf{x}\, dt$. Thus

$$\frac{dN}{dV} \sim \frac{1}{dV} \sim \frac{d^4x}{dV} \sim dt,$$

showing that the photon density transforms like the time-like component of a four-vector. Note that the same argument may be applied to obtain the transformation laws for charge density.

Combining Eqs. (1) through (4), we obtain

$$\Gamma_0 = 4\pi^2 \frac{(2s_\mu + 1)}{(2s_e + 1)} \frac{M^2}{(M^2 - m^2)^2} \frac{U(\omega_0)}{\omega_0} \frac{1}{\tau}.$$

However, this is the rate in the center-of-mass frame, and must be corrected for time dilation to obtain the rate in the lab. Hence

$$\Gamma_{\text{lab}} = \frac{\Gamma_0}{\gamma} = \frac{8\pi^2(2s_\mu + 1)M^3 m}{(2s_e + 1)(M^2 - m^2)^2(M^2 + m^2)} \frac{U(\omega_0)}{\omega_0} \frac{1}{\tau}.$$

For $s_\mu = s_e = \frac{1}{2}$, we have $(2s_\mu + 1)/(2s_e + 1) = 1$; when $s_\mu = \frac{3}{2}$, $s_e = \frac{1}{2}$, then $(2s_\mu + 1)/(2s_e + 1) = 2$.

When this problem was proposed, the problem was of interest because of the conceivable decay $\mu \longrightarrow e + \gamma$. This decay is now known not to occur. The problem is still of interest today, however, in connection with the photo-production of high-spin ($\geqslant 3/2$) baryonic resonances.

35. Let n be the neutron density and Ω the total volume of the pile. Then in time dt there will be $n\Omega(v\, dt/\lambda)$ collisions and hence a net increase in the number of neutrons by $n\, \Omega v(k - 1)/\lambda N$ per unit time due to collisions. However, in this time $nvA\, dt/4$ neutrons strike the surface and escape (A is the surface area of the pile). Thus the equation for the time rate of change of the neutron density is

$$\frac{dn}{dt} = nv\left[\frac{k - 1}{\lambda N} - \frac{A}{4\Omega}\right].$$

The pile is critical when $dn/dt > 0$; this condition occurs when

$$\Omega > A\, \lambda N/4(k - 1).$$

For a cube $\Omega = L^3$ and $A = 6L^2$. Therefore we need

$$L > \frac{6 \times 10 \times 100}{4 \times 0.04} = 3.75 \times 10^4 \text{ cm.}$$

36. (a) Consider a neutron scattered at an angle θ in the center-of-mass frame (nonrelativistic). The originally stationary proton has a recoil energy

$$\frac{m}{2}\left[\frac{v^2}{4}\cos^2\theta + \frac{v^2}{4}(1 - \cos\theta)^2\right].$$

By averaging over θ we obtain the average energy loss in a single collision of a neutron of energy E with a stationary proton as $E/2$. Therefore after n collisions

$$\langle E_n \rangle = \frac{\langle E_{n-1} \rangle}{2}, \quad \langle E_{n-1} \rangle = \frac{\langle E_{n-2} \rangle}{2}, \quad \cdots, \quad \langle E_1 \rangle = \frac{E_0}{2}.$$

Hence $\langle E_n \rangle = (\tfrac{1}{2})^n E_0$.

To calculate $\rho_n(E)$ we first note that in a *single* collision a neutron of energy E is scattered isotropically in the center-of-mass frame, i.e. the probability that a neutron is scattered into a solid angle $d\Omega$ is $d\Omega/4\pi$; hence the probability $\rho(E)\,dE$ that a scattered neutron has an energy between E and $(E + dE)$ is dE/E', where $0 \leq E \leq E'$. This may be proved by differentiating the expression $E = E'(1 - \cos\theta)/2$. After n collisions the probability of obtaining a neutron of energy E is found from

$$\rho_n(E)dE = \frac{dE}{E_0}\int \frac{dE_{n-1}}{E_{n-1}} \cdots \frac{dE_2}{E_2}\frac{dE_1}{E_1},$$

where the integration is over the region $E_n \leq E_{n-1} \leq \cdots \leq E_1 \leq E_0$. However the integral is a symmetric function of $E_{n-1}, E_{n-2}, \ldots, E_1$, and the limits of each integration may be taken as $E_n \leq E \leq E_0$ after division by $(n - 1)!$, which is the number of ways the sequence $E_{n-1}, E_{n-2}, \ldots, E_1$ may be arranged. Thus

$$\rho_n(E) = \frac{1}{(n-1)!}\int_E^{E_0}\frac{dE_{n-1}}{E_{n-1}}\int_E^{E_0}\frac{dE_{n-2}}{E_{n-2}} \cdots \int_E^{E_0}\frac{dE_1}{E_1} = \frac{1}{(n-1)!}\left\{\ln\left(\frac{E_0}{E}\right)\right\}^{(n-1)}.$$

(b) In the steady state the number of neutrons per unit energy is independent of time, thus the number entering the interval dE per unit time must equal the number leaving:

$$Nn(E_0)v_0\sigma_s(E_0)\frac{dE}{E_0} + N\int_E^{E_0}f(E')\sigma_s(E')dE'\frac{dE}{E'} = Nf(E)[\sigma_s(E) + \sigma_a(E)]dE.$$

$$(1)$$

In this expression N is the total density of scatterers; $n(E_0)$ is the density of neutrons at energy E_0; $f(E)dE$ is the flux of neutrons that are in the interval $(E, E + dE)$. The problem is to calculate this flux per unit energy, given that neutrons of energy E_0 are produced at the rate q/cm^3 sec.

Differentiating Eq. (1) with respect to E, we obtain the differential equation

$$\frac{d}{dE}\{f(E)\sigma_T(E)\} + \frac{f(E)\sigma_s(E)}{E} = 0, \qquad \text{where} \qquad \sigma_T = \sigma_s + \sigma_a. \qquad (2)$$

The solution is

$$f(E) = \frac{A}{\sigma_T(E)} \exp\left[-\int_{E_0}^E \frac{\sigma_s}{\sigma_T}\frac{dE'}{E'}\right] = \frac{AE_0}{E\sigma_\tau(E)} \exp\left[+\int_{E_0}^E \frac{\sigma_a}{\sigma_T}\frac{dE'}{E}\right].$$

The constant of integration is found by substituting in Eq. (1), where it is found that $A = n(E_0)\, v_0\, \sigma_s(E_0)/E_0$. It only remains to relate this expression for A to q. This is done by the requirement that the number of neutrons produced per unit time, of energy E_0, equal the number that leave, i.e.

$$q = Nn(E_0)\, v_0[\sigma_s(E_0) + \sigma_a(E_0)].$$

Therefore $A = q\, \sigma_s(E_0)/E_0 N\sigma_T(E_0)$.

37.
$$\frac{1}{\tau_+} = \frac{1}{\tau_0}; \qquad \frac{1}{\tau_-} = \frac{1}{\tau_0} + \frac{1}{\tau_Z}.$$

Assume that each proton acts independently, and that the nucleus is point-like; then in calculating the ground-state muon wave function, we find $1/\tau_Z = Z^4/\tau_4$. A factor of Z^3 comes from the normalization of the muon wave function, and one factor of Z for the total number of protons with which the muon may interact. Then

$$1/\tau_- = 1/\tau_0 + Z^4/\tau_4,$$

while $\tau_+ = \tau_0$ since the μ^+ is not absorbed.

In zinc, $\tau_+ = 2.1 \times 10^{-6}$ sec, $\tau_- = 2.45 \times 10^{-12}$ sec. A more accurate calculation would include the finite size of the nucleus by using a Coulomb force for $r > R$ and a harmonic force inside the nucleus.

38. In the limit of zero range, the deuteron wave function is $e^{-\gamma r}/r$, where $\gamma^2 = mB$ (B is the binding energy of the deuteron and m is the neutron mass). In the rest frame of the deuteron this wave function gives a distribution function for the neutron momentum proportional to $d^3p/(p^2 + \gamma^2)^2$. For $\gamma \ll p_0$, where $p_0 = [4m(200 \text{ MeV})]^{1/2}$ is the deuteron momentum, the angle which the neutron makes with the original beam direction is small. If p_t is the transverse momentum which the neutron possesses when the proton is removed, then this angle is given by $\theta \approx p_t/p_0$.

From the distribution function for the neutron momentum, $d^3p/(p^2 + \gamma^2)^2$, the distribution in transverse momentum is found by integration over the longitudinal momentum. In this manner the distribution in transverse momentum is found to be $p_t\, dp_t/(p_t^2 + \gamma^2)^{3/2}$, or, in terms of the angle θ, the normalized distribution is $(\gamma/p_0)\, \theta\, d\theta/(\theta^2 + \gamma^2/p_0^2)^{3/2}$. In the small angle approximation used here, $\theta \approx \sin\theta$ and thus $\theta\, d\theta \approx d\Omega/2\pi$, where $d\Omega$ is the increment of solid angle. Finally the probability that a neutron goes into the solid angle $d\Omega$ is

$$dP = \left(\frac{\gamma}{2\pi p_0}\right)\frac{d\Omega}{(\theta^2 + \gamma^2/p_0^2)^{3/2}}.$$

From the formula it is seen that neutrons are confined to a cone in the forward direction of opening angle $\delta \approx \gamma/p_0$. In this problem $\gamma/p_0 \approx \frac{1}{20}$.

39. In Problem (2–35), we studied the free oscillations of a liquid drop, under the influence of its own gravitational attraction. In this problem, the Coulomb energy is obtained from the gravitational potential energy of Solution (2–35) by the substitution $-G\rho_0^2 \rightarrow (3Ze/4\pi R^3)^2$. In addition, the surface tension T gives rise to a potential energy, which must be estimated.

When the drop undergoes a deformation $h(\hat{n})$, in the notation of Solution (2–35), the area changes by a small amount dA, and the potential energy of the deformation is $u = \int T \, dA$.

We proceed to calculate the change in area resulting from a small deformation. We begin by noting the relation between solid angle and area, $dA = r^2 \, d\Omega/(\mathbf{n}\cdot\hat{\mathbf{r}})$. The unit normal $\mathbf{n}$ is obtained from the equation of the surface,

$$f(r, \theta, \phi) = \text{const.}$$

Here it is most convenient to use spherical coordinates. Then, using the fact that θ and ϕ may be regarded as the independent variables, with $r = r(\theta, \phi)$, we write

$$\nabla f = \frac{\partial f}{\partial r}\hat{\mathbf{r}} + \frac{1}{r}\frac{\partial f}{\partial \theta}\hat{\boldsymbol{\theta}} + \frac{1}{r\sin\theta}\frac{\partial f}{\partial \phi}\hat{\boldsymbol{\phi}}$$

$$= \frac{\partial f}{\partial r}\left\{\hat{\mathbf{r}} - \frac{1}{r}\left(\frac{\partial r}{\partial \theta}\right)_\phi \hat{\boldsymbol{\theta}} - \frac{1}{r\sin\theta}\left(\frac{\partial r}{\partial \phi}\right)_\theta \hat{\boldsymbol{\phi}}\right\}$$

and

$$|\nabla f| = \left|\frac{\partial f}{\partial r}\right|\left[1 + \frac{1}{r^2}\left(\frac{\partial r}{\partial \theta}\right)^2 + \frac{1}{r^2\sin^2\theta}\left(\frac{\partial r}{\partial \phi}\right)^2\right]^{1/2}.$$

Hence

$$\mathbf{n}\cdot\hat{\mathbf{r}} = \frac{1}{\left[1 + \frac{1}{r^2}\left(\frac{\partial r}{\partial \theta}\right)^2 + \frac{1}{r^2\sin^2\theta}\left(\frac{\partial r}{\partial \phi}\right)^2\right]^{1/2}}$$

and

$$U = \int Td\Omega(R + h(\mathbf{n}))^2\sqrt{1 + \frac{1}{r^2}\left(\frac{\partial r}{\partial \theta}\right)^2 + \frac{1}{r^2\sin^2\theta}\left(\frac{\partial r}{\partial \phi}\right)^2}.$$

Because $(\partial r/\partial\theta)^2 = (\partial h/\partial\theta)^2$ is already second-order in h, we may take $(\partial r/\partial\theta)^2/r^2 = (\partial h/\partial\theta)^2/R^2$, and perform a Taylor expansion of the square root. Then, to second order in h,

$$A = A_1 + A_2 + A_3,$$

where

$$A_1 = \int d\Omega\, R^2 \left\{ 1 + \frac{1}{2R^2}\left[\left(\frac{\partial h}{\partial \theta}\right)^2 + \frac{1}{\sin^2 \theta}\left(\frac{\partial h}{\partial \phi}\right)^2 \right] \right\}$$

and

$$A_2 + A_3 = \int d\Omega (2Rh + h^2).$$

Nuclei may be regarded as incompressible. Hence

$$\int d\Omega\, Rh = -\int h^2\, d\Omega,$$

as in Solution 2–35. Therefore

$$A_2 + A_3 = -\int d\Omega\, h^2.$$

Collecting terms, we have

$$\delta A = \int \frac{d\Omega}{2}\left[\left(\frac{\partial h}{\partial \theta}\right)^2 + \frac{1}{\sin^2 \theta}\left(\frac{\partial h}{\partial \phi}\right)^2 - 2h^2 \right].$$

We integrate by parts to obtain

$$\delta A = \frac{d\Omega}{2}[+h\,\mathbf{L}^2\, h - 2h^2],$$

where

$$-\mathbf{L}^2 = \frac{1}{\sin\theta}\frac{\partial}{\partial\theta}\left(\sin\theta\frac{\partial}{\partial\theta}\right) + \frac{1}{\sin^2\theta}\frac{\partial^2}{\partial\phi^2}.$$

This operator has the familiar property $\mathbf{L}^2\, Y_{l,m} = l(l+1)\, Y_{l,m}$. Then, since $h = R \sum_{l,m} A_{l,m}\, Y_{l,m}$, we have

$$U = \frac{R^2 T}{2} \sum_{l,m} [l(l+1) - 2] A_{l,m} A_{l,m}^* = \frac{TR^2}{2} \sum_{l,m} (l-1)(l+2) A_{l,m} A_{l,m}^*.$$

In writing the Hamiltonian, we use the kinetic energy as given in solution (2–35) to obtain

$$H = \sum_{l,m} \left\{ \left(\frac{\rho_0 R^5}{2l}\right)\dot{A}_{l,m}^* \dot{A}_{l,m} + \left[\frac{TR^2}{2}(l-1)(l+2) - \frac{3(Ze)^2(l-1)}{4\pi(2l+1)R}\right] A_{l,m}^* A_{l,m} \right\}.$$

The frequency of the mode $A_{l,m}$ is thus

$$\omega_l^2 = \frac{4\pi l(l-1)}{3Am_p}\left[T(l+2) - \frac{3(Ze)^2}{2\pi(2l+1)R^3} \right].$$

Since, for $l = 2$, m ranges from -2 to $+2$, taking integral values, the lowest mode has five degrees of freedom. The energy for the $l = 2$ mode is therefore

$$E = \hbar\omega_2 (n_1 + \cdots + n_5 + \tfrac{5}{2}) = \hbar\omega_2 (n + \tfrac{5}{2}).$$

That is, the $(l = 2)$-mode behaves like a five-dimensional harmonic oscillator, with level spacing

$$\hbar\omega_2 = \hbar\left\{\frac{8\pi}{3Am_p}\left[4T - \frac{3(Ze)^2}{10\pi R^3}\right]\right\}^{1/2}$$

The oscillations will remain bounded so long as ω is real. However, we notice that when

$$T(l + 2) < \frac{3(Ze)^2}{2\pi}\frac{1}{(2l + 1)R^3},$$

ω becomes imaginary, corresponding to unbounded oscillations, or fission. If we use the incompressibility condition to write

$$R = R_0 A^{1/3} \qquad \text{with} \ \ R_0 = 1.2 \times 10^{-13} \ \text{cm},$$

and also write

$$T = u_0/4\pi R_0^2,$$

the condition for fission becomes $3(Ze)^2/10R_0 > u_0 A$. Upon using the data given, one finds stability provided by $Z^2/A < 39$. For an unstable nucleus such as $_{92}U^{236}$, $Z^2/A \approx 36$. This is good agreement, considering the simplicity of the model.

USEFUL INFORMATION

List of Notations

Gravitational constant	$G = 6.67 \times 10^{-8}$ dyn cm^2 gm^{-2}
Speed of light	$c = 3 \times 10^{10}$ cm sec^{-1}
Electronic charge	$e = 4.8 \times 10^{-10}$ esu $= 1.6 \times 10^{-19}$ C
Planck's constant	$h = 6.625 \times 10^{-27}$ erg sec
	$\hbar = h/2\pi = 1.05 \times 10^{-27}$ erg sec
Fine-structure constant	$\alpha = e^2/\hbar c = 1/137$
Electron mass	$m_e = 9.1 \times 10^{-28}$ gm $= 0.51$ MeV
Muon mass	$m(\mu) = 207\ m_e = 106$ MeV
Pion mass	$m(\pi) = 270\ m_e = 140$ MeV
Proton mass	$M = 1.6 \times 10^{-24}$ gm $= 938$ MeV
Bohr radius	$a_0 = \hbar^2/me^2 = 0.53 \times 10^{-8}$ cm
Classical electron radius	$r_c = e^2/mc^2 = 2.82 \times 10^{-13}$ cm
Bohr magneton	$\mu_B = e\hbar/2m_e c = 0.927 \times 10^{-20}$ erg gauss^{-1}
Nuclear magneton	$\mu_N = e\hbar/2Mc = 0.505 \times 10^{-23}$ erg gauss^{-1}
Boltzmann constant	$k = 1.38 \times 10^{-16}$ erg ($^\circ$K)$^{-1}$
Stephan-Boltzmann constant	$\sigma = (\pi^2/60)(k^4/\hbar^3 c^2)$
	$= 5.7 \times 10^{-5}$ erg cm^{-2} sec$^{-1} \times$ ($^\circ$K)$^{-4}$
Avogadro's number	$N_0 = 6.02 \times 10^{23}$ mole^{-1}
Gas constant	$R = N_0 k = 2$ cal-mole^{-1} $^\circ$K^{-1}
	$= 8.3 \times 10^7$ ergs·mole^{-1} $^\circ$K^{-1}
Proton magnetic moment	$\mu_p = 2.79$ nm
Neutron magnetic moment	$\mu_n = -1.91$ nm

Units of measurement:

1 Angstrom $= 1$Å $= 10^{-8}$ cm

1 year $= \pi \times 10^7$ sec

1 electron volt $= 1.6 \times 10^{-12}$ erg

1 calorie $= 4.2$ J

1 atmosphere $= 14.7$ lb in^{-2} $= 10^6$ dyn cm^{-2}

Stirling's Formula

$$n! \approx \sqrt{2\pi n}\ (n/e)^n \text{ for large } n$$

Spherical Harmonics (for $L = 0, 1, 2$)

$$Y_0 = \sqrt{\frac{1}{4\pi}}, \qquad\qquad Y_1^0 = \sqrt{\frac{3}{4\pi}} \cos\theta,$$

$$Y_1^{\pm 1} = \mp\sqrt{\frac{3}{8\pi}} \sin\theta\, e^{\pm i\phi} \qquad\qquad Y_2^{\pm 1} = \mp\sqrt{\frac{15}{8\pi}} \cos\theta \sin\theta\, e^{\pm i\phi}$$

$$Y_2^0 = \sqrt{\frac{5}{16\pi}} (2\cos^2\theta - \sin^2\theta) \qquad Y_2^{\pm 2} = \sqrt{\frac{15}{32\pi}} \sin^2\theta\, e^{\pm 2i\phi}$$

Hydrogen Atom Wave Functions

The normalized wave functions are defined by $\psi_{n,l,m} = R_{nl}(r)\, Y_l^m$. The lowest levels are described by:

$$R_{10}(r) = \left(\frac{Z}{a_0}\right)^{3/2} 2e^{-Zr/a_0},$$

$$R_{20}(r) = \left(\frac{Z}{2a_0}\right)^{3/2} \left(2 - \frac{Zr}{a_0}\right) e^{-Zr/2a_0},$$

$$R_{21}(r) = \left(\frac{Z}{2a_0}\right)^{3/2} \left(\frac{Zr}{a_0\sqrt{3}}\right) e^{-Zr/2a_0}.$$

Vector Identities

$$\nabla(\mathbf{F}\cdot\mathbf{G}) = (\mathbf{F}\cdot\nabla)\mathbf{G} + (\mathbf{G}\cdot\nabla)\mathbf{F} + \mathbf{F}\times(\nabla\times\mathbf{G}) + \mathbf{G}\times(\nabla\times\mathbf{F}),$$
$$\nabla\cdot(\phi\,\mathbf{F}) = \phi\nabla\cdot\mathbf{F} + \mathbf{F}\cdot\nabla\phi,$$
$$\nabla\cdot(\mathbf{F}\times\mathbf{G}) = \mathbf{G}\cdot(\nabla\times\mathbf{F}) - \mathbf{F}\cdot(\nabla\times\mathbf{G}),$$
$$\nabla\times(\phi\,\mathbf{F}) = \phi\nabla\times\mathbf{F} + \nabla\phi\times\mathbf{F},$$
$$\nabla\times(\mathbf{F}\times\mathbf{G}) = \mathbf{F}(\nabla\cdot\mathbf{G}) - \mathbf{G}(\nabla\cdot\mathbf{F}) + (\mathbf{G}\cdot\nabla)\mathbf{F} - (\mathbf{F}\cdot\nabla)\mathbf{G},$$
$$\nabla\times(\nabla\times\mathbf{F}) = \nabla(\nabla\cdot\mathbf{F}) - \nabla^2\mathbf{F},$$

$$\oint \mathbf{F}\cdot d\mathbf{l} = \oint (\nabla\times\mathbf{F})\cdot\mathbf{n}\, dA)$$

$$\int \frac{\partial F}{\partial x_i} dV = \int n_i\, F\, dA.$$

Various familiar identities may be obtained from the above formula by an appropriate choice of F. For example, if $F = V_i$ is a vector, it follows that

$$\int \nabla\cdot\mathbf{V}\, dV = \int \mathbf{V}\cdot\mathbf{n}\, dA.$$

Another identity is obtained by choosing $F = \mathbf{e}_l \epsilon_{lij} V_j$; then the above formula yields

$$\int \nabla\times\mathbf{V}\, dV = \int \mathbf{n}\times\mathbf{V}\, dA.$$

Units for Electromagnetism

There is considerable arbitrariness permissable in the choice of units for electromagnetism; as a result, a variety of systems is in use. We have used

Gaussian units or rationalized MKS units, whichever seemed most suitable to the problem at hand. In these two systems of units, the fundamental equations of the theory are

System	ϵ_0	μ_0	D, H	Maxwell's equations	Lorentz force
Gaussian	1	1	$D = E + 4\pi P$ $H = B - 4\pi M$	$\nabla \cdot \mathbf{D} = 4\pi\rho$ $\nabla \times \mathbf{H} = \dfrac{4\pi\mathbf{J}}{c} + \dfrac{1}{c}\dfrac{\partial\mathbf{D}}{\partial t}$ $\nabla \cdot \mathbf{B} = 0$ $\nabla \times \mathbf{E} + \dfrac{1}{c}\dfrac{\partial\mathbf{B}}{\partial t} = 0$	$q\left(\mathbf{E} + \dfrac{\mathbf{v}}{c} \times \mathbf{B}\right)$
Rationalized MKS	$\dfrac{10^{-9}}{36\pi}$	$4\pi \times 10^{-7}$	$D = \epsilon_0 E + P$ $H = \dfrac{B}{\mu_0} - M$	$\nabla \cdot \mathbf{D} = \rho$ $\nabla \times \mathbf{H} = \mathbf{J} + \dfrac{\partial\mathbf{D}}{\partial t}$ $\nabla \cdot \mathbf{B} = 0$ $\nabla \times \mathbf{E} + \dfrac{\partial\mathbf{B}}{\partial t} = 0$	$q(\mathbf{E} + \mathbf{v} \times \mathbf{B})$

One may convert physical quantities from one set of units to the other through the use of the following table, which expresses unit amount of the quantity measured in MKS units, in the equivalent amount of Gaussian units.

Quantity	MKS units	Gaussian units
Charge	1 coulomb (C)	3×10^9 esu
Potential	1 volt (V)	1/300
Magnetic flux	1 weber (Wb)	10^8 maxwells (Mx)
Magnetic intensity (B)	1 weber/m^2	10^4 gauss (G)
Magnetic field (H)	1 amp-turns/m	$4\pi \times 10^{-3}$ oersteds (Oe)
Capacity	1 farad (F)	9×10^{-11} cm